Industrial Chemistry Library, Volume 2

Calcium Magnesium Acetate

An Emerging Bulk Chemical for Environmental Applications

Industrial Chemistry Library

Industrial Chemistry Library, Volume 2

Calcium Magnesium Acetate

An Emerging Bulk Chemical for Environmental Applications

Edited by

D.L. Wise, Y.A. Levendis and M. Metghalchi

Department of Chemical Engineering, Northeastern University, 342 Snell Engineering Center, Boston, MA 02115 (U.S.A.)

ELSEVIER   **Amsterdam — Oxford — New York — Tokyo**   **1991**

ELSEVIER SCIENCE PUBLISHERS B.V.
Sara Burgerhartstraat 25
P.O. Box 211, 1000 AE Amsterdam, The Netherlands

Distributors for the United States and Canada:

ELSEVIER SCIENCE PUBLISHING COMPANY INC.
655, Avenue of the Americas
New York, NY 10010, U.S.A.

Library of Congress Cataloging-in-Publication Data

Calcium magnesium acetate, an emerging bulk chemical for environmental
   applications / edited by D.L. Wise, Y.A. Levendis, and M.
   Metghalchi.
        p.   cm. -- (Industrial chemistry library ; v. 2)
      Includes bibliographical references and index.
      ISBN 0-444-88511-0
      1. Calcium magnesium acetate.   I. Wise, Donald L. (Donald Lee),
   1937-   . II. Levendis, Y. A. (Yiannis A.) III. Metghalchi, M.
   (Mohamad) IV. Series.
   TP248.A17C35   1991
   625.7'63--dc20   90-19536
                                                             CIP

ISBN 0-444-88511-0

TABLE OF CONTENTS

v

vi

PREFACE

"CMA" or Calcium Magnesium Acetate is a new bulk chemical emerging on the world marketplace. Two major uses for this chemical identify it as being used in multi-million ton quantities in the near future. One use is as an organic biodegradable non-corrosive deicing salt, already established by the U.S. Federal Highway Administration as the road salt selected to replace sodium chloride. Production and use of up to 100 million tons per year are projected within the next decade. The second major use for CMA is as an additive to coal-fired combustion units such as used by electrical utilities. In this application the CMA has a dual role: in one, the calcium acts as a catalyst to facilitate combustion; thus, more coal can be burned in the same size unit and more electrical power result without increased capital expenditure. In the second combustion application of CMA, the calcium acts as a "grabber" for sulfur in the coal. The CMA reacts with the sulfur forming the solid particle calcium sulfate, which is recovered from the stack gases; the acetate is burned. As a result, the use of CMA in this manner substantially reduces "acid rain." The prevention of "acid rain" is believed by many to be a major societal goal within this decade. Thus, calcium magnesium acetate holds promise as becoming a major bulk chemical by the end of this decade. Production at pilot plant levels now exist, where it is prepared from limestone (the source of the Ca/Mg) and either petroleum or natural gas derivatives (the source of the acetate). On the other hand, production of CMA from renewable woody biomass is being investigated, as is production of CMA from industrial and municipal wastes.

The editors have undertaken the assembly of this reference text with the sincere opinion that the background and potential for CMA will result in this material having a surprisingly large positive environmental impact. All chemical companies, energy companies, and municipal/government authorities will soon be needing to know more about this new material. Just as nylon was unknown in the 1930's and silicon "chips" were unknown later, so CMA is now moving from the unknown into major world significance. We expect and hope that this text will be welcomed.

Donald L. Wise
Yiannis A. Levendis
Mohamad Metghalchi

BIOGRAPHICAL SKETCHES

Background on Northeastern University

All three editors are with Northeastern University, Boston, MA. Northeastern University is one of the largest private universities in the United States. It is perhaps most well-known for its cooperative education program in which students alternate between on-the-job training and educational course work. The College of Engineering is particularly well-known in the Boston area for providing many of the technical staff that has fueled the economic expansion of this area of the United States. Graduate programs at Northeastern University are becoming especially strong; every attempt is being made to increase the strength of the graduate program while continuing to maintain and grow the undergraduate programs.

Editors

Donald L. Wise, Ph.D., Cabot Professor of Chemical Engineering, Northeastern University, has had a career in biotechnology research and development including novel applications to fossil fuels. Dr. Wise has been primarily responsible for the initiation of development work on peat, lignite, and sub-bituminous coal to gaseous fuel, liquid fuels, and organic chemicals; he also originated work on the biconversion of coal gasifier product gases to these products. Dr. Wise has worked in the area of biotechnology research and development for two decades and has approximately fifty publications in the field. He is an Associate Editor of <u>Solar Energy</u> responsible for the review of manuscripts in the biomass/bioconversion area. He is on the Editorial Board of the international journal <u>Resources, Conservation, and Recycle</u> and is an editor of several texts in the areas of advanced biomaterials development and bioconversion, (i.e., <u>Energy Recovery from Lignin, Peat and Lower Rank Coals</u> by Elsevier Science Publishers, The Netherlands, 1989). Dr. Wise is also the editor of the reference text entitled <u>Bioprocessing and Biotreatment of Coal</u> to be published by Marcel Dekker, Inc., 1990. Dr. Wise holds the B.S. (magna cum laude), M.S., and Ph.D. degrees in Chemical Engineering from the University of Pittsburgh.

Yiannis A. Levendis, Ph.D., Assistant Professor of Mechanical Engineering, Northeastern University, holds a B.S. and an M.S. in Mechanical Engineering from the University of Michigan and an Ph.D. in Environmental Engineering from the California Institute of Technology (Caltech). He was a postdoctoral research engineer at Caltech from October 1987 to July 1988 and joined the Northeastern University faculty in August 1988. Dr. Levendis' technical interests include energy related problems, combustion of liquid and solid fuels, catalyzed

x

combustion, internal and external combustion engines, turbomachinery, incineration of hazardous wastes, atmospheric pollution, solar energy, high temperature materials, and pyrometry. His doctoral dissertation dealt with fundamentals of coal combustion and pollutant formation. In particular he developed model synthetic carbonaceous materials of optimal properties for experiments aimed at developing and evaluating models of coal combustion and investigating the effects of physical factors, such as porosity, pore structure and density on the combustion rate of carbon. In addition, he studied the effects of chemical factors, such as atomic structure of the material and effects of catalysis on the oxidation of carbon. To accomplish such studies Dr. Levendis constructed pertinent apparatus including a high temperature furnace, aerosol generators, a three-stage thermal reactor, an optical pyrometer for automated recording of temperature and associated data acquisition systems. Moreover, he took part in the construction of a system used for gas adsorption studies where density and porosity data on solids was rapidly deduced, as well as in the development of a system for small-angle scattering. Dr. Levendis is a member of various engineering societies such as TBII, IITΣ, ASME, SAE, SME.

Mohamad Metghalchi, Sc.D., Associate Professor of Mechanical Engineering, Northeastern University, has been involved with combustion research since 1975. As a graduate student at Massachusetts Institute of Technology he was in charge of design and construction of a spherical combustion bomb to study the properties of laminar flames. He measured laminar velocity of fuel/air mixtures at different conditions. His work at M.I.T. was funded by the Department of Transportation, General Motors, and the Army Research Office. He was associated with professor J.C. Keck and J. Heywood in this work. He cooperated closely with the Sloan Automotive Laboratory at M.I.T. in several programs.

Professor Metghalchi joined the Department of Mechanical Engineering at Northeastern University in 1979; he has been involved in combustion research both in experimental and theoretical areas. He was primarily responsible for the construction of an Internal combustion Engine Laboratory and in setting up a Laser Doppler Anemometry system to be used with combustion and biotechnology research. He was awarded a National Science Foundation grants for the purchase of equipment to set up these laboratories. Currently, Professor Metghalchi supervises graduate students doing research in combustion. His research topics are: (i) modeling of coal combustion in a diesel engine, (ii) determination of turbulent intensity in a reactor using laser doppler velocimetry, (iii) experimental investigation of cycle to cycle variation of a spark ignition engine, (iv) combustion stability analysis of a four-cylinder engine, and (v) determination of ignition delay of fuel/air mixtures.

Chapter 1

THE INVOLVEMENT OF THE FEDERAL HIGHWAY ADMINISTRATION WITH CALCIUM MAGNESIUM ACETATE

Brian H. Chollar[1], Douglas L. Smith[2], and Joseph A. Zenewitz[1]
[1]Research Chemist, Office of Engineering and Highway Operations R&D, Federal Highway Administration, McLean, Virginia 22101-2296
[2]Ecologist, Office of Engineering and Highway Operations R&D, Federal Highway Administration, McLean, Virginia 22101-2296

1.1   INTRODUCTION

Salt has been and presently continues as the main deicing agent for highway use in the United States. The use of salt mixed with abrasives began in the 1930's. By the late 1940's and early 1950's, highway agencies began to adopt a "bare pavement" policy as a standard for winter pavement conditions. The "bare pavement" policy developed in response to a number of factors including a rapid increase in the number of motor vehicles, increasing highway speeds, and an increasing reliance on cars and trucks for commuting and commerce. "Bare pavement" has also become a useful concept for maintenance because it is a simple and self-evident guideline for highway crews.

The most commonly used salts are sodium chloride (NaCl) and calcium chloride ($CaCl_2$). Sodium chloride quickly became the most common salt and deicing chemical because of its overall effectiveness, relative abundance, and low cost. Salt use for highway deicing increased from 4 million tons per year in the early 1960's to 9 million tons by 1970, and eventually rose to 12 million tons by the late 1970's.[1] Current use of salt for deicing is approximately 10 million tons of sodium salts and 0.3 million tons of calcium salts annually. In addition, about 11 million tons of abrasives are used annually on the nation's roads and highways. Salting rates are approximately 200 to 500 pounds of salt per lane mile of highway per application. Actual rates vary considerably among jurisdictions.

Concern about the effects of salt on the natural environment began in the late 1950's with the evidence of damage to roadside sugar maples (*Acer saccharum*) in New England. During the mid-1970's several Federal-aid studies were conducted which investigated the impact of salt on roadside vegetation.[2, 3, 4] These studies also identified plant species which were able to tolerate higher concentrations of salt. By the early 1960's, concern arose over contamination of drinking water from wells adjacent to unprotected salt storage areas.[5] At about the same time automobiles began to pit and "rust out," and bridge decks were beginning to show signs of corrosion.[6]

In the early 1970's the Federal Highway Administration (FHWA) began an extensive research program designed to minimize or eliminate damage caused by salt use on highways and bridges. Internally sealed concrete[7] and polymer concrete[8], used to slow down the penetration of Cl⁻ through portland cement concrete (PCC) bridge decks, were developed as a result of the program. FHWA also sponsored research into the development of epoxy-coated reinforcing bars (rebars)[9] and the use of cathodic protection[10] technology to minimize the corrosion of rebars by Cl⁻. As a result, new bridges constructed today have a much better chance of lasting 40 to 70 years. In addition, corrosion of older bridges has been slowed down considerably when cathodic protection has been installed.

Measures to reduce the use of chemical deicers have been relatively successful. These include measures to improve the performance of the deicer by providing methods to get the chemical to the proper location in the proper amount, and keeping it there. Prewetting of salt, normally with liquid calcium chloride, has resulted in quicker melting and less salt loss due to bounce and scatter.[11] The use of salt brine applicators has achieved similar results, but required modifications of equipment and application practices have inhibited implementation of this technology.[12] However, a few states with suitable sources of "waste brines" from oil and gas development are pursuing this technology.[13]

A key component of the spreader technology is the "optimum application rate."[14] When used in conjunction with a calibrated spreader[15] significant reductions in salt can be achieved. FHWA, state highway agencies, and the Salt Institute have sponsored research to find the most effective concentrations of salt for deicing. State Highway Agencies in cooperation with Salt Institute have instituted a program, "Sensible Salt Use."[16]

The economics associated with snow and ice control were also of great concern. In addition to investigating ways of reducing the amount of salt applied to highways, research also focused on the economics of using salt for highway deicing. FHWA and 11 state highway agencies sponsored a cooperative research study to investigate the state-of-the-art practices for assessing the economic impact of highway snow and ice removal.[17, 18, 19] The study addressed factors such as maintenance, traffic and safety, environment, structural damage, and vehicle corrosion. Table 1.1 highlights some of the conclusions from that study.

A study by Abt Associates Inc. for the U.S. Environmental Protection Agency (EPA) investigated the economics of the specific use of salt.[20] This study covered corrosion of bridge materials; environmental effects to plants, animals, and water supplies; and effects on other highway and vehicle components by salt.

The conclusions from this study are summarized in Table 1.2.

The Institute of Safety Analysis investigated the beneficial and adverse effects of using salt (NaCl) in highway deicing.[21] They reported that the principal benefits of road salting were: reduced winter driving accidents; reduced dollar losses accompanying reductions in mobility delay; and reduced delay in emergency transportation services to aid traffic accident victims, heart attack and other emergency medical victims, and public safety including fires and crime. The principal adverse effects of road salting were vehicle corrosion, highway bridge deck deterioration, increase of salinity of water supplies and its adverse effect on health, vegetation deterioration, and utilities corrosion. They reported quantitative data for each of these effects and found the economic benefits of road salting outweigh the economic losses of the adverse effects. They concluded road salt should be continued to be used in keeping roads open during winter storms.

These studies did much to identify and estimate the true costs paid for bridges and the environment when salt is used for deicing pavements. FHWA and the state highway agencies began to realize there were additional "hidden" costs to the use of salt for deicing and initiated research into alternative deicing chemicals. An alternative deicing chemical must have an effective melting range similar to salt, lack detrimental effects, and be cost-comparable to be an acceptable, implementable, and affordable substitute for salt. In 1973, the California Department of Transportation (Caltrans) began looking at the effectiveness and drawbacks of 16 alternative deicers.[22, 23] In 1976, FHWA conducted a survey of the use of alternative deicers in the United States and in several countries around the world.[24] These studies by Caltrans and FHWA showed that deicers such as urea, formamide, a urea-formamide mixture, ethylene glycol, ammonium acetate, or tetrapotassium pyrophosphate were being used and were less corrosive to metals than salt, but the costs were prohibitive as compared with the price of salt. These alternatives also had some detrimental environmental impacts.

1.2 DISCOVERY OF CALCIUM MAGNESIUM ACETATE

By 1976, the time had come to look at whether deicing practices could be changed to reduce the corrosion of bridges and the environmental and health impacts to vegetation, streams, lakes, and groundwater due to salt contamination. The FHWA initiated a research program, "Improved Traffic Operations During Adverse Conditions," to try to find alternative ways to physically and chemically prevent or dislodge and remove snow and ice from the roadway.

A contract for the study of chemical alternatives to salt deicers was awarded to Bjorksten Research Laboratories, Inc. under the new research program.

4

TABLE 1.1

Summary of Conclusions from FHWA Study on the "Economic Impact of Snow and Ice Control" [17,18,19]

- Older drivers find present levels of maintenance more acceptable than younger drivers. Problems may result in any attempt to reduce maintenance levels because the public has become accustom to present levels.

- Accident rates are significantly higher when road surfaces are wet or snow-packed than when they are dry.

- Drivers significantly reduce speeds during snow storms.

- There is a significant range in the business losses (other than absenteeism and tardiness) dependent upon a business' ability to reschedule production. Very few businesses actually benefit from a snow storm. Losses due to tardiness, as well as comfort and convenience, are complex.

- Volume reductions on a highway depend upon the availability of alternate routes with better maintenance. The time of day of the storm also has a great effect on volume reduction.

- Vehicles were more corrosion resistant prior to the reduction of sheet metal thickness (which accompanied the introduction of unibody construction from 1955 - 1958). Reduction in sheet metal thickness contributes more to the decreased time required for perforation by corrosion than does salt.

- Substantial increases in salt usage between 1953 and 1965 did not result in the increased corrosion that might have been expected.

- Air pollution has been shown to as much as double corrosion rates.

- Structural deterioration is increasing faster than the funds necessary for repair, rehabilitation, or replacement.

- The technology exists to evaluate and contain corrosion that causes structural deterioration. Lack of funds has resulted in limited implementation.

TABLE 1.2

Summary of Conclusions from EPA Sponsored Study "An Economic Analysis of the Environmental Impact of Highway Deicing"[20]

- Several states have experienced significant increases of salt in groundwater and surface drinking water supplies that have been directly linked to the use of deicing salts.

- In particular cases, the levels exceeded Public Health Services safety standards set in 1962 and in most cases the levels exceeded the standards set by leading researchers, heart specialists, and the American Heart Association.

- The cost in terms of permanent health degradation is extremely difficult to measure, but is likely to be high.

- The cost of actual damage to vehicles, highways and structures, utilities, and vegetation are immense. Conservatively, annual damage costs approach $3 billion. this "hidden" cost is almost 15 times the annual national budget for the purchase and application of road salt, and about 6 times the entire annual national budget for snow and ice removal.

The main objective of the research was to attempt to develop an alternative deicer that is non-corrosive to bridge steels, does not affect other highway components, has minimal or no environmental detriments, and can be produced inexpensively using materials readily available in the U.S. The study was approached from a chemist's viewpoint, eliminating various elements in the periodic table due to various detriments such as toxicity, cost, or availability. As a result of this elimination process, twelve elements remained which, in various combinations, provided many candidate deicing materials.

Preliminary literature searches and laboratory tests further eliminated most of the remaining elements due to deicing limitations, toxicity, cost, or environmental concerns. From this process two candidate deicers remained, methanol and calcium magnesium acetate (CMA). Further laboratory studies and field trials showed that both methanol and CMA melted snow and ice at lower temperatures than salt. These laboratory studies also suggested that both chemicals would be non-corrosive to metals, non-toxic to fish and animals, and non-destructive to concrete and other highway materials. Methanol, however, was eliminated from further consideration due to its flammability, solvent nature, difficulty in dispensing with most currently available equipment, and evidence that it "attacks" asphalt concrete. This left only one chemical that was felt suitable for further consideration.[25]

CMA was the only acceptable chemical that met the criteria of being non-corrosive to metals, non-destructive to other highway materials, environmentally safe, and could be produced by readily available materials. CMA is produced by the direct reaction of acetic acid and dolomitic lime. The research, however, also showed that the possible production cost of CMA would be 5 to 10 times that of NaCl.[25] Later estimates showed the cost of CMA was 10-20 times that of NaCl. At that cost, States would not be able to use CMA continuously except in those specific areas where NaCl cannot be used due to excessive corrosion, impacts to vegetation or water supply, or surface or groundwater contamination by the deicing salt. Thus, the FHWA thought that, if CMA was as promising as this preliminary study indicated, it would only be used for deicing in those designated areas where salt could not be used.

1.3 FHWA EVALUATION OF CMA

The next phase for FHWA was to begin a comprehensive project for the complete evaluation of CMA. The objective was to develop methods for economically producing CMA, to determine its environmental acceptability and technical merits and to develop warrants for its use.

This project encompassed all phases of developing the technology for producing, evaluating, and using CMA for snow and ice control. An economically viable production method was developed and demonstrated. CMA, in both pure and commercial grades, was evaluated for its environmental acceptability and fate. The technical merits of CMA (such as its deicing properties and corrosiveness to highway structures and vehicles) were measured. An assessment was made to determine the conditions under which it would be most effective to use CMA and how to use it. Table 1.3 lists the specific tasks and objectives in the FHWA project on CMA.

1.4 CMA ENVIRONMENTAL CONSIDERATIONS

In 1982, FHWA awarded a contract to Caltrans to conduct an extensive literature search and laboratory analyses to further identify potential environmental concerns with the use of CMA. Caltrans used reagent grade materials to make the CMA used for these evaluations.

Caltrans found that CMA had no significant detriments relating to air, vegetation, water quality, and public and occupational health. The research did find that CMA was more toxic to certain algae than NaCl at high concentrations. Overall, CMA at concentrations likely to be generated in snow and ice control would be less environmentally damaging than NaCl. Caltrans' investigation of the effects of CMA on soil had indeterminate results.[26]

In 1985, a study on the "Environmental Monitoring and Evaluation of CMA," funded through the National Cooperative Highway Research Program, was initiated with the University of Washington.[27] This research contained two phases:

(1) the laboratory screening of potential environmental effects, and (2) an in situ environmental evaluation of CMA. Phase 1 used CMA made from acetic acid prepared by reacting *Clostridium thermoaceticum*, an anaerobic bacterium, with corn starch. The second phase, using a commercial grade of CMA, focused on the concerns identified under Phase 1 and the above Caltrans study.[26]

Results from this study showed that CMA has little or no impact on soil chemistry or physical characteristics. CMA also has minimal impact on aquatic organisms or terrestrial vegetation. In greenhouse experiments, most of the woody and herbaceous plants tested grew well in CMA concentrations of up to 2,500 ppm, with some stimulation of growth. However, CMA suppressed the growth of seedlings at higher concentrations, and at concentrations of 5,000 ppm, seedlings died if planted immediately. If planting was delayed 2 days, most seedlings survived and grew well. The study concluded that normal CMA concentrations expected in runoff were neutral or beneficial to plant growth.[27]

Research conducted by the Chevron Chemical Company investigated the effects of CMA through acute oral toxicity, acute inhalation toxicity, acute dermal toxicity, eye and skin irritation, and subchronic oral effects of CMA on rats. The research concluded that CMA will not affect human health any more than would similar use of common salt (NaCl).[28]

TABLE 1.3

Description of Tasks under FHWA's Research Program on CMA

Environmental Acceptability of CMA - This task assessed the environmental consequences of producing, storing, handling and using CMA. It addressed the question of the fate of the components of CMA in the receiving environment, their benefits and their deficiencies. Analysis of the effect of CMA on water, soil, vegetation, and aquatic life lead to guidelines on acceptable techniques for its use. Any problems with production, storage, and handling were examined to determine proper techniques and guidelines.

Development of Manufacturing Technology - This task developed an economically viable method for producing CMA from available raw materials. Starting from literature and laboratory studies, procedures were developed and optimized. The manufacturing procedure was scaled up to the pilot plant size. Production from this plant was used for extensive field evaluations.

Technical and Economic Evaluation of CMA - This task evaluated the technical merits of CMA and developed guidelines to determine when it is advantageous to use CMA. Technical evaluations of ice melting ability and corrosiveness to highway structural materials were performed, as well as field evaluations of CMA's deicing ability.

1.5   CMA PRODUCTION TECHNOLOGY

In 1982, FHWA awarded a contract with SRI International to investigate the process development for the production of CMA.   The study was a survey and evaluation of existing production technologies using nationally available and inexpensive raw materials.   After examining many production technologies, the researchers concluded that the microbiological production of acetic acid from hydrolysed corn grain, the further in situ reaction with dolomitic lime in water, and then drying would obtain a CMA product at the least expense.   They recommended using *Clostridium thermoaceticum* for the production of acetic acid. The estimated cost of CMA produced by this process based on a 1,000 ton/day plant output at 90 percent capacity was $376 per ton.[29]

The second phase of the contract had originally called for the production of 200 tons of CMA using the above recommended process. However, that process could not be run because the improved strain of *Clostridium thermoaceticum* needed for this production process had not been developed.   Therefore, the 200 tons of CMA was manufactured using a combination of dolomitic lime and purchased acetic acid.   This phase of the research was conducted by RAD Industries, a subcontractor for SRI International, and supported through a cooperative study sponsored by FHWA and 22 states.

A two-step process was employed in the production of CMA for the study. In the first step, magnesium oxide, a by-product of step 2, was dissolved in excess acetic acid. In step two, the acid solution was mixed with dolomitic lime to a pH of 9 and, after the solids settled, the solution was decanted and dried to produce a CMA compound of approximately 1:1 molar ratio of Calcium acetate: Magnesium acetate.   The material was between 25 and 28 percent pure CMA.   The advantages of this process were that all starting materials were used with very little waste, and an optimal reaction between acetic acid and dolomitic lime was achieved.[30]

In 1983, FHWA awarded another contract with the University of Georgia Research Foundation, Inc. for the development of new strains of the bacterium *Clostridium thermoaceticum* (CT) to produce acetic acid from corn or corn wastes for the production of CMA.   This study afforded durable CT strains that were reactive in producing acetic acid from corn in solutions of differing pH and acetate concentrations.   The study was able to develop strains of CT for efficient large scale production of CMA. Results from the study demonstrated that commercial CMA could be produced at an estimated cost of $400/ton (20¢/lb), excluding transportation costs, by using this bacteriological process which uses CT to convert corn and corn products into acetate. One disadvantage with the process was that it produced CMA solutions with a range of only 2 to 6 percent solid CMA. These solutions had to be dried to obtain the desired CMA product.[31]

The next step was to develop an effective drying procedure using diluted CMA solutions to obtain solid CMA with the best properties. In 1984, another contract was awarded with this objective. The study, conducted by Energy and Minerals Research Company, recommended a fluid bed dryer process for drying large quantities of diluted CMA solutions. The CMA pellets obtained from this process had good deicing properties and had no dusting or abrasion problems. The size, shape, and durability of these pellets were compatible with existing field application equipment.[32]

In 1984, FHWA awarded a contract to Gancy Chemical Company for the development of their technology to produce large-scale quantities of CMA with desired deicing properties. Four laboratory versions of industrial pelletizers/agglomerators were evaluated. Of these, the drum pelletizer was selected. Over 200 lb (91 kg) of CMA were produced. The CMA was comparable to the best laboratory CMA in most respects. It was compatible in size, shape and strength for use in conventional application equipment. In addition, the CMA was less friable than laboratory produced CMA and had no dust problems. The main advantage of this process was that it produced a solid CMA product; therefore, drying of solutions was not necessary. The estimated cost of commercially produced CMA by this process was $400/ton (20¢/lb), excluding transportation costs.[33]

This production research sparked the interest of several private organizations and state concerns. Chevron Chemical Company began commercial production of CMA in 1985 and is now one of the main producers of CMA. The state of Iowa produced and evaluated a CMA coated sand product containing about 25 percent CMA.[34, 35, 36, 37, 38] The cost of producing this coated material was approximately $100/ton or roughly one fourth of the estimated cost to produce pure CMA. The University of Iowa has also demonstrated that acetic acid can be made from several different low cost by-products of the wet milling process of corn such as clarified starch, pericarp starch, and corn steep liquor using several microbial organisms. The study found that several alternatives to CT exist to convert these substrates to acetic acid.[39] Further development of these processes, if successful, may significantly reduce the cost to produce acetic acid and, subsequently, CMA. Verdugt, Inc., a Netherlands company, has commercially produced CMA since 1985. They have supplied CMA for deicing use to Sweden and California. Alaska, as part of a federally aided research program, also produced CMA solutions from dolomitic lime and acetic acid. Cost analysis data from the research indicate that CMA can be commercially produced for $400 to $420/ton (20 to 21¢/lb) using petroleum derived, low grade, low cost, acetic acid from Alaskan sources.[40, 41] Dynatech Scientific, Inc. conducted research sponsored by the New York State Energy Research and Development Authority

(NYSERDA) to produce CMA by bacterial action from low cost waste materials such as cheese whey. They found that a process to convert this low-cost waste feedstock to CMA was technically and economically feasible using bacterial action.[42] For an optimized pilot plant batch type reaction they obtained yields of 75 - 90 percent low molecular weight fatty acids, over half of which was acetic acid. Concentrations of these resulting acids in solution were lower (2 - 3 percent) than those (2 - 6 percent) resulting acids from higher cost raw materials (i.e.,corn). Since the costs of drying the end product from solution may outweigh raw material costs, the cost of producing CMA from corn may be lower than from a cheese whey feedstock.

1.6 PROPERTIES OF CMA

One of the first priorities of FHWA's program in CMA was to find out the characteristics and the deicing properties of CMA, both in solution and in solid form. In 1983, FHWA initiated research with Bjorksten Research Laboratories, Inc. to study the colligative properties of CMA made by a mixture of reagent grade calcium and magnesium acetate at various Calcium Acetate: Magnesium Acetate molar ratios. Table 1.4 summarizes some of the findings of this research. As a result of these findings, the study concluded that CMA produced with a Calcium Acetate: Magnesium Acetate molar ratio of between 0.43 and 0.25 and a solution pH range of 7 to 9 would have optimal desired characteristics and would be the material of choice for deicing use.[43]

Prior to this research, CMA had been produced with a Calcium Acetate: Magnesium Acetate molar ratio of 1. As a result of this study, FHWA recommended that further field studies be conducted using CMA with a Calcium Acetate: Magnesium Acetate molar ratio of 0.43. In response to the research and FHWA's recommendation, Chevron Research Co., Verdugt Inc., and Gancy Chemical Co. changed their CMA material formulations. Gancy Chemical Co. also began an investigation of CMA materials with Calcium Acetate: Magnesium Acetate molar ratios less than 0.43.

FHWA was also interested in identifying the effects of CMA on all metallic and nonmetallic components of highways and on all components of transportation vehicles. A contract was awarded to the University of Oklahoma in 1983 to investigate the effects of CMA on metallic components of all highway structures. A laboratory study utilizing spray and dip immersion, electrochemical, and concrete ponding techniques was employed to identify corrosion differences among CMA, NaCl, and $CaCl_2$ solutions on highway metals. Results showed that CMA solutions are from 10 to 15 times less corrosive to all highway metals than NaCl.[44] Results of the effects of CMA on reforcing steel in NaCl contaminated concrete slabs were not conclusive.

TABLE 1.4

Summary of the Results and Conclusions from the FHWA Sponsored Research to Identify the Colligative Properties of CMA

- Scaling of mortar samples was reduced for CMA solutions above pH 7.

- Scaling of mortar samples was reduced when the Calcium Acetate: Magnesium Acetate molar ratio was less than one.

- The rate of dissolution for solid CMA on ice at temperature between 0° and 3°F (-18° to -1°C) was approximately one half that of salt. The fastest rate of dissolution for solid CMA was at 23°F (-5°C).

- The solubility of CMA was constant for Calcium Acetate: Magnesium Acetate molar ratios >1.

- The solubility of CMA increased as the Calcium Acetate: Magnesium Acetate molar ratio decreased until optimal solubility was achieved at a Calcium Acetate: Magnesium Acetate ratio of 0.25. At Calcium Acetate: Magnesium Acetate molar ratios less than 0.25 the solubility of CMA decreased.

- The heat of solution for CMA increased as the Magnesium Acetate: Calcium Acetate molar ratio increased.

- The freezing point depression of CMA solutions were constant for Calcium Acetate: Magnesium Acetate molar ratios >1.

- The freezing point depression of CMA decreased as the Calcium Acetate: Magnesium Acetate molar ratio decrease below one until a minimum was reached at a Calcium Acetate: Magnesium Acetate molar ratio of 0.43. This effect was prevalent for CMA concentrations ranging from 0.23 molar to a saturated (2.5 to 3.5 molar) solution.

In March 1988, FHWA reported on a staff research study designed to determine the effects of CMA solutions on reinforcing bars in uncontaminated bridge deck concrete.[45] In the study, CMA and NaCl solutions were ponded on 5-ft by 8-ft concrete slabs imbedded with rebars, and changes in the electric potential of the rebars were monitored as a means of analyzing rebar corrosion. The results showed that the electric potentials of rebars in slabs ponded only with NaCl solutions moved into the corrosion range within 3 months of ponding. The electric potentials of rebars in slabs ponded only with CMA solutions had not changed appreciably from starting potentials after 4 years of intermittent ponding to simulate annual deicer application and exposure to outside weather.

Another staff laboratory study was begun in early 1989 by FHWA to determine the effects of CMA solutions on reinforcing bars in bridge decks already contaminated with salt. Mortar samples ladened with sodium chloride and other deicing salts and imbedded with reinforcing bars were immersed in CMA solutions. Electrical potentials and corrosion currents of these mortar samples are being monitored. To date results are not available.

The FHWA awarded another contract to Daedalean Inc. in 1984 to determine the effects of CMA solutions on all non-metallic highway components and on automobile components. The study conducted ponding, electrochemical, and immersion tests, following ASTM standards. The study concluded that CMA is at least 15 times less corrosive to metallic automobile parts than NaCl, and that it is not harmful to highway or automobile non-metallic components.[46]

Other organizations also have sponsored research on the properties of CMA. The Province of Ontario conducted studies of the scaling effect of CMA on different types of portland cement concretes used in pavements in North America.[47] Their ponding studies showed that damage, spalling, and scaling from CMA for most concrete types were significantly less than for salt. Maine conducted corrosion studies with CMA and salt. They reported that CMA solutions do not corrode steel panels as much as salt solutions.[46] The Naval Air Development Center, Warminster, Pennsylvania, also conducted corrosion experiments with CMA according to Aerospace Recommended Practice (ARP) specifications for aircraft. They reported that CMA met all ARP specifications and could be used for deicing airport runways.[49] Testing by the Airport Authority Group of Transport Canada has also found CMA to meet most of their specification requirements.[50] The Research and Development Committee for the International Aviation Snow Symposium[1] is now working on revised airport runway

---

[1]The Research and Development Committee for the International Aviation Snow Symposium is a committee comprised of representatives from local, state, and Federal governments; local port authorities; deicing chemical manufacturers; and airline and airport associations who are interested in airport snow and ice control.

deicing specifications for CMA.

Michigan conducted corrosion studies with mixtures of NaCl and CMA solutions on metals.[51] Their data indicated that solutions of CMA and salt caused less corrosion of metals than a pure salt solution and approximately the same amount of corrosion as CMA alone. Iowa investigated the corrosion of reinforcing bars embedded in PCC by CMA solutions and found that CMA does not cause as much corrosion as salt solutions.[52]

1.7   CMA FIELD EVALUATION

As part of its overall effort to evaluate CMA as an alternative deicer, FHWA worked with 22 state highway agencies in a cooperative study to investigate the use and deicing properties of CMA in the field. As was discussed earlier, 200 tons of CMA were produced for use in field evaluations. The participating states agreed to have Michigan and Washington field test the material. Field tests were conducted in both states over the winters of 1983 - 1984 and 1984-1985.[53, 54] Table 1.5 summarizes the results of these evaluations.

Many of the participating states in the cooperative research effort were also concerned about roadway friction characteristics of CMA. As a result, Pennsylvania and Washington conducted skid resistance tests with CMA solutions on asphaltic and portland cement concrete test tracts.[55, 56] They reported that the friction characteristics of CMA solutions are almost the same as those of salt solutions. Also, as discussed earlier, Iowa funded research to evaluate the production and use of CMA containing sand to improve the friction characteristics and increase the specific gravity of CMA. The researchers found that abrasion of traffic was needed to initiate the melting action of the CMA deicer. This was also true for the sand/salt mixture, but to a lesser degree.[35]

In 1986 a second cooperative study was funded by 22 states to test a new CMA product produced by Chevron Chemical Company containing a 7:3 molecular ratio of Magnesium Acetate: Calcium Acetate. The objective of the research was to evaluate in quantitative terms how the improved formulation of CMA performed as compared with salt. For this study, Wisconsin, Michigan, and Minnesota field tested the material. Results of the effort were mixed.

During the winter of 1986 - 1987, Wisconsin found that the CMA could be easily applied and was as effective as salt in deicing pavements, although it

TABLE 1.5

Summary of results from field trials of CMA in the states of Michigan and Washington.[53,54]

- CMA performed as well as NaCl in deicing roadways. Salt was faster than CMA in achieving equal deicing effectiveness but, from a practical standpoint, the difference was not significant.

- Both Michigan and Washington had problems with CMA sticking to the truck bed, clumping in the bed, and sticking to the material dispersing spinner. These problems were caused by moisture from wet snow condition during the tests. They combated these problems by covering truck beds and spinner with tarps during snow storms, substituting dispensing chutes for spinners, or knocking the coated CMA off the spinner or chute after accumulation.

- Washington found that CMA flowed freely over the spinner after a CMA coating on the wet spinner had built up. They were able to continue to apply CMA easily using the CMA coated spinner.

- Both Michigan and Washington found CMA to be very dusty and had to use protective masks and clothing to handle the material.

- Washington found that when CMA was applied to the roadway before a storm, the amount of CMA used during that storm was lower than that used without the initial preapplication. Salt preapplication did not decrease the amounts of salt used during the storm.

- Both Michigan and Washington found that CMA did not corrode the deicing trucks as much as salt.

---

is somewhat slower acting. An evaluation of the various deicers by Wisconsin showed that it took 1.33 times as much CMA to be as effective as salt when tested over a 14-mile stretch of state highway.[57, 58] The study also showed that it took 2.2 times as much CMA, in a 1:1 CMA/sand mixture, as salt, in a 1:1 salt/sand mixture, to be equally effective. In a similar evaluation, it took only 0.71 times as much CMA, in a mixture of 1 part CMA coated on 3 parts sand, to be as effective as salt, in a 1:1 salt/sand mixture. Michigan found that CMA was an effective deicer but was slower and less effective in deicing on the average than salt, even when applied at a rate two times that of salt. They also noted that CMA failed to create a brine. This resulted in the need for removal equipment in many situations for efficient deicing. Minnesota started using CMA during the winter of 1988 - 1989. They found that CMA was not as effective as salt in deicing pavements. They also found that it was not effective under their winter conditions at temperatures below 23 ° F(-5 °C).[59]

The results of the field study also lead to the conclusion that trucks dispensing CMA materials should be covered to prevent CMA wind scatter and loss.

The CMA material is also subject to bouncing off the pavement when applied. A more angular, less spherical, CMA material is recommended to alleviate this problem. It also was noted that appreciable amounts of CMA built up on the spreader wheel during application.[57, 58]

Several states have also field-tested CMA on their own over the last several years. During the winter of 1985 - 1986, California field-tested CMA produced by Verdugt, Inc., Tiel, Netherlands.[60] They found that CMA worked well to help in clearing roadways in combination with cinders. Snow-plowing was easier when CMA was used rather than salt. California, Massachusetts, Michigan (Zeewaukee Bridge), and Ontario have also used CMA produced by Chevron Chemical Co. over the last three winters (1986 -1989).[57, 61, 62, 63, 64] Results from these evaluations show that CMA is as effective at deicing pavements as salt, but requires a longer period of time depending upon conditions. At warm temperatures, CMA application rates of 1.4 to 1.6 times that of salt were required for equal deicing. However, fewer applications of CMA were needed than salt. Thus, only 1.1 to 1.2 times more CMA than salt, by weight, was used in most storms (California has experienced the need for even less applications than this ratio; Michigan has experienced the need for greater).[64, 65] The studies also conclude that trucks must be covered to protect the CMA from precipitation. However, in drier areas such as California, trucks do not need to be covered. The CMA dust problems were minimized by use of protective masks in its handling.

1.8   CURRENT FHWA INVOLVEMENT

The FHWA pioneered the development of CMA in cooperation with the state highway agencies. As a result of this effort, a material with superior environmental qualities and corrosion resistance compared with salt was developed. In the last 2 years, however, FHWA has not funded any administrative research in CMA technology. The FHWA believes that it is now time to turn the responsibility for the further development and manufacturing of CMA over to the private sector and the state highway agencies. Industry, universities, and states are continuing to conduct research. They are continuing to search for more economical ways of manufacturing CMA, to provide a manufactured material with better deicing properties, and to conduct expanded field trials of CMA.

There are a few lingering problems that have not been fully addressed in which FHWA will continue to have a role. As discussed earlier, there have been attempts to conduct cost benefit analyses comparing the use of CMA with salt.[20, 21, 66, 67, 68] Most of these considered only corrosion costs or environmental concerns. The FHWA has now been directed by Congress to conduct a cost benefit analysis to include all aspects of deicing properties and effects of both CMA and salt on all materials and the environment.[69]

The FHWA is conducting a staff effort to investigate the effects of CMA on rebars under simulated conditions of bridge decks with and without salt. The FHWA will also continue to monitor CMA field studies and use by states and CMA manufacturing techniques and properties by industry and universities as part of its Nationally Coordinated Program (NCP) for Research and Technology, Program E6, "Snow and Ice Control."

1.9 OUTLOOK FOR CMA

Due to the FHWA research program, more knowledge has been accumulated for CMA than for any deicing chemical besides NaCl, $CaCl_2$, or urea. CMA is now being used in several states on an experimental basis, and in several states as an alternative to salt in areas where NaCl cannot be used. CMA is now one of the best alternatives to salt for use in preserving environmentally sensitive areas where NaCl cannot be used.

The major drawback to CMA is its cost. Currently, CMA costs 10 to 20 times as much as NaCl. Thus, in order for states to achieve the benefits of reduced bridge and automobile corrosion, pavement damage, and environmental impacts, a material with a higher cost has to be used. The states have realized this and have started planning larger budgets for the use of alternative deicers such as CMA. Companies, seeing this change, have started developing other alternative materials. CMA is a first generation material. Both Chevron Chemical Co. and Gancy Chemical Co. are working on developing improved CMA type materials which will have the same environmental qualities but better deicing properties at lower temperatures.

The future for CMA and related materials, for use in environmentally sensitive areas and for situations where the control of corrosion of highway structures is necessary, is promising.

REFERENCES

1   "Survey of Salt, Calcium Chloride and Abrasive Use in the United States and Canada," Salt Institute, Alexandria, VA, 1964-65, 1973-74, 1978-79, 1982-83.

2   F.R. Drysdale and D.K. Benner, "The Suitability of Salt Tolerant Species for Revegetation of Saline Areas Along Selected Ohio Highways," Ohio Department of Transportation, Report No. OHIO-DOT-21-73, Columbus, Ohio, August 1973.

3   E. Sucoff, "Effects of Deicing Salts on Woody Vegetation Along Minnesota Roads," Minnesota Agricultural Experiment Station, Technical Bulletin 303, 1975.

4   R.E. Hanes, et al., "Effects of Deicing Salts on Plant Biota and Soil." National Cooperative Highway Research Program, Report No. 170, Washington, DC, 1976.

5   B.N. Lord, "Program to Reduce Deicing Chemical Usage," in the Proceedings of an Engineering Foundation Conference on Current Practice and Design Criteria for Urban Quality Control, Design of Urban Runoff Quality Controls (Roesner, Larry A., et al., editors) (pp. 421 - 435), American Society of Civil Engineers, New York, N.Y., 1989.

6   "Solving Corrosion Problems of Bridge Surfaces Could Save Billions," Report to Congress of the United States by the Comptroller General, PSAD-79-10, United States General Accounting Office, Washington, D.C., January 19, 1979.

7   K.C. Clear, and S. W. Forster, "Internally Sealed Concrete: Material Characterization and Heat Treating Studies," Federal Highway Administration, Report No. FHWA-RD-77-16, Washington, D.C., March 1977.

8   T.J. Pasko, Y. P. Virmani, and W. R. Jones, "Polymer Concrete Used in Redecking a Major Bridge," Public Roads, Vol. 49, No. 3, December 1985.

9   J.R. Clifton, H. F. Beeghly, and R. G. Mathey, "Non-metallic Coatings for Concrete Reinforcing Bars," Federal Highway Administration, Report No. FHWA-RD-74-018, Washington, D.C., February 1974.

10 D.R. Jackson, "Cathodic Protection for Reinforced Concrete Bridge Decks," Interim Report, Federal Highway Administration, Report No. FHWA-DP-34-2, Washington, D.C., October 1982.

11 F. Bozarth, and C. Huisman, "Implementation Package for Use of Liquid Calcium Chloride to Improve Deicing and Snow Removal Operations," Federal Highway Administration, Report No. FHWA-IP-73-2, Washington, D.C., April 1973.

12 M.M. Kasenskas, "Evaluation of the Use of Salt Brine for Deicing Purposes," Connecticut Department of Transportation, Report No. 3, Hartford, Ct., 1978.

13 R. Eck, and W. Sack, "Determining Feasibility of West Virginia Oil and Gas Brines as Highway Deicing Agents," West Virginia Department of Highways, Morgantown, West Virginia, January 1987.

14 L. Minsk, "Optimizing Deicing Chemical Application Rates," Federal Highway Administration, Washington, D.C., August 1982.

15 W. Besselievre, "Automatic Controllers for Hydraulically Powered Deicing Chemical Spreaders," Federal Highway Administration, Report No. FHWA-RD-76-505, Washington, D.C., August 1976.

16 "The Snowfighter's Handbook," Salt Institute, Alexandria, VA, 1977.

17 B.H. Welch, et al., "Economic Impact of Highways Snow and Ice Control," Federal Highway Administration, FHWA-RD-77-20, Washington, DC, September 1976.

18 J.C. McBride, et al., "Economic Impact of Highway Snow and Ice Control, Final Report," Federal Highway Administration, FHWA-RD-77-95, Washington, D.C., December 1977.

19 J.C. McBride, et al., "Economic Impact of Highway Snow and Ice Control, User's Manual," Federal Highway Administration, FHWA-RD-77-96, December 1977.

18

20 D. Murray, "An Economic Analysis of the Environmental Impact of Highway Deicing," U.S. Environmental Protection Agency, EPA-600/2-76-105, Washington, D.C., May 1976.

21 R. Brenner, and J. Mashman, "Benefits and Costs in the Use of Salt to Deice Highways," The Institute for Safety Analysis, Washington, D.C., November 1976.

22 R.F. Stratfull, D. L. Spellman, and J. A. Halterman, "Report on the investigation of Alternative Deicing Chemicals," Presented at 53rd Annual Meeting of TRB, Washington, D.C., January 1974.

23 R.F. Stratfull, D. L. Spellman, and J. A. Halterman, "Further Evaluation of Deicing Chemicals," State of California Research Report CA-DOT-TL-5197-2-74-01, January 1974.

24 J.A. Zenewitz, "Survey of Alternatives to the Use of Chlorides for Highway Deicing," Federal Highway Administration, FHWA-RD-77-52, Washington, D.C., May 1977.

25 S. Dunn, and R. Schenk, "Alternative Highway Deicing Chemicals," Federal Highway Administration, FHWA-RD-79-108, Washington, D.C., October 1979.

26 G.R. Winters, J. L. Gidley, and H. Hunt, "Environmental Evaluation of Calcium Magnesium Acetate," Federal Highway Administration, FHWA/RD-84/094, Washington, D.C., June 1985.

27 R.R. Horner, "Environmental Monitoring and Evaluation of Calcium Magnesium Acetate (CMA)", Transportation Research Board, National Cooperative Highway Research Program Report 305, Washington, D.C., April 1988.

28 G.F.S. Hiatt, et al., "Calcium Magnesium Acetate: Comparative Toxicity Tests and An Industrial Hygiene Site Investigation," in Transportation Research Record 1157, Deicing Chemicals and Snow Control (pp. 20 - 26), Transportation Research Board, National Research Council, Washington, D.C., 1988.

29 C. Marynowski, "Process Development for Production of Calcium Magnesium Acetate (CMA)," Federal Highway Administration, FHWA/RD-82/145, Washington, D.C., March 1983.

30 C. Marynowski, "Production of Calcium Magnesium Acetate (CMA) for Field Trials," Federal Highway Administration, FHWA/RD-83/062, Washington, D.C., April 1984.

31 L.G. Lungdahl, et al., "CMA Manufacture (II): Improved Bacterial Strain for Acetate Production," Federal Highway Administration, FHWA/RD-86/117, Washington, D.C., June 1986.

32 J. Solash, "Preferred Drying Methods of Calcium Magnesium Acetate Solutions," Federal Highway Administration, FHWA/RD-87/045, Washington, D.C. November 1986.

33 A. Gancy, "Preparation of High Quality Calcium Magnesium Acetate Using a Pilot Plant Process," Federal Highway Administration, FHWA/RD-86/006, Washington, D.C., January 1986.

34 M. Sheeler, and W. Rippie, "Production and Evaluation of Calcium Magnesium Acetate," Final Report, Iowa Highway Research Board, Research Project HR-243, Ames, Iowa, October, 1982.

35 M. Sheeler, "Experimental Use of Calcium Magnesium Acetate," Final Report, Iowa Highway Research Board, Research Project HR-253, Ames, Iowa, June 1983.

36 Ibid, Addendum to the Final Report, September 1984.

37 M. Sheeler, "Continuous Production of Calcium Magnesium Acetate/Sand Deicer," Progress Report, Iowa Highway Research Board, Research Project HR-253, Ames, Iowa, April 1985.

38 M. Sheeler, "Evaluation of Calcium Magnesium Acetate Deicer in Scott County," Progress Report, Iowa Highway Research Board, Research Project HR-253, Ames, Iowa, March 1987.

39 V.J. Marks, "Production of Acetic Acid for CMA Deicer," Final Report, Iowa Highway Research Board, Research Project HR-304, Ames, Iowa, June 1988.

40 M. Economides, and R. Osterman, "Preliminary Design and Feasibility Study for a Calcium Magnesium Acetate Unit," Final Report, Department of Transportation and Public Facilities, Report No. AK-RD-83-16, Fairbanks, Alaska, July 1982.

41 M. Econimides, and R. Osterman, "Pilot Plant Studies and Process Design for the Production of Calcium Magnesium Acetate," Final Report, Department of Transportation and Public Facilities, Report No. AK-RD-86-24, Fairbanks, Alaska, November 1985.

42 A. Leuschner, et al., "Calcium Magnesium Acetate Production and Cost Reduction," Dynatech Scientific, Inc., Report 888-ERER-ER-86, Cambridge, MA, February 1988.

43 R.V. Schenk, "Ice-Melting Characteristics of Calcium Magnesium Acetate," Final Report, Federal Highway Administration, FHWA/RD-86/005, Washington, D.C., February 1985.

44 C.E. Locke, and K. J. Kennelley, "Corrosion of Highway and Bridge Structural Metals by CMA," Federal Highway Administration, FHWA/RD-86/064, Washington, D.C., June 1986.

45 B.H. Chollar, and Y. P. Virmani, "Effects of Calcium Magnesium Acetate on Reinforced Steel Concrete," Public Roads, Vol. 51, No. 4, March 1988.

46 D.S. Slick, "Effect of Calcium Magnesium Acetate (CMA) on Pavements and Motor Vehicles," Federal Highway Administration, FHWA/RD-87/037, Washington, D.C., April 1987.

47 F. Pianca, K. Carter, and H. Sedlak, "A Comparison of Concrete Scaling Caused by Calcium Magnesium Acetate and Sodium Chloride in Laboratory Tests," Ontario Ministry of Transportation and Communications, MI-108, Downswille, Ontario, March 1987.

48 M.T. Hsu, "Production and Testing of Calcium Magnesium Acetate in Maine," Materials and Research Division, Maine Department of Transportation, Technical Paper 83-1, Augusta, Maine, July 1983.

49 K. Clark, Naval Air Development Center, Warminster, Pennsylvania, February 21, 1984 (private communication).

50 R.W. Elliot, "Aircraft Compatibility Test on Calcium Magnesium Acetate," Airports Authority Group, Transport Canada, Report No. CS 128-86, Ottawa, Ontario, November 1986.

51 R.L. McCrum, "Corrosion Evaluation of Calcium Magnesium Acetate (CMA), Salt (NaCl), and CMA/NaCl Solutions," Materials and Technology Division, Michigan Department of Transportation, Report No. R-1295, Lansing, Michigan, December 1988.

52 M. Callahan, "Deicing Salt Corrosion With and Without Inhibitors," Final Report, Iowa Department of Transportation, Research Project MLR-87-8, Ames, Iowa, January 1989.

53 D.D. Ernst, et al., "CMA Research in Washington State," Washington State Department of Transportation, Olympia, Washington, 1984.

54 J.H. DeFoe, "Evaluation of Calcium Magnesium Acetate as an Ice Control Agent," Testing and Research Division, Michigan Department of Transportation, Report No. R-1248, Lansing, Michigan, June 1984.

55 J.C. Wambold, "Evaluation of Wet Skid Resistance Using Four Deicing Salts," The Pennsylvania Transportation Institute, Report No. PTI 8316, University Park, Pennsylvania, July 1983.

56 R.L. Schultz, "Evaluation of Wet Skid Resistance Using Various Concentrations of Two Deicing Chemicals," Materials Office, Washington State Department of Transportation, Report No. 188, Olympia, Washington, March 1984.

57 B.H. Chollar, "Field Evaluation of Calcium Magnesium Acetate During the Winter of 1986-87," Public Roads, Vol. 52, No. 1, June 1988.

58 R.V. Schenk, "Field Deicing Tests of High Quality Calcium Magnesium Acetate (CMA): Road Deicing Tests on USH 14, Dane County, Wisconsin, Winter 1986-87," Final Report, Bjorkstin Research Laboratories, Inc., Madison, Wisconsin, June 1987.

59   M. Martolla, Minnesota Department of Transportation, St. Paul, Mn., August 1989 (private communication).

60   J.L. Gidley, "Preliminary Evaluation of Calcium Magnesium Acetate for Use as a Highway Deicer in California," Office of Transportation Laboratory, California Department of Transportation, Project No. E86TL63, Sacramento, California, September 1986.

61   "Reduced Salt Experiments: 1986-87," Highway Maintenance Division, Commonwealth of Massachusetts Department of Public Works, 1987.

62   D. G. Manning, and L. W. Crowder, "A Comparative Field Study of Calcium Magnesium Acetate and Rock Salt During the Winter of 1986-87," Research and Development Branch, Ontario Ministry of Transportation and Communications, Report No. ME-87-16, September 1987.

63   D.G. Manning, and L. W. Crowder, "A Comparative Field Study of Calcium Magnesium Acetate and Rock Salt During the Winters of 1986-87 and 1987-88," Research and Development Branch, Ontario Ministry of Transportation an Communications, Report No. MAT-88-06, September 1988.

64   Symposium on Calcium Magnesium Acetate (CMA), Sponsored by the Materials and Technology Division, Michigan Department of Transportation, Lansing, Michigan, May 1989.

65   J.L. Gidley, Office of Transportation Laboratory, California Department of Transportation, Sacramento, Ca., July 14, 1989 (personal communication).

66   D. Nottingham, et al., "Costs to the Public Due to the Use of Corrosive Deicing Chemicals and a Comparison to Alternate Winter Road Maintenance Procedures," State of Alaska Department of Transportation and Public Facilities, Report No. AK-RD-84-14, Fairbanks, Alaska, December 1983.

67   L. Hudson, "Calcium Magnesium Acetate (CMA) From Low Grade Biomass," New York State Energy and Research Development Authority, New York, New York, 1986.

68   A. Bacchus, "Financial Implications of Salt vs. CMA as a Deicing Agent," Research and Development Branch, Ministry of Transportation of Ontario, Report No. ME-87-20, December 1987.

69   Department of Transportation and Related Agencies Appropriations Bill (Supplemental Appropriations Bill), Senate Report 100-411, July 6, 1988.

Chapter 2

PHYSICAL AND CHEMICAL PROPERTIES OF CALCIUM MAGNESIUM ACETATE

Roy U. Schenk, Ph. D.
Bioenergetics, Inc. PO Box 9141, Madison, WI 53715

2.1   INTRODUCTION

In 1982, Bjorksten Research Laboratories was commissioned by the Federal Highway Administration to investigate the physical and chemical properties of Calcium Magnesium Acetate (CMA) with varying ratios of calcium and magnesium and to evaluate the effects of pH on the damage done to concrete by CMA. (1)   The objective was to determine the ice-melting characteristics of the various ratios and so determine which ratios would be preferable as road deicers.

The heat of solution, solubilities, liquidus/solidus points, and ice melting rates of several ratios were determined.(2) For simplicity, a shorthand notation is utilized for CMA of various ratios as follows: CMA 6/4 = CMA with a 0.6 mole fraction fo calcium and a 0.4 mole fraction of magnesium; CMA 3/7 = CMA with a 0.3 mole fraction of calcium and a 0.7 mole fraction of magnesium.

2.2   PHYSICAL PROPERTIES OF CMA

Heats of Solution: Heats of solution of various ratios of CMA were determined with 100 ml. samples of water containing the experimental acetate salt at a salt-to-water mole ratio of 1 to 420 $\pm$ 20.   Except for magnesium acetate tetrahydrate, the salts tested were dried overnight in a vacuum oven at 100°C. Calcium acetate was dried from the monohydrate, whereas, the CMA 2/1 and CMA 1/2 were dried from concentrated aqueous solutions.   Titration for calcium, magnesium, and alkalinity showed that the dried salts were not totally anhydrous as was expected.   The salts had the following hydration levels: $CaAc_2.1/3H_2O$; $Ca_2MgAc6.1/3H_2O$; $CaMg_2Ac6.1/6H_2O$.

The calculated heats of solution of the salts are shown in Table 2.1. Literature values for calcium acetate, for sodium chloride, and for calcium chloride and its hydrates are shown in Table 2.2.

There appears to be a slightly higher heat of solution at 16°C than at other temperatures for calcium acetate and CMA. However, the differences may well be within experimental error.   It is evident that the anhydrous salts of both calcium and magnesium acetate have strong positive heats of solution. The data indicate that the anhydrous magnesium acetate has a higher heat of solution than does calcium acetate. In fact, by extrapolation anhydrous magnesium acetate appears to have a heat of solution of 17 to 18 Kcal/mole. This is comparable to anhydrous calcium chloride (@ 18Kcal/mole).

   Total dehydration of magnesium acetate was not attempted because earlier drying attempts at 140°C had produced substantial decomposition with the loss of substantial acetate. Adding calcium acetate appeared to stabilize the magnesium acetate. Table 2.3 shows that a calcium acetate level greater than 50 percent was needed to prevent significant acetate loss from the powdered mixed salts dried at 140°C.

   The drying results indicate that vacuum drying at 100°C may attain essentially anhydrous CMA, and probably magnesium acetate also, without significant decomposition since the monocalcium dimagnesium

TABLE 2.1

Heats of Solution of Calcium Magnesium Acetates

Substance	Temp, °C	Mole Ratio (H_2O:1 Solute)	Kcal/mole
Calcium acetate 1/3 H_2O or CMA10/0.13H_2O	8.2	419	6.40
	12.1	416	6.07
	16.7	435	6.67
	19.9	421	6.38
	25.3	407	6.29
CMA 6.7/3.3 .1/9H_2O	8.0	419	9.13
	12.4	424	8.97
	16.5	408	9.28
	19.2	440	9.40
	25.2	427	9.18
CMA 3.3/6.7 .1/18H_2O	7.6	422	13.70
	12.4	414	13.91
	15.6	422	14.33
	20.4	407	13.95
	25.2	425	13.78

TABLE 2.2

Literature Data - Heats of Solution

Substance	Temp. °C	Mole Ratio (H$_2$O:1 mole Solute)	Heat of Solution (Kcal/mole)
CaAc2 (3)	16	440	+6.93
CaAc2.H2O (3)	17	600	+5.85
NaCl (4)	25	400	-1.02
	18	200	-1.28
	20	315	-1.08
CaCl2 (5)	18	300	+17.41
	25	555	+19.82
CaCl2 (6)	18	400	+17.99
CaCl2.H2O (6)	18	300	+11.71
CaCl2.2H2O (6)	18	400	+10.04
CaCl2.4H2O (6)	18	400	+2.2
CaCl2.6H2O (6)	18	400	-4.56

acetate sample was dried to a very low moisture level with no evident decomposition. Furthermore, the hydration remaining in the CMA salts appeared to be associated with the calcium acetate, since the amount of hydration halved as the calcium acetate concentration halved in the two mixed salts which were vacuum dried.

TABLE 2.3

Extent of Acetate Loss From CMA Dried at 140°C for 18 hours.

CMA Composition	% Acetate Loss
CMA 10/0	0
CMA 6.7/3.3	0
CMA 5/5	2.8
CMA 0/10	20.2

Solubility: Weighed quantities of various ratios of calcium acetate and magnesium acetate were mixed with water in amounts slightly greater than saturation and equilibrated at 20, 0, and -18°C. The calcium and magnesium concentrations of the solution were then analyzed and CMA solubility was calculated. An additional solubility test was conducted on the pure calcium acetate and magnesium acetate at -14.7°C.

The solubility at 20°C. of approximately CMA 2/8 was essentially identical to that at 0°C. It therefore appears that there is only a very slow change in

24

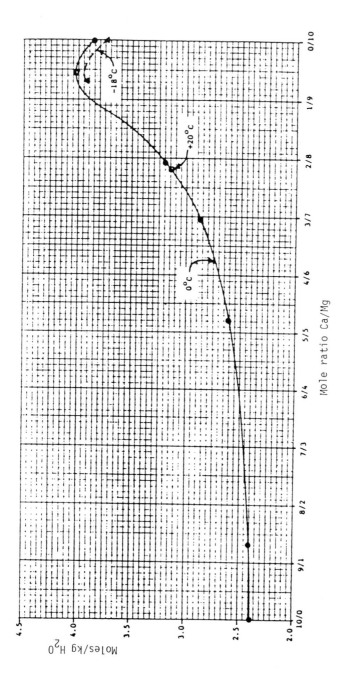

Fig. 1. Solubility vs. temperature of CMA (moles/kg water).

solubility with changes of temperature for high magnesium CMA. At 0°C, as the concentration of magnesium vs. calcium increased, a gradually accelerating increase in molality was observed. The maximum osmolality was at approximately 95% magnesium acetate. (Figure 1). A similar increase in osmolality was evident at -18°C with the actual solubility being only slightly less than at 0°C.

The solubilities of pure calcium acetate and magnesium acetate were obtained from the literature (7, 8) and additional data were collected to extend the results. These results are shown in figures 2 and 3. Calcium acetate increases in solubility substantially as its temperature is lowered below about 40°C until it reaches its eutectic at -14.7°C. Magnesium acetate solubility declines slowly below 0°C. Rivett (8) extrapolated his data to project an eutectic of -29°C. However supercooling was observed in saturated solutions. This was followed by heat evolution and solidification at temperatures as high as -8.5°C. Analysis of the solid showed a composition of magnesium acetate octahydrate. This solid dissolved in its own water of hydration when placed on the finger, i.e. about 34°C.

Liquidus-Solidus: Calcium and magnesium acetates were dissolved in water to produce concentrations of 0.23 molal (equivalent to 2.0% NaCl), 0.5, 1.0, and 1.5 molal solutions with 11 Ca/Mg ratios ranging from 10/0, 9/1, 8/2 to 1/9, and 0/10. In addition, solutions with these ratios were prepared at concentrations close to saturated.

Approximately 10 ml of solution was frozen in an insulated tube and the sample was permitted to warm slowly. The solidus was considered to be the temperature at which the metallic thermometer probe was released from firm retention in the sample. The liquidus was determined as the temperature at which all of the solid dissolved. These values were reproducible to 0.1°C for liquidus and about 1°C for solidus.

Figure 4 shows the liquidus data. At low concentrations, the liquidus changed very little with changes in the Ca/Mg ratio. As the solution concentration increased, the decrease in liquidus temperature was greater than proportional to the increase in concentration down thru 1.5 molal. In the near saturated solutions the liquidus decreased substantially with the increased proportion of magnesium cation up to the 2/8 ratio, due to increasing solutility as the magnesium content increased. Beyond that point the liquidus rapidly increased to higher than 0°C due to the formation of a solid that was insoluble at 0°C. Analysis of the insoluble solid from the pure magnesium acetate sample indicated that it was the octahydrate of magnesium acetate.

Figure 5 shows the solidus data. At the lowest concentration tested, the solidus temperature increased slowly with increasing magnesium composition. With the more concentrated solutions, the solidus temperature declined as the magnesium content increased, particularly above the CMA 5/5 composition. In the 1.0 and 1.5 molal solutions the solidus temperature continued to decline. In the 0.5 molal and saturated solutions, the CMA 3/7 composition represented a minimum solidus temperature. Very likely this is dependent on whether the octahydrate formed. When this occurred, the solidus temperature increased appreciably.

In the near saturated solutions with ratios between CMA 2/8 and CMA 0/10, supercooling occurred regularly. When crystalization finally occurred, a substantial heat evolution was observed with the solution temperature increasing as much as 20°C. When this occurred, the solidus temperature also increased.

Chemical Stability of CMA: CMA 5/5 samples were dissolved in water at concentrations of 1%, 5%, 10%, and 30% and were stored at ambient temperature (@ 20°C) or -18°C in either open or closed containers for 9 months. The closed samples lost little weight during storage, though the ambient temperature samples developed considerable mold growth. In spite of this, the samples retained over 89% of the original acetate composition.

The ambient temperature stored, open samples - which dried out over the first six months to a composition resembling calcium acetate monohydrate/magnesium acetate tetrahydrate - lost at least as much acetate. The pH of the solutions declined from pH 10 to approximately pH 7.5, apparently due to the absorption of carbon dioxide since gas evolved when hydrochloric acid was added to the samples.

The -18°C samples had only very small weight changes and the acetate losses were modest for both the open and closed samples.

Chemical analysis of the crystals from the dry, ambient, open samples revealed some relatively pure calcium acetate and magnesium acetate crystals. Other crystals had calcium to magnesium ratios suggestive of 1:1, 2:1 and 3:1 salts. X-ray crystalography gave a mixture of hydrated calcium acetate, magnesium acetate and additional lines which did not fit the patterns of any calcium or magnesium acetate or acetate hydrate in the Powder Diffraction File. This appears to confirm the presence of intermediate compositions of complex salts.

Rates of Ice Melting: Eleven ratios of hydrated CMA, from 10/0 to 0/10, were tested for the rate of ice melting in comparison with rock salt crystals (NaCl) and hydrated calcium chloride at 0, -2, -5, -10, and -20 C, using cubed pellets of 0.125 or 0.25 inch.

With either sized pellet, the penetration by the sodium chloride and calcium chloride pellets was more rapid than was the penetration by any of the CMA pellets. With a few exceptions calcium chloride penetrated more rapidly than did sodium chloride even though the calcium chloride was hydrated. There was no evident pattern of ice melting rates favoring any one ratio in the CMA ratio studies, except that at $-20^{\circ}C$ the high calcium pellets often did not melt any ice.

The slower melting rate by CMA in comparison to sodium chloride and calcium chloride appears to be a result of the lower density and of the high level of hydration of the CMA. The high hydration reduced any heat of solution effects, as also occurred with the calcium chloride. The low density caused the pellets to float higher above the ice surface, thus reducing the effective ionic concentration at the ice surface.

Scaling Potential of CMA: One hundred and twenty Portland cement concrete cylinders (3"D x 6"H) were used in this test. Cylinders were exposed in duplicate to 2% sodium chloride (pH7), to the osmolal equivalent with calcium chloride (pH7), a commercial CMA 1/1, and with 11 Ca/Mg ratios of CMA at 5 pH's. The pH's selected were 6.0, 7.0, 8.0, 8.5, and 9.0 Higher pH's (10 and 11) were initially proposed, but those pH's could not be achieved without a substantial excess of hydroxide. Controls were deionized water and air.

The sample cylinders were lowered into the specified solutions for a period of 8 hours on Monday, Wednesday and Friday and were then removed and suspended in the air to dry. Approximately a circular half of the cylinder was immersed in the solution. The immersion/drying test was conducted for 12 months. At the end of the study the cylinders were evaluated on the basis of the degree of scaling of the concrete, using a reinement of the ASTM rating system (ASTM C672).

None of the rated cylinders had scaling greater than a rating of 1 by the ASTM Method, so the region from 0 to 1 was subdivided. A longer testing period would have been desirable as differences were just becoming apparent at the end of 12 months. Summations of the ratings, presented in Table 2.8, show that pH6, and to a lesser extent pH7, was detrimental to concrete. In addition for CMA, a Ca/Mg ratio of 5/5 or lower appeared to have less effect on the concrete. None of the CMA ratios appeared more detrimental to concrete than was sodium chloride when the CMA solution had a pH of 7 or greater.

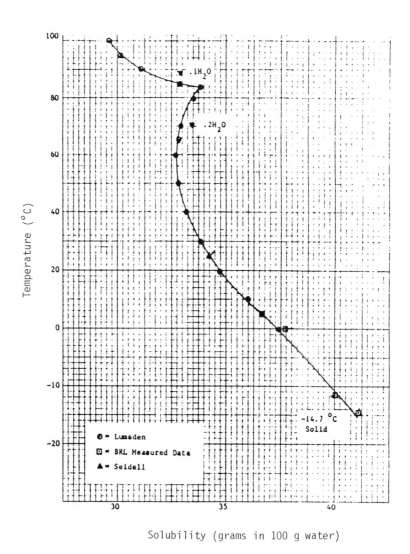

Fig. 2. Solubility vs. temperature of calcium acetate, $Ca(OAc)_2$.

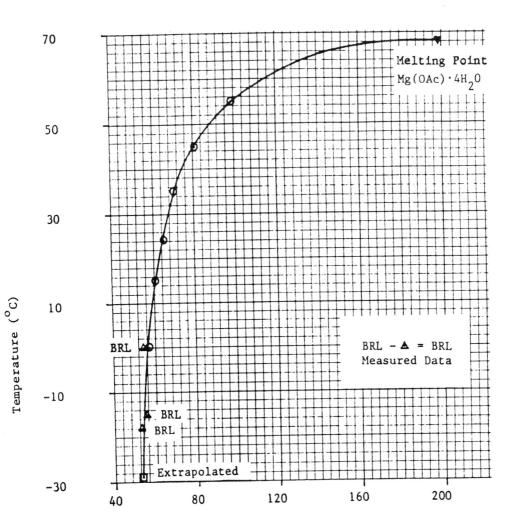

Solubility (grams in 100 grams water)

Fig. 3. Solubility vs. temperature of magnesium acetate.[2]

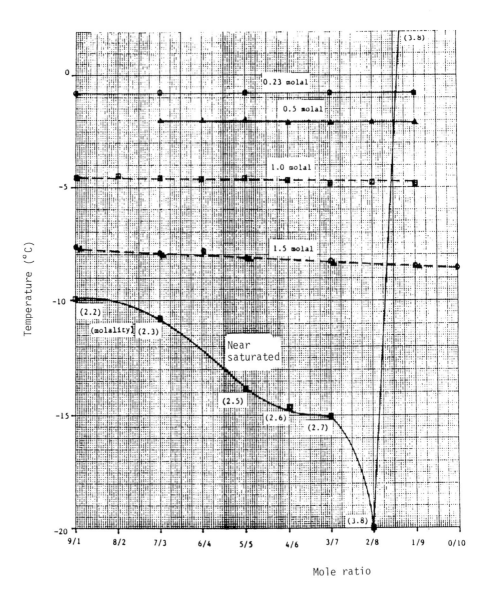

Fig. 4. Liquidous vs. mole ratio.

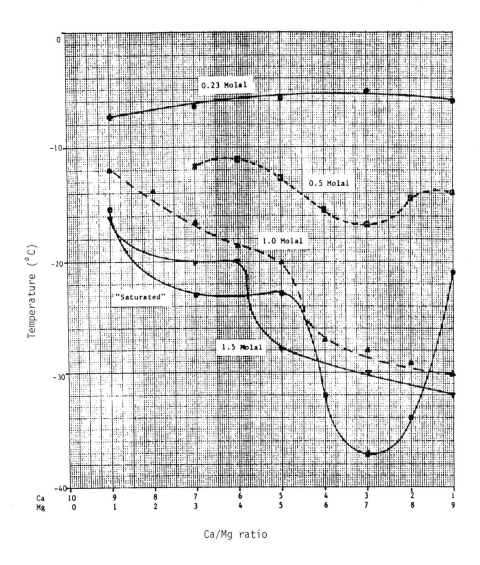

Fig. 5. Solidus vs. mole ratio.

TABLE 2.4

Summation of Immersion/Drying Ratings.

CMA	Average Ratings	Ratio Ca	Mg	Average pH 7-9	Ratings pH 6
pH 6	0.77	10	0	0.28	0.80
pH 7	0.29	9	1	0.21	0.80
pH 8	0.17	8	2	0.26	1.00
pH 8.5	0.23	7	3	0.36	0.55
pH 9	0.20	6	4	0.26	0.79
		5	5	0.29	0.85
Sodium Chloride	0.35	4	6	0.19	0.75
Calcium Chloride	0.10	3	7	0.20	0.90
Commercial CMA5/5	0.40	2	8	0.14	0.65
Deionized Water	0.20	1	9	0.11	0.75
		0	10	0.13	0.75

The most likely explanation as to why the higher magnesium CMAs were less detrimental to the concrete is because the solution pH's were higher for the higher magnesium samples (Table 2.5). Whether the initial pH was 6.0 or 9.0, the final pH's were very similar. Further tests (Table 2.6) showed that the pH change occurred mostly in the first few immersions. If the higher pH's of the solutions does indeed explain the lesser damage by the high magnesium CMAs, this would have no relevence under conditions encountered on highways since the exposure periods would be so much shorter that the pH's would not change significantly during the contact period.

TABLE 2.5

pH of CMA Immersion Solutions after One Month

Original pH	CMA 10/0	Final pH CMA 7/3	CMA 5/5	CMA 3/7	CMA0/10
6.0	7.70	7.73	7.79	7.89	8.30
7.0	7.69	7.76	7.55	7.88	8.24
8.0	7.76	7.74	7.79	7.87	8.50
8.5	7.72	7.69	7.74	7.82	8.45
9.0	7.72	7.69	7.72	7.80	8.16
Average	7.72	7.72	7.72	7.85	8.34

Other Samples (original pH)	Final pH
Water (pH5.2)	8.92
Sodium Chloride (pH7.0)	9.86
Calcium Chloride (pH7.0)	7.22

TABLE 2.6

Change of pH 6.0 Solutions during Nine Immersions

Sample	Original pH	Final pH Number of Immersions			
		1	2	3	9
Water	5.2	7.5	8.7	8.4	8.4
CMA 10/0	6.0	6.3	6.8	7.1	7.6
CMA 7.3	6.0	6.4	6.9	7.4	7.7
CMA 5/5	6.0	6.5	7.2	7.6	7.8
CMA 3/7	6.0	6.7	7.4	7.7	7.9
CMA 0/10	6.0	7.1	8.3	8.3	8.3

Samples of several immersion solutions were analyzed after one month of useage for calcium, magnesium and acetate. As can be seen in Table 2.7, the concentration of calcium in the solutions was somewhat increased over the initial solution composition. Thus it appears that there was a gradual extraction of calcium from the concrete. The acetate recovery varied from 78 to 100 percent of the original amounts. The greatest variability occured with the CMA 5/5 samples. There was substantial mold growth on some samples, and this may have accounted for much of the acetate loss.

TABLE   2.7

Compositional Changes of Immersion Solutions after One Month

pH	Mole Percent of Magnesium in Solution		
	CMA 10/0	CMA 5/5	CMA 0/10
6	99.95	52.4	4.1
7	99.95	51.9	9.7
8	99.95	51.4	3.0
8.5	99.95	54.5	3.2
9.0	99.95	56.0	5.2
Average	99.95	53.2	5.2
	Percent of Original Acetate in Solution		
	CMA 10/0	CMA 5/5	CMA 0/10
6	95	79	85
7	89	78	90
8	89	100	87
8.5	91	100	85
9	87	92	89

2.3   SUMMARY AND CONCLUSIONS

In the final report of the FHWA project, it was recommended that a pH of 8.0 or somewhat higher was the optimum pH for CMA when it is used as a highway deicer.   This recommendation still seems appropriate.   In fact, the highest reasonable pH attainable up to pH 9 increasingly appears to be the most desirable.

In addition in the FHWA final report it was reported that the bulk of the evidence supported a Ca/Mg ratio of 3/7 for CMA road deicer.   Subsequent field observations of the CMA 3/7 revealed that with many road conditions this material tended to be much slower in its ice melting rate than was the CMA 5/5. (9)   This appears to be due to the higher viscosity, even gelation, of magnesium acetate at high concentrations in water.   This property in combination with the higher cost of the magnesium ion component of CMA make this high magnesium CMA less desirable.

In combination with the subsequent recognition that the reduced concrete damage by the higher magnesium CMA was undoubt-edly simply a pH effect, it therefore now appears that there can be no justification for this continued recommendation.   At this time it appears that CMA 5/5 can be recommended as the optimum calcium to magnesium ratio.   Actually, the differences between the various ratios are so modest that a rather wide ratio of these two cations is likely to be found to produce a satisfactory deicer.

REFERENCES

1   FHWA Contract DTFH61-83-C00041. 1983.
2   R.U. Schenk, Ice-Melting Characteristics of Calcium Magnesium Acetate. Report No. FHWA/RD-86/005. 1986.
3   Handbook of Chemistry & Physics, 39th Edition. 1957. Chemical Rubber Publ. Co., Cleveland, OH.
4   International Critical Tables. Vol. 5. 1929.
5   Calculated from Kaufman, Sodium Chloride: The Production and Properties of Salt and Brine. ACS Monograph 145. 1960. American Chemical Society. Washington, D. C.
6   N.A. Lange, Handbook of Chemistry. 1952.
7   J.S. Lumsden, Solubilities of the Calcium Salts of the Acids of the Acetic Series. J. Chem. Soc 81:350-361. 1902.
8   A.C.D. Rivett, The Constitution of Magnesium Acetate Solutions. J. Chem. Soc. 129:1063-70. 1926.
9   R.U. Schenk, Field Deicing Tests of High Quality Calcium Magnesium Acetate (CMA). Final Report to FHWA and Wisc. Dept of Transportation. 1987.

Chapter 3

ALTERNATIVE ROAD DEICER

Roy U. Schenk, Ph.D.
Bioenergetics, Inc., PO Box 9141, Madison, WI 53715

3.1   INTRODUCTION

In 1976, Bjorksten Research Laboratories was commissioned by the Federal Highway Administration to develop an alternative highway deicer.[1] The material was to do minimal environmental damage, be relatively inexpensive and be produceable from solid wastes. The impetus for this search stemmed from numerous drawbacks associated with the prevalent use of NaCl (rock salt) as a road deicer.

All types of chemical compounds were reviewed. Selections were made on the basis of criteria such as water solubility and freezing point lowering, corrosion, toxicity, relative cost or cost potential, effect on soils, plants and water supplies, flammability, etc. The search was a continuous process of elimination based on the growing depth of information on a constantly diminishing number of deicer compounds. Finally two prime candidate deicers remained.

The low molecular weight organic acid salts of dolomitic limestone (Calcium-magnesium carboneate) were promptly recognized as promising candidates. The low solubility of the calcium and magnesium formates made them unattractive, so the more soluble acetate salts were initially proposed.

A search of the periodic table was subsequently pursued, and a total of nine elements were tentatively identified as suitable components of a deicer. These elements are Hydrogen, Carbon, Nitrogen, Oxygen, Sodium, Magnesium, Phosphorus, Potassium and Calcium.

Compounds containing these nine elements were evaluated for solubility and eutectic temperature, solution pH, and estimated cost. However, ultimately, compounds containing nitrogen and phosphorus were eliminated because of their serious eutrohic effects, and sodium was eliminated because of its damaging effects in drinking water and to plants and animals. The high cost of potassium made it considerably less desireable, though the carbonate/bicarbonate mixed salt showed some promise except for a relatively high alkalinity and modest solubility.

Calcium/Magnesium acetate (CMA) and methyl alcohol were the deicers selected for further testing in the laboratory and on the road. These tests showed that methyl alcohol's deicing was rapid but lacking in persistence; whereas CMA had deicing properties very similar to those of rock salt.

3.1.1 Magnitude of Deicer Usage

Over ten million tons (9.1×10^9 1kg) of rock salt are used for deicing annually in the United States.[2]   On an equiosmolal basis, this would require 16 million tons of CMA.

On a smaller scale, current spreading rates vary from 200 to 800 pounds of salt per lane mile (presumed 10 ft wide).[3,4] This corresponds to 2.4 acres of road surface and, therefore involves 83 to 330 pounds of salt per acre per application. Repeated applications during a winter result in the spreading of five to 25 tons or more of salt per lane mile depending upo whether usage is light, moderate or heavy.[5,6] This corresponds to a total expenditure of 4,100 to 20,600 pounds or more per acre of road surface per year.

3.1.2 Dissipation of Deicers

Once the deicers are applied, they dissipate either through aqueous solution and dilution, chemical reaction, microbial breakdown, volatilization or a combination of these processes. For one of the deicers considered, acetone, flammability was a potential problem since solutions as dilute as 20% in water have a flash point below $10^{\circ}C$, thus presenting an explosion hazard thru accumulation in underground structures. On the road, the low temperature and high dilution would minimize this risk.

A factor which has received only minimal attention, but which may be of some importance, is that of traffic generated spray of sodium chloride solutions.[7]   This spray has a serious effect on shrubs, trees and grass along the roadside, especially on the downwind side and on surfaces not covered with snow. Thi damage is mainly a specific ion effect of the sodium and/or chloride ions.[8]

3.2   PRELIMINARY SELECTION OF CANDIDATE DEICERS

3.2.1 Initial Elimination of Unsuitable Deicer Candidates

Since the primary requisite for a deicing chemical is water solubility, there are many substances to choose from. Selection of the best deicing chemical thus became a process of elimination. The first round of judgements was based on a general knowledge of the elements in the Periodic Table, including scarcity, costs and hazardous properties. The following groups were eliminated in the first screening:

1.   Transuranium elements (elements with atomic numbers greater than that of uranium).
2.   Actinide series (heavy radioactive metals).
3.   Lanthanide series (rare earth metals).
4.   Elements in Periods 4, 5, and 6, excepting potassium (K), calcium

(CA), barium (Ba), manganese (Mn), iron (Fe), zinc (Zn), and bromine (Br).

5.   The Noble gases of Group VIIIA.

6.   The miscellaneous group, beryllium (Be), fluorine (F), and sulfur (S in the sulfide form).

The remaining chemical elements were considered under two headings, Organic Compounds and Inorganic and Mixed Salts. Organic compounds are those which contain carbon atoms, with the exception of the carbonates. Inorganic Salts include the carbonates and other non-organic compounds. The Mixed Salts contain an inorganic component and an organic component. Included in this group is CMA.

In evaluating the potential deicing cost, it is necessary to take into account the unit weight cost of the material in comparison to rock salt, as well as its effectiveness per unit weight relative to rock salt. For first round cost evaluations, the effectiveness of the various deicing candidates was calculated on the basis of freezing point depression theory, assuming a figure of 1.86°C for the freezing point depression of water per unit osmolal concentration of solute. The osmotic effects of ions and of molecular solutes was assumed to be equivalent. This is approximately correct according to osmotic theory, which points out further that freezing point depression by a solute is a function of the number of particles in solution rather than a function of their nature.

3.2.2 Further Evaluations of Elements

Cost evaluations induced the elimination of several more chemical elements. These were barium, manganese and zinc. Partly for reasons of cost and partly for reasons of corrosion, chlorine and bromine were eliminated as a components of inorganic deicers. Sulfur was eliminated as a component for reasons of toxicity (sulfide) and corrosion (sulfate - the only other stable form). Several elements were eliminated because they required excessively high acidity or alkalinity to be soluble. Silicon, as silicate, and iron and aluminum were eliminated on this basis. Lithium was found to be much too expensive. Boron, in the form of borates, was also too costly and too toxic to consider. The nitrate combination of nitrogen and oxygen was found to be intolerably corrosive.

3.2.3 Tentatively Acceptable Elements

At this stage, only nine elements were left for serious considerations as constituents of deicers. These are Hydrogen (H), Carbon (C), Nitrogen (N), Oxygen (O), Sodium (Na), Magnesium (Mg), Phosphorus (P), Potassium (K) and Calcium (Ca).

Four of the nine elements are metallic (Na, Mg, K, and Ca) and exist in solution primarily as ions. Only the salts of these compounds would be of interest as the oxides and hydroxides when soluble are far too alkaline. The other elements can exist as components of ionic or non-ionic compounds. The non-ionic compounds would normally be organic.

(i) Sodium Containing Salts

The damage inflicted by sodium ion on plants and soils, as well as the indication of increased hypertension in human beings via high sodium levels in water supplies, have been well documented.[6,9,10] Because of these environmental drawbacks it is questionable whether sodium salts should be used for deicing except in special circumstances such as, for example, in locations where the runoff flows directly into bodies of salt water.

(ii) Phosphorus Containing Salts

Only dibasic or mixed mono- and dibasic potassium phosphate had sufficient solubility to warrant consideration, and even this is marginal.[11] Besides this, the eutrophic effects of phosphorus are serious enough to make such compounds of questionable value.

(iii) Potassium Containing Salts

Neither the carbonate nor the bicarbonate salt of potassium alone is satisfactory as a deicer. The carbonate is too alkaline (pH = 12) and the bicarbonate is insufficiently soluble (Eutectic = -8.8 C). In combination the bicarbonate could mitigate the alkalinity of the carbonate and the carbonate could enhance the solubility of the bicarbonate, though a pH of 10 or higher is still anticipated when adequate solubility is attained; and the solution is highly buffered and so on dilution it will not decrease substantially in pH.

Potassium formate may be worth consideration in the future since the deleterious effects of the formate ion on concrete may well be pH related. CMA also damaged concrete in early tests until the test solutions were maintained moderately alkaline (about pH 9). Moderate alkalinity may well eliminate concrete damage for the formate also.

Potassium is an essential fertilizer element, as are nitrogen and phosphorus. However, it does not appear to affect eutrophication as do the other two, probably because it is generally not the limiting element. Since the health effects of the sodium ion in water appear to be related to the Na/K ratio, potassium salts could actually have a beneficial health effect.[10] Consequently, the higher cost of the potassium salts needs to be balanced against the benefits they may confer.

(iv) Ammonium Salts

Nitrogen bonded to hydrogen produces ammonia which in solution forms the ammonium ion. Initially several ammonium salts appeared to be useful. The carbonate salts in particular were attractive. However, the high level of toxicity of the ammonium ion to aquatic biota made these salts unacceptable.[12]

In addition, the conversion of ammonia to nitrate can be rapid under certain conditions, and this creates a risk of serious corrosivity. Nitrogen is also a serious contributor to eutrophication. Consequently, it seems unlikely that a nitrogen containing compound can be justified as a major deicer.

(v) Calcium and Magnesium Salts

A great many of these salts are insoluble or only moderately soluble. The carbonates, for example, are the components of dolomitic limestone. The salts of low molecular weight organic acids do tend to be soluble. The formate salts, however are not sufficiently soluble to function as effective deicers. The acetate salts in combination do suply enough solubility to make them promising candidates. The cost of the organic component is the major drawback to these compounds.

In the late nineteenth Century, technology was developed to convert cellulosic materials to predominately acetic acid.[13] This process of alkaline fusion was initially performed with sodium hydroxide. However, preliminary tests at Bjorksten Research Laboratories showed that Calcium/Magnesium Hydroxides could be substituted. Subsequently, the Dakota School of Mines obtained a patent on a method for carrying out this reaction.[14] In their process they also obtain lactate and glycolate salts which can enhance the solubility of the resultant deicer. This method has the potential to permit small scale production of CMA at scattered sites utilizing solid wastes.

A conservative cost estimate for producing CMA via the alkaline fusion process indicated that the material could be produced for approximately $.05 per pound.[15] This compares with current costs of approximately $.30 per pound utilizing commercially synthesized acetic acid. Success of alkaline fusion production of CMA would require development of the appropriate equipment and probably governmental subsidies to encourage installation of the equipment. However, the process once underway could dramatically reduce solid waste accumulations in the northern half of the country.

Other low molecular weight organic acid salts were also investigated. Glycolate, propionate and lactate salts have significant solubility. However, synthesis costs would be greater for these than for acetate if produced in the pure state. As expected, [13], modest amounts of these acids occur in the alkaline fusion process for producing the acetates.[14] Their presence would actually enhance the ice-melting properties of the product, so there would be no incentive to remove them.

CMA does not contribute significantly to eutrophication. The acid anion is indeed organic but it is broken down rather slowly at the low temperatures experienced in deicing situations.[16] There are only rare conditions on a very small scale where concentrations would be so high as to cause serious effects on aquatic biota either through concentration effects or through reductions in the oxygen level of the water.

Soils in the eastern half of the United States, where most deicing is done, are deficient in calcium and magnesium. Restoration of these ions to the soil in these areas via a road deicer could be beneficial.[17,18] These divalent cations tend to improve the soil structure whereas, in contrast, sodium, and to a lesser degree other monavalent cations (e.g. potassium and ammonium), tend to cause the breakdown of soil structure.[9,19] Such breakdown results in a decreased permeability for both water and air, often a serious agricultural problem.[20]

(vi)   Organic Compounds

A number of low molecular weight organic compounds have quite high water solubilities. Consequently, they have potential as alternative deicers. These include alcohols, ketones and several nitrogen containing compounds.

The compound with the lowest molecular weight and a high water solubility is methanol. Methanol exhibits a eutectic with water (eutectic = $-125°C$) far below that of rock salt. It is neutral and non-corrosive. It contains no nitrogen or phosphorus and so contributes nothing to eutrophication problems. It has been used as an antifreeze for gasoline engines, but was replaced by glycols because of its relatively high evaporation rate. This high evaporation rate and its low viscosity, which causes it to run down into the cracks on the road and so lose its contact with the ice on the road, are the two major disadvantages of this material as a deicer.

The flash point of methanol, ($60°F$) is substantially above the temperatures encountered when deicing roads. Its toxicity limits are relatively high, having been cleared as a food additive. [21]   No ill effects have been found upon prolonged exposure to low concentrations or to short intermittent exposures at high concentrations of its vapors.[22] The purchase cost of methanol is also one of the lowest for potential deicers.

Other organic compounds evaluated include Ethanol and Isopropanol and Acetone. These are relatively more expensive than methanol. Acetone is substantially more flammable, but all three have very low eutectics, are neutral, noncorrosive, contain no nitrogen or phosphorus, and are only moderately toxic. Ethanol has the disadvantage that it reacts with chlorine in water to produce chloroform, a significant water pollutant. These materials are substantially more expensive than is methanol, and produce substantially less melting effect per pound.

Urea, Formamide and Urethane were also tested. They all contain nitrogen and so contribute to eutrophication. The only one which has an adequate eutectic is formamide ($-45°C$). Urea has a moderately low eutectic ($-13°C$) and is used on a limited basis as a deicer. However, it has been found to cause serious corrosion in some tests, probably due to conversion to nitric acid. Also, it can be broken down to ammonia which is highly toxic to aquatic biota. None of these compounds seem worthy of serious consideration as a major alternative deicer.

3.3 FURTHER EVALUATION OF CMA DEICER

On the basis of the information cited above, CMA and Methanol were selected for further evaluation in comparison to rock salt. Testing included traction, skidding friction, field performance, compatibility with cement, asphalt and road paint, and corrosion of the commonly used metals associated with vehicular and roadway construction.

3.3.1 Traction and Skid Friction

Tests of CMA in comparison to rock salt on traction and skid friction were conducted using a British Portable Tester and a braking wheel constructed specifically for the tests.

The results showed that CMA performed at least as well as salt in the retention of traction.[13] In fact, in some tests a crude product from the alkaline fusion test, containing 9% CMA produced substantially improved traction. This was due to the insoluble components of the reaction. However, it suggests that the reaction product of the alkaline fusion process might be useable without major purification. This would constitute another significant cost reduction.

3.3.2 Corrosion Testing

(i) Reinforcing Bars (Rebars)

Two reinforcing bars were embedded in concrete slabs, 14" x 10" x 3". Solutions of deicers were maintained on the surface of the slabs for a period of 9 months. One hundred and fifty milliliters of solution was placed on the surface and this dried out overnight. Subsequently, 150 mls of deionized water was added to the CMA and salt treated slabs, and methanol solution was added to the methanol treated slabs daily. Concentrations used were 25% and 50% saturation for salt and for CMA, and 25% and 50% concentration for methanol. Triplicate samples of each test were used and deionized water was used as a control. A thin steel wire (1.2 mm D) was embedded between the rebars to measure corrosion effects during the test.

Resistance readings were taken over a period of 9 months. At 5 months the sodium chloride treated sample of one of the three replicates at each concentration began to show increased resistance. By the end of the nine months, the resistance had increased dramatically on one sample indicating a complete rusting through of this wire.

At the end of the test period, the slabs were broken open and the rebars and wire were examined. The rebars of the sodium chloride treatments showed substantial rust extending into the slabs for several inches. The piano wires of one replicate came out in two pieces due to the extensive rusting.

By contrast, none of the wires and only one end of one bar of the CMA treatments had any evidence of rust. Scrapings on the one rusted area, in comparison to a comparable unrusted section, showed the rusted section to contain over ten times as much chloride ion. Reference to the sample layout showed that this latter block had been directly across from a NaCl treated block in the adjacent row. So the corrosion was likely a result of contamination from that treatment.

The near absence of corrosion in the CMA sample was not due simply to a lack of contact between CMA and steel. White crystalline deposits of CMA, probably from spillage of overflow, were founded on or directly adjacent to the projecting ends of several of the rebars.

There was no corrosion evident on the rebars or on the steel wire for either the methanol or the water samples.

(ii) Metal Corrosion

Triplicate samples of A-36 hot rolled sheet steel were cleaned and exposed for a total of five months to aqueous solutions of NaCl, CaCl2 and CMA. Following exposure the samples were cleaned and the weight losses were determined. The solutions were equi-osmolal to 2% NaCl.

The acetate samples, with one exception, showed dramatically lower weight loss at the end of the test period in comparison to the chloride solutions. The one CMA treated sample had very extensive corrosion. This is likely the result of the growth of adventitious iron oxidizing bacteria since these bacteria corrode iron more in the presence of organic materials.[23]

A series of tests were run in triplicate on zinc, rolled and cast aluminum. NaCl again produced serious corrosion with the zinc. CMA and methanol showed substantially less corrosion, but still more than occured with water alone. The NaCl also caused the most severe corrosion of the cast aluminum (A-3560); but surprisingly, the methanol showed the highest rate of attack on the rolled aluminum (6061-T3).

A recent test of CMA combined with NaCl found that the mixture produced only "slightly more total corrosion than CMA alone for mixtures down to at least the 0.46 CMA/NaCl weight ratio."[24] However, a solution containing a 0.50 ratio of CMA/NaCl corroded concrete-embedded reinforcing bars whereas pure CMA solutions did not.[25]

In later highway trials, observations of trucks which delivered CMA have confirmed the near total exclusion of corrosion to steel by CMA.

3.3.3 Deicer Compatibility with Concrete

Observation of the surfaces of the Portland cement concrete slabs of the 9 month study described in B-1 above, revealed that the surfaces of the slabs were essentially unaffected by the water and the methanol treatments. All of the sodium chloride treatments produced a generalized roughening to the surface of the concrete, exposing a substantial amount of aggregate surface.

The CMA treatments produced deeper scaling of the concrete surface in most instances. The CMA solutions had been adjusted to pH 7 before application. It was later realized that this produced a substantial amount of free acetic acid, and this evidently produces serious damage to the surface. Later tests were performed with CMA solutions which were moderately alkaline (about pH 9) and these did not produce any comparable scaling. In fact, further tests have shown that CMA does substantially lower scaling damage than does salt when the pH is maintained at a moderately alkaline condition.[26]

3.3.4 Freeze-Thaw Deicer Treatments of Portland Cement and Asphaltic Concrete

Slabs of portland cement (pcc) concrete and asphaltic concrete were treated in triplicate with two concentrations each of methanol, CMA and NaCl, as well as water. These slabs were then subjected to 50 cycles of freezing and thawing.

None of the deicers caused serious disruptions of the pcc surface. There were a substantial amount of very small chips on the methanol treated concrete surfaces and one NaCl treated surface lost a moderate sized chunk.

At the termination of the exposures, the methanol treated asphaltic concrete slabs had many uncoated aggregate surfaces exposed, whereas NaCl showed essentially no uncoated aggregates. CMA was intermediate in behavior.

3.3.5 Field Testing of Candidate Deicers

CMA, methanol and NaCl were applied to stretches of two separate roads in the winter of 1978-9 in the first actual field tests of these alternate deicers. The one road was a two lane, minimally travelled road and the deicers were applied on 1,000 foot sections. The second road was a four-lane, limited access highway, and the deicers were applied to four bridges and 1,000 feet of road leading up to the bridges. Methanol was tested first in contrast to NaCl, and subsequently CMA was tested in comparison to NaCl.

In these tests, it was found that equi-osmolal applications of CMA behave similarly to NaCl both in rate and degree of effectiveness. Methanol takes effect much more rapidly than either CMA or NaCl but is less persistent. It is effective, however, at temperatures well below the operating temperature of either of the other two de-icers. CMA was shown to be readily dispensed from equipment similar to that used for NaCl. In later field studies, it was found that CMA may build up seriously on the spreader wheels during application.[27] This may require the use of Teflon spreader wheels, or perhaps merely requires protecting the wheels from water spray. Methanol requires fluid handling equipment and it seems important to apply it directly on snow or ice and not to a cleared surface.

3.3.6 Subsequent Testing of CMA

Numerous subsequent laboratory and field tests of CMA have been conducted.[e.g. 27, 28, 29, 30] Generally, CMA has been found to react somewhat more slowly than NaCl on highway application. Approximately equi-osmolal quantities of the CMA produce effects similar to NaCl. This requires about 1.6 times the weight of CMA, though in some trials substantilly less than this was required. Damage to concrete has generally been found to be substantially less by CMA than by NaCl; and corrosion to metals has generally been about an order of magnitude less.[26,31]

Effects of CMA on vegetation are substantially less than those of salt.[16,32] Although a modest amount of aluminum and iron were found to be extracted from at least one soil treated with CMA, no comparison was made with NaCl so it is unknown wheter the effect is greater than would have occured with salt. Damage to aquatic biota occur at only high concentrations of CMA; and it is unlikely that these concentrations would occur except in very small and rare conditions.

3.4   CONCLUSION

Results of research to date indicate that CMA offers a desirable alternative to NaCl for deicing roads, except for the current high cost of the material. The use of alkaline fusion or the production of acetate from less expensive sources such as cellulose or solid waste are the most promising routes for achieving a more cost effective product.

The vision that this researcher has had for many years, is the development of simple technology which will permit the installation of converters in every major center of solid waste production so that the cellulosic wastes, which constitute about 50% of solid wastes, could be convereted to CMA and be spread on the local highways. This would dramatically reduce landfill useage; but even more, by eliminating most of the carbon going into the landfills, it would alsmost eliminate methane production, a current bane of landfills.

CMA spread on the highways would also serve to neutralize a significant amount of the acidity from automobile exhausts and so reduce the intensity of acid rain. CMA in runoff would also neutralize some of the acidity of the lakes and streams into which the runoff flows. The extent of this benefit still needs to be assessed, but it promises to be significant.

In summary, CMA continues to offer promise as an ecologically sound alternative to NaCl, and only awaits the development of methodology which will lower the production costs of the acetate.

Much of the material presented in this chapter was published first in the final report to the Federal Highway Administration, Report No. FHWA-RD-79-108, March 1980.

REFERENCES

1   FHWA Contract no. DOT-FH-11-9100. 1976.
2   Passaglia, E. and R.A. Haines, The National Cost of Automobile Corrosion, in NACE: Automotive Corrosion by Deicing Salts, R, Baboian, Ed. Houston, 1981.
3   Manual for Deicing Chemicals. Application Practices EPA-670/2-74-045. 1974.
4   Road Research. Motor Vehicle Corrosion and Influence of De-Icing Chemicals. 1969.
5   Survey of Salt, Calcium Chloride and Abrasive Use in the United States and Canada for 1973-74, Salt Institute, Alexandria, VA.
6   Murray, D.M. and U.F.W. Ernst, An Economic Analysis of the Environmental Impact of Highway Deicing. EPA-600/2-76-105. 1976.
7   Lumis, G.P. et al. Salt Damage to Roadside Plants, Ontario Ministry of Ag & Food Reprint. 1971.
8   Encyclopedia of Science and Technology, McGraw-Hill, NYC. 1971.
9   Russell, E.W., Soil Conditions and Plant Growth. Longmans, Green & Co., London, 8th Ed. 1950.
10 Meneely, G.R., Toxic Effects of Dietary Sodium Chloride and the Protective Effects of Potassium in Toxicants Occuring Naturally in Foods. 2nd Ed. National Acad. Sci., D.C. 1973.
11 Dunn, S.A. and R.U. Schenk, Alternative Highway Deicing Chemicals in Snow Removal and Ice Control Research. Special Rpt. of NAS Transp. Res. Board (TRB). 1979.
12 Quality Criteria for Water. EPA EP1.2: W29/34/976-2. 1976.
13 Dunn, S.A. and R.U. Schenk, Alternative Deicing Chemicals. Final Report, FHWA-RD-79-108. 1980.
14 Sandvig, R.L. et al. De-Icing Chemicals and their Preparation from Polysaccharide Sources. U.S. Patent 4,430,240. 1984.
15 Dunn, S.A. and R.U. Schenk, Alternative Deicing Chemicals. Final Report, Appendix D. FHWA-RD-79-108. 1980.
16 Horner, R.R. Environmental Monitoring and Evaluation of Calcium Magnesium Acetate (CMA). TRB Washington, D.C. 1988.
17 SOIL. 1957 Yearbook of Agriculture. USDA
18 Bown, M.J.M., Trace Elements in Biochemistry. Academic Press. NYC. 1966.
19 Joffe, J.S., Pedology. Pedology Publications. New Brunswick, N.J., 1948.
20 Bayer, L.D., Soil Physics, 2nd Ed. J. Wiley & Sons, NYC. 1943.
21 Sax, N.I. Dangerous Properties of Industrial Materials, 4th Ed. Reinhold, NYC. 1975.

48

22   U.S. Bureau of Mines Information Cir. 6415. 1930.20.
23   Ehrlich, H.L. The Geomicrobiology of Iron, in GEOMICROBIOLOGY. Marcel
      Dekker, Inc., NYC. 1981.
24   McCrum, R.L. Corrosion Evaluation of Calcium Magnesium Acetate (CMA), Salt
      (NaCl), and CMA/Salt Solutions. Michigan Transp. Com. Research Rpt. No.
      R-1295. 1988.
25   Callahan, M. Deicing Salt Corrosion with and without Inhibitors. TRB,
      68th Annual Mtg. 1989.
26   Nadezhdin, A. et al.   The Effect of Deicing Chemicals on Reinforced
      Concrete. TRB. 67th Annual Meeting. 1988.
27   Schenk, R.U.   Field Deicing Tests of High Quality Calcium Magnesium
      Acetate (CMA).   Final Report to FHWA and Wisconsin Dept. of
      Transportation. 1987.
28   McElroy, A.D. et al. Comparitive Study of Chemical Deicers. TRB, 67th
      Annual Meeting. 1988.
29   Hamilton, G.B. et al. 1987-88 City of Ottawa, Ontario, Canade De-Icer
      Field Trials. TRB, 68th Annual Mtg. 1989.
30   Manning, D.G. and L.W. Crowder.   A Comparative Field Study of the
      Operational Characteristics of Calcium Magnesium Acetate and Rock Salt.
      TRB. 68th Annual Mtg. 1989.
31   Slick, D.S. Effects of Calcium Magnesium Acetate (CMA) on Pavements and
      Motor Vehicles. TRB, 66th Annual Mtg. 1987.
32   Winters, G.R. et al.   Environmental Evaluation of Calcium-Magnesium
      Acetate (CMA). Report no. FHWA/RD-84/094.

Chapter 4

EFFECTS OF CALCIUM MAGNESIUM ACETATE (CMA) ON PAVEMENTS AND MOTOR VEHICLES

David Straup Slick, P.E.
Daedalean, Incorporated, Oakland Ridge Industrial Center, 9017 Red Branch Road, Columbia,
Maryland 21045

4.1   INTRODUCTION
4.1.1 Program Abstract

The corrosion and/or deterioration of highway-related and automotive-related materials by chloride-containing deicing chemicals has become a major economical problem in the United States. The Federal Highway Administration has proposed the use of calcium magnesium acetate (CMA) as an alternative to these chloride-containing chemicals (specifically, sodium chloride). This report describes the comparative effects of CMA and sodium chloride on various highway-related and automotive-related materials. The results of various exposure techniques, followed by extensive additional testing, indicate that CMA is much less deleterious to highway-related and automotive-related materials than sodium chloride.

4.1.2 Program Definition

Each year, millions of tax dollars are spent to reconstruct or repair damaged highways and associated highway structures, and millions of private dollars are spent to repair or replace corroded motor vehicles. Additionally, large sums of money and time are used to assess and correct environmental problems which have resulted from the use of sodium chloride as a deicing chemical.

The Federal Highway Administration (FHWA) has recognized this problem, and has determined that calcium magnesium acetate (CMA) is a suitable alternative deicing chemical. CMA is a mixture of calcium acetate and magnesium acetate, manufactured by reacting acetic acid with dolomitic lime or limestone.

The FHWA wanted a deicing chemical that had similar deicing capabilities to those of sodium chloride, but had less deleterious effects on highway-related and automotive-related materials than sodium chloride. The FHWA also wanted to substantiate previous research that indicated that CMA was less deleterious to highway-related and automotive-related materials than sodium chloride. This research could possibly also indicate previously unknown effects of CMA on the aforementioned materials. This was the basis for the institution of this program.

At this time, there are no plans for CMA to replace sodium chloride as the United States' primary road deicing chemical. However, it is necessary to have a suitable, effective roadway deicing alternative that can be used in areas where sodium chloride may have significant effects on highway-related and automotive-related materials, or great environmental impact.

4.1.3 Basic Technical Approach

The purpose of this program was to determine the comparative effects of sodium chloride and calcium magnesium acetate on highway-related and automotive-related materials.

Test material categories for this program were chosen because they include materials that are commonly used in highway-related and automotive-related applications, and also are representative of other products made from similar materials. 2240 specimens were tested from 81 categories including portland cement concrete, asphalt, traffic marking paint, plastic, and tape, reflective and ceramic pavement markers, pavement marker adhesives, sign reflective sheetings and paints, joint sealant materials, plastic and wood stanchions, bridge bearing materials, drainage pipe, automotive paints, primers, and undercoatings, automotive exterior adhesives, automotive plastics and elastomers, automotive tires and rubber, brake lining materials, automotive steels, hydraulic brake line tubings, automotive aluminum alloys and stainless steels, and automotive combined metals such as chrome, aluminized steel, galvanized steel, terne coated steel, and nickel-zinc coated steel. The specimens were made from new materials to insure accuracy in testing and reproducibility. It must be noted that the purpose of this program was not to evaluate the test materials' susceptibility to corrosion. Rather, the program sought to evaluate the comparative effects of sodium chloride and calcium magnesium acetate on the various materials tested.

The types of tests performed on the specimens were performed as specified by the American Society for Testing and Materials (ASTM). 6048 tests were performed on the aforementioned specimens, and included salt fog exposure, concrete compression, abrasion, and scaling resistance testing, static loading, impact testing, testing of rubber properties, testing of light-related properties, paint evaluations, shear testing, thickness testing, indentation hardness testing, environmental testing, friction testing, adhesion testing, immersion testing, exfoliation testing, and electrochemical testing.

The specimens were exposed to a control solution and four test solutions. The control solution was deionized water and the test solutions were 4 percent by weight sodium chloride acquired from the Maryland State Highway Administration (4% NaCl), 1 percent by weight (1% CMA $_{(FHWA)}$) and 6.8 percent by weight (6.8% CMA

(FHWA)) calcium magnesium acetate (having a one-to-one molar ratio of calcium and magnesium ions) supplied by the FHWA, and 6.8 percent by weight calcium magnesium acetate made by the contractor by mixing pure calcium acetate and pure magnesium acetate, 4-hydrate (both purchased from American Hoechst Corporation) in a one-to-one molar ratio (6.8% CMA (DAI)), all in aqueous solution.

Not all of the aforementioned tests were performed on each type of material specimen. Certain tests were performed on certain types of materials.

In the following Discussion of Results of Testing, specimens are comparatively described as more or less affected by the aforementioned test solutions. It would be unfair to attempt to be more specific about the test results, since each test was only repeated two, three, or four times.

4.2 DISCUSSION OF RESULTS OF TESTING

This section discusses the highlights of the results of testing performed during this program.

4.2.1 Discussion of Results of Testing of Highway-Related Materials

i. Pavement Materials

The compressive strengths of Portland cement concrete and asphalt products were comparatively unaffected by sodium chloride or calcium magnesium acetate solutions. Additionally, scaling did not occur on any asphalt products.

However, some scaling did occur on Portland cement concrete. In general, the scaling was severe on specimens exposed to sodium chloride solutions, moderate on specimens exposed to deionized water, and minimal on specimens exposed to calcium magnesium acetate solutions.

The same magnitudes of scaling occurred on joint sealant material specimens, but the joint sealant materials themselves appeared to be comparatively unaffected.

ii. Guidance Materials

Road marking materials such as paints, plastic, and tape were affected by both sodium chloride and calcium magnesium acetate solutions. Road marking paint specimens exposed to solutions of both compounds exhibited chalking, checking, erosion and flaking. In all cases, the sodium chloride solutions were either more deleterious than the calcium magnesium acetate solutions, or so severely scaled the specimen substrates, that no paint remained to be examined. Plastic and tape specimens were similarly affected. Hardness properties appeared to be uneffected, while whiteness of specimens exposed to deionized water and calcium magnesium acetate solutions appeared to be enhanced.

Similarly, hardness properties of pavement markers appeared to be unaffected. Light-related properties were difficult to analyze, since the specimen sizes and configurations inhibited consistent analyses. Although some pavement marker adhesives exhibited increases and/or decreases in shear strength and impact resistance over exposure for one year, these increases and/or decreases were slight, and it appeared that neither solutions of sodium chloride nor solutions of calcium magnesium acetate affected the adhesives.

Sheetings and paints used in directional and informational highway signs that were exposed to sodium chloride and calcium magnesium acetate solutions all appeared to retain adhesion to their aluminum substrates. Light-related properties of the sign materials were either relatively unaffected, or similarly affected. Physical degradation of sign paints did not occur.

The light-related properties of plastic road delineators, or stanchions, were either relatively unaffected, or similarly affected by the sodium chloride or calcium magnesium acetate solutions. Again, specimen sizes and configurations inhibited consistent analyses. Likewise, hardness properties were also unaffected. Impact resistance of plastic stanchions, and shear strengths of wood stanchions appeared to be unaffected.

iii. Construction Materials

Hardness properties and impact resistance of bridge bearing materials were unaffected by solutions of sodium chloride or calcium magnesium acetate.

Crushing strengths of concrete, bituminous coated corrugated metal, and plastic pipes were also unaffected by solutions of sodium chloride or calcium magnesium acetate.

4.2.2 Discussion of Results of Testing of Automotive-Related Materials

i. Automotive Paints, Coatings, and Adhesives

The adhesive properties, light-related properties, and hardness properties of automotive paints, primers, and undercoatings appeared to be relatively unaffected by solutions of sodium chloride or calcium magnesium acetate. Additionally, no blistering occurred on any of these specimens.

Some paint, primer, and undercoating breakdown, and subsequent rusting occurred on both scribed and unscribed specimens. In all cases, however, sodium chloride caused more coating breakdown, and more severe rusting, than solutions of calcium magnesium acetate.

Some automotive exterior adhesives exhibited increases and/ ordecreases in shear strength over exposure for one year, but these increases and/or decreases were slight, and it appeared that neither solutions of sodium chloride nor solutions of calcium magnesium acetate affected the adhesives.

ii. Automotive Plastics and Rubber

Solutions of sodium chloride and calcium magnesium acetate caused slight changes in the light-related properties of automotive hard plastics, elastomers, and transparent lens materials, but these changes usually occurred consistently for all materials. One exception was the haze exhibited by the Automotive Transparent Lens Material 1 specimen exposed to the solution of 6.8% CMA $_{(FHWA)}$. However, since the same respective specimens exposed to the other calcium magnesium acetate solutions appeared to be unaffected, it was determined that the subject specimen had most likely been damaged prior to the performance of the Test for Haze and Luminous Transmittance of Transparent Plastics (ASTM D1003). Additionally, with the exception of transparent lens materials, most automotive hard plastics and elastomers on automobiles are either painted, or are not subjected to light. The specimens were therefore considered comparatively unaffected. Accordingly, the hardness properties and impact resistances of these specimens were also considered comparatively unaffected.

The hardness properties of automotive tire materials and rubber compounds exposed to solutions of sodium chloride and calcium magnesium acetate appeared to be unaffected. Compression sets of all specimens increased, while ultimate tensile strengths remained unaffected. The 300-percent moduli and tear resistances of the same specimens were only slightly affected. Some surface cracking appeared on these specimens, but remained consistent throughout the specimen type. Surface cracking therefore was not attributable to the exposure to solutions of sodium chloride or calcium magnesium acetate, but rather to the specimen composition.

iii. <u>Automotive Brake Systems</u>

Solutions of sodium chloride and calcium magnesium acetate appeared to have no effect on the hardness properties of the automotive brake lining materials. However, specimens exposed to solutions of sodium chloride consistently exhibited lower static and kinetic coefficients of friction.

In general, automotive hydraulic brake line tubings exposed to solutions of sodium chloride exhibited greater weight losses due to corrosion and more rusting and deterioration, than the same specimens exposed to solutions of calcium magnesium acetate. No pitting corrosion was observed on any specimens.

iv. <u>Automotive Metals</u>

Automotive steels exposed to solutions of sodium chloride consistently exhibited greater weight losses due to corrosion, more pitting corrosion, and more general corrosion than the same respective specimens exposed to calcium magnesium acetate. Specimens exposed to solutions of sodium chloride were also susceptible to crevice corrosion, while the same respective specimens exposed to solutions of calcium magnesium acetate were not.

Automotive aluminum alloys exposed to solutions of sodium chloride consistently exhibited greater weight losses due to corrosion, and more pitting corrosion than the same respective specimens exposed to calcium magnesium acetate. General corrosion of specimens was either minimal for all solutions, or slightly enhanced by solutions of sodium chloride. Susceptibility to crevice corrosion varied from specimen to specimen, and solution to solution.

Automotive stainless steels exposed to all solutions exhibited insignificant weight losses and minimal differences in general corrosion. Pitting corrosion was also minimal but did occur in some specimens exposed to solutions of sodium chloride. Specimens exposed to solutions of sodium chloride were also susceptible to crevice corrosion, while the same respective specimens exposed to solutions of calcium magnesium acetate were not.

Automotive combined metals exposed to all solutions exhibited minimal pitting corrosion. However, greater resistances to pitting corrosion were prevalent in specimens exposed to the calcium magnesium acetate solutions, than in specimens exposed to the sodium chloride solutions. Susceptibility to crevice corrosion varied from specimen to specimen, and solution to solution. Chrome-plated steel and stainless steel alloys were only slightly affected by all solutions. Aluminized steel exposed to solutions of sodium chloride exhibited greater weight losses than the same respective specimens exposed to the calcium magnesium acetate solutions. Galvanized steel and terne-coated stainless steel were equally affected by all solutions. Results varied for exposures of nickel-zinc alloy coated steel.

It must be noted that the electrochemical tests were immer-sion-type tests which were conducted in a nitrogen environment. Thus, the resulting susceptibilities to general corrosion and corrosion resistance were of limited importance, since, under actual highway conditions, automotive metals are in contact with air-saturated environments. It must also be noted that in cases where no general corrosion was detected, as was the case for many metals exposed to deionized water and solutions of sodium chloride, the rate of corrosion would increase exponentially with the degree of aeration. The principle main danger to automotive-related materials under actual highway conditions is their susceptibility to localized corrosion (pitting and crevice corrosion). As previously stated, pitting and crevice corrosion were generally more prevalent in specimens exposed to solutions of sodium chloride, than in specimens exposed to solutions of calcium magnesium acetate.

4.3 CONCLUSIONS

Asphalts, plastics, elastomers, ceramics, wood, sign sheetings and paints, rubber compounds, sealers, and adhesives appeared to be either unaffected by solutions of sodium chloride or calcium magnesium acetate, or similarly affected.

Road marking paints were affected by both sodium chloride and calcium magnesium acetate solutions. However, it was obvious that sodium chloride solutions were much more deleterious. Road marking plastics and tapes were affected by both solutions because of the solutions' effects on the Portland cement concrete substrates. Again, specimens exposed to sodium chloride solutions were more severely damaged.

Portland cement concrete was affected by both sodium chloride and calcium magnesium acetate solutions. While the calcium magnesium acetate solutions did minimally affect Portland cement concrete, the sodium chloride solutions' effects were much more severe.

Automotive paints and coatings exposed to calcium magnesium acetate solutions exhibited some breakdown, but those same respective specimens exposed to sodium chloride solutions exhibited far greater breakdowns.

Automotive brake lining materials exposed to sodium chloride solutions consistently exhibited slightly lower static and kinetic coefficients of friction than the same respective specimens exposed to calcium magnesium acetate solutions, while automotive hydraulic brake line tubings exposed to sodium chloride solutions consistently exhibited more severe corrosion than the same respective specimens exposed to calcium magnesium acetate solutions.

In general, solutions of sodium chloride cause more severe corrosion in automotive steels, aluminum alloys, stainless steels, and combined metals, than solutions of calcium magnesium acetate. This is true for general corrosion, as well as localized corrosion.

Also, regarding general corrosion, under nitrogen conditions, specimens exposed to solutions of sodium chloride exhibited no passivation, while specimens exposed to solutions of calcium magnesium acetate did. In actual highway (aerated) exposures, since most sodium chloride solutions exhibit no passivation, corrosion rates of metals caused by exposure to sodium chloride solutions will increase much more rapidly than corrosion rates caused by solutions of calcium magnesium acetate.

Regarding localized corrosion, it can be concluded that in most metals, with the exception of aluminum, solutions of sodium chloride caused more pitting and crevice corrosion than solutions of calcium magnesium acetate.

4.4   RECOMMENDATIONS

Results of testing of highway-related materials indicated that the effects of corrosion by salts of sodium chloride and calcium magnesium acetate could be relatively accurately predicted. This condition is probably not as true for automotive-related metallic materials, since new technology is constantly providing new alloys with enhanced and varying properties. Periodic studies of the susceptibility to corrosion of these new alloys must be performed to ensure the use of the best corrosion resistant products.

Additionally, further analyses of the affects of sodium chloride and calcium magnesium acetate on the light-reflectance properties of sign sheetings and paints is recommended.

As previously stated, calcium magnesium acetate was never intended to replace sodium chloride as the United States' primary road deicing chemical. However, it must be noted that, in light of new manufacturing techniques of calcium magnesium acetate leading to lower production costs, and its much lower corrosive tendencies, CMA has the potential for significant impact on the commercial and private consumer deicing markets.

REFERENCES

1   D.S. Slick, P.E., "Effects of Calcium Magnesium Acetate (CMA) on Pavements and Motor Vehicles," U.S. Department of Transportation Federal Highway Administration Report Number FHWA/RD-87/037 (April 1987).
2   D.S. Slick, P.E., "Effects of Calcium Magnesium Acetate (CMA) on Pavements and Motor Vehicles," National Research Council Transportation Research Board Transportation Research Record 1157 (1988), 27-30.

Chapter 5

ENVIRONMENTAL EVALUATION OF CALCIUM MAGNESIUM ACETATE

Richard R. Horner, Mark V. Brenner, Richard B. Walker, Richard H. Wagner
University of Washington, Seattle, Washington, 98195

5.1 ABSTRACT

Laboratory and controlled field plot experiments were performed to determine environmental transport, fate, and effects of calcium magnesium acetate (CMA), a potential highway deicing agent. These experiments concerned CMA movements, reactions, and effects in soil, vegetation, surface water, and groundwater. They involved both testing in the laboratory and at controlled field plot/pond systems.

Preliminary modeling as a basis for experimental planning estimated typical concentrations in highway runoff or spray at order of magnitude 10^2 ppm CMA, while maximum concentration could be an order of magnitude higher. Laboratory adsorption/desorption experiments on three soils indicated a tendency for mobilization of native copper, zinc, and aluminum through ion exchange with the applied major cations. These experiments also predicted that the majority of acetate reaching soils could be mobile rather than bound. Observations in field plot soil failed to confirm metal mobilization, but some elevated metals were measured in shallow groundwater. Acetate appeared in both soil water and groundwater, but at far lower than applied concentrations. CMA decomposition rates in soils were studied in laboratory incubations at 2 and 10 C. A high level of decomposition was achieved within two weeks at the higher temperature, but at 2 C up to four weeks were required to degrade the material fully. In greenhouse pot tests, various herbaceous and woody plant species withstood root-zone CMA concentrations up to 2500 ppm. The osmotic stress of larger concentrations killed seedlings, unless planting was delayed for a day or two, while cations were adsorbed in the soil. Field applications by spray and flood did not affect the yield, cover, vigor, rooting, or external damage symptoms of grasses or tree seedlings and saplings. Oxygen depletion due to CMA addition to water was demonstrated in both the laboratory and the field. Approximately 10 ppm CMA temporarily reduced saturated dissolved oxygen levels in field ponds by about half. This ability for relatively low CMA concentration to deplete dissolved oxygen is the leading potential environmental impact of concern identified by the research. Laboratory bioassays showed CMA to enhance bacterial in preference to algal growth, but suppression of the phytoplankton community was not observed in the field ponds. Rainbow trout hatching and survival in aerated laboratory bioassay chambers were high up to CMA concentrations in excess of 1000 ppm, but fish could not survive removal of the air supply if substantial CMA remained in the vessels. Despite some depletion, oxygen was always high enough in the field ponds to maintain bluegill sunfish and fathead minnow populations. Conclusions of the research were applied to formulate interim environmental protection guidelines for CMA use.

5.2   INTRODUCTION

Evidence has gathered over the years that highway deicing salts, primarily sodium chloride (NaCl), negatively impact the natural environment and constructed facilities in a variety of ways. Roadside trees and other vegetation have been damaged by direct impaction and soil accumulation of chlorides (e.g., 1-6). Soil property changes also have accompanied salt accumulation (7, 8). Natural waters may be affected by density stratification, the promotion of nuisance blue-green algal blooms, and ion exchange with toxic mercury sequestered in the sediments (4, 5, 9). A sufficient increase in the ionic concentration can alter the ionic balance in aquatic organisms. High chloride concentration imparts unpleasant taste and odor to drinking water, while sodium (Na) endangers hypertensive human consumers (10, 11). Chloride accelerates metal corrosion, causing substantial economic damage to public property and private vehicle owners (12). Bridge decks and underground utilities are especially vulnerable. Portland cement also exhibits relatively poor resistance to chloride salts (13).

Recognition of the numerous drawbacks of continued use of sodium and calcium chlorides for highway deicing has stimulated substantial thought and effort to identify alternatives. Dunn and Schenk (14) reported on a comprehensive investigation to identify alternative highway deicing agents. They screened numerous compounds on the basis of properties, cost, availability, toxicity, and potential to affect the natural and man-made environments. Two candidates, methanol and calcium magnesium acetate (CMA), were selected. The latter has received most of the attention, probably because of its greater persistence.

5.3   PREVIOUS INVESTIGATIONS OF CALCIUM MAGNESIUM ACETATE

With identification of CMA as a promising alternative to chloride salts, the Federal Highway Administration (FHWA) sponsored several research efforts to document its environmental acceptability, manufacturing technology, and technical and economic feasibility. Investigation of CMA manufacture established that the most feasible method is to react acetic acid with dolomitic lime (15). For performance evaluation in the field, small lots of CMA were distributed among 24 states and in larger volumes to the States of Michigan and Washington for highway deicing evaluations under actual conditions. Results reported by several states agree that CMA is capable of achieving a degree of bare pavement equal to that realized with sodium chloride but that the latter acts slightly faster (15).

The California Department of Transportation (CalTrans) performed the most extensive investigation of CMA environmental effects prior to the research reported in this paper. Using CMA produced from reagents, the CalTrans work included aquatic bioassays on fish, zooplankton, and phytoplankton; assessment of the effects of irrigation and foliar CMA applications to terrestrial plants; and an evaluation of CMA effects on soils (16). Static bioassays on fathead minnow (Pimephales promelas) and rainbow trout (Salmo gairdneri) demonstrated that calcium acetate alone and a mixture of acetates were less lethal than NaCl. In chronic tests with rainbow trout, a continuously maintained 5000 mg/l concentration of equimolar CMA slightly delayed hatching and was associated with some hatching difficulty in a small number of cases but did not reduce its ultimate success. Lower CMA concentrations caused no significant hatching delays or difficulty. In acute and chronic bioassays on the zooplankter Daphnia magna, observable effects occurred between 125 and 250 mg/l of CMA but not until approximately 2500 mg/l of NaCl. Separate calcium and magnesium acetates and an equimolar mixture all induced significant inhibition of the phytoplankters Selenastrum capricornutum and Anabaena Flos-aquae at concentrations above 83.55 mg/l, whereas NaCl did not significantly reduce growth at any concentration up to 1000 mg/l. In irrigation experiments exposing 18 species of terrestrial plants to CMA and NaCl, nine species were more severely damaged by NaCl and one by CMA. When sprayed on foliage, the NaCl application caused greater damage than CMA to 17 of the 18 species. CalTrans exposed soils from various parts of the United States to CMA solution in laboratory lysimeters. The principal result of this study was that CMA is capable of extracting substantial amounts of iron, aluminum, sodium, hydrolyzable orthophosphate, and potassium from soil.

Elliott and Linn (17) assessed CMA impact on soil copper (Cu) and zinc (Zn) mobility in batch and flow-through column leaching experiments. In batch soil suspensions maintained at several pH levels in the range 3-7, CMA solutions substantially enhanced Cu and Zn desorption from a metal-enriched soil. In the columns, bound metal ions were initially displaced, but increasing base saturation and pH ultimately resulted in net suppression of metal efflux. The investigators speculated that CMA application might stimulate metal translocation at the outset, but that sustained use should reduce metal flux.

5.4 RESEARCH OBJECTIVES AND GENERAL APPROACH

5.4.1 Objectives

The CalTrans research was performed using CMA produced from reagent-grade chemicals. It was desired to subject a potential production-run grade of CMA to laboratory testing of environmental effects. No previous work investigated the environmental transport and fate of CMA, nor its effects on natural communities of organisms. Therefore, it was also desired to experiment on natural media and communities. The specific research objectives involved the transport and transformations of CMA in the environment and its effects on soil and soil water properties, plants, ground and surface water quality, and aquatic biota. The results were intended to serve as a basis for developing interim guidelines for using CMA in actual highway practice and for preparing environmental impact assessments regarding CMA use. To fulfill the objectives, a program of laboratory and controlled field plot studies was outlined. This paper presents representative results and conclusions, while the project completion report (18) contains full details.

5.4.2 Laboratory studies

Laboratory studies, for the most part, were assays conducted according to standard procedures. The philosophy in laboratory assays was to create a range of CMA exposures on soils, vegetation, and aquatic biota that would produce effects ranging from none to severe. As a guide for experimental planning, preliminary modeling was used to predict typical and worst-case exposures in actual highway applications. All laboratory experiments were conducted with untreated controls, as well as CMA-treated vessels, and were replicated.

Table 5.1 lists constituents measured in water samples and the methods and equipment employed. Volatile fatty acids (VFA) include acetic, propionic, butyric, and other acids of higher carbon number. VFA were determined by direct aqueous injection of acidified samples into the Hewlett-Packard gas chromatograph with flame ionization (FID) detector. Acidification with formic acid to pH < 2 drives the chemical equilibrium toward the acid species of the salts actually present in samples for detection by gas chromatograph.

5.4.3 Controlled Field Plots

Controlled field plots were small terrestrial and aquatic communities established in the natural environment, but with a greater degree of control than would be afforded by full-scale systems at highway locations. One purpose of experimentation at controlled field plots was to verify laboratory findings in natural conditions. The plots also permitted extending the study to investigate environmental transport of CMA on the surface and through the soil, and the interrelationships of organisms living in diverse communities.

Controlled plots were located at the University of Washington's Pack Forest in the Cascade Mountain foothills approximately 90 km southeast of Seattle, Washington. At elevation 366 m and receiving cold air drainage from nearby Mount Rainier, the site experiences considerably colder winter conditions than the Puget Sound lowlands.

Controlled plot experiments were designed to simulate distribution of CMA from highway surfaces to adjacent soils, vegetation, and water. The plots were treated to distinguish the effects of spray versus flood (runoff) treatment, as well as those of CMA solution versus water not containing CMA (experimental controls). In order to provide duplicate plot systems for the various treatments, eight areas 10 m^2 in area were prepared as identically as possible on slopes of 8 percent and about 7 m apart. Each area consisted of a plot for terrestrial experiments, a pond for aquatic experiments, and a trench system to collect runoff from the plot and channel it to the pond. Figure 1 illustrates a controlled plot schematically.

Controlled plots were seeded in June, 1985, with red fescue (Festuca rubra) and Dutch white clover (Trifolium repens). Also planted were seedlings of the following species: buckhorn plantain (Plantago lanceolata), buttercup (Ranunculus acris), Douglas fir (Pseudotsuga menziesii), and red maple (Acer rubra).

Suction lysimeters (Soil Moisture Equipment Corporation 1900L series) and a well point (perforated cast iron pipe) were installed in each plot to sample soil water and groundwater, respectively. Lysimeters were placed for sampling at 30 and 60 cm depths, while well points were driven 3 m deep.

Ponds were excavated approximately 3.6 m in diameter and 1 m deep to provide 10 m^3 water volumes. They were lined with 19 mil thick reinforced polyvinyl chloride to provide a seal. Mixed-size gravel and cobbles were placed on the beds of the ponds to provide a natural substrate for aquatic organisms. The ponds were initially stocked with several species each of algae, zooplankton, and invertebrates obtained from laboratory sources, as well as bluegill sunfish (Lepomis macrochirus) and fathead minnows (Pimephales promelas) from hatcheries. Smaller life forms also colonized from the supply pond and the surrounding environment.

Water supply for ponds and site treatments came by gravity from a former farm pond. At the site, water could be directed to the plots (for irrigation), the ponds, or into fiberglass CMA solution and water supply tanks through a valving system.

Near the field plots, 0.5 m^2 duplicate boxes were planted with Douglas fir, balsam fir (Abies balsamina), and red maple seedlings and treated by spray with water (controls) and CMA during the second winter of experiments. These boxes were necessitated when poor tree seedling survival on plots resulted from shading and competition by vigorously growing grasses.

To provide observations of CMA effects on larger tree specimens, two sets of four Douglas firs each about 1 m in height were transplanted to the field site. Similar sets of red alder (Alnus rubra) about 3 m high growing near the field plots were tagged. These sets were sprayed with CMA solution or water during the second winter.

5.5   FORECAST OF CMA TRANSPORT FROM HIGHWAYS

Lacking measurements of CMA in actual highway drainage, an initial predictive analysis of its transport in runoff was made to aid in planning experiments. As a basis for modeling typical annual CMA application to a multi-lane road, the following conditions were assumed: (1)   250 lb/lane-mile/storm (69 kg/lane-km/storm), (2) 20 storms/year, and (3) 4 lanes in one direction of travel draining to the shoulder.

The annual loading equivalent to these conditions would be 10 tons per linear mile of highway (5.5 metric tons/km), or about 4 lb per linear foot (6 kg/m). It was assumed that this entire loading would be removed from the highway within the annual period.

In addition to cumulative annual mass loading, there was interest in modeling CMA concentration (mass per unit volume) in highway spray or runoff, in order to represent conditions that may affect receptor organisms at any given time in a typical winter or early spring period. Assuming minimum winter precipitation is 12 inches (30.5 cm), water equivalent, on a 50-foot (15.2 m) wide right-of-way zone, the water volume available to dilute CMA would be 50 ft^3 per linear foot of highway (4.6 m^3/m). Dilution of 4 lb of CMA in 50 ft^3 of water would result in a concentration of 1282 mg/l.

This CMA concentration estimate implicitly assumes that the maximum annual application in a relatively arid area is washed off in a single melting event. Lower concentrations would result from different assumptions about dilution volume and melting pattern. More typical concentrations likely would be an order of magnitude less than the estimate above. On the other hand, washoff of a large CMA application by a very intense small storm could result in a concentration of perhaps 5000 ppm. Therefore, it was concluded that typical CMA concentrations to expect in actual highway runoff or spray are order of magnitude 10^2 ppm, while a normal upper limit would be order of magnitude 10^3 ppm and a worst case condition would be 5000 ppm.

The estimated concentrations were used to establish ranges of exposure in laboratory assays. The estimated annual mass loading is equivalent to 450 g

CMA/m^2 on a controlled field plot (total 4.5 kg/plot), which was the loading rate used in the first of two winters of application (1985-86). This quantity was delivered to the CMA treatment plots in five equal applications totaling 1500 liters of water, equal to runoff from 50-80 inches (127-203 cm) of snow (concentration was 3000 mg CMA/1). Controls received equal applications of plain water. To investigate a higher loading condition, the mass of CMA applied was doubled (to 900 g/m^2) in the second winter by doubling the hydraulic loading (concentration remained 3000 mg/1). The soil boxes containing tree seedlings and the larger tree stands received CMA concentrations and areal and hydraulic loadings equal to the plots.

5.6   CHARACTERISTICS OF CALCIUM MAGNESIUM ACETATE

The CMA used in this research program was produced by the Center for Biological Resource Recovery at the University of Georgia under a Federal Highway Administration contract.   It was made from corn starch hydrolysate by fermentation using Clostridium thermoaceticum, with the addition of a dolomitic limestone (dolime) slurry to the fermentor. At the end of fermentation, more dolime was added to bring pH to approximately 8.   CMA was shipped to the University of Washington in liquid form in new, plastic 55-gallon drums, either during a period of low temperatures or in a refrigerated truck. Upon receipt, it was placed in cold storage until it was dried.

Drying was performed to provide a material for experiments like that expected to be marketed eventually, as well as to realize several experimental advantages. Drying involved holding CMA solution in a shallow, open pan on a 55 C hot plate in a greenhouse during the summer. After a nearly dry product was obtained, it was placed in a 70 C oven for one hour to drive off any remnant moisture. This procedure avoided any volatilization and loss of acetate. After drying, the material was immediately placed in close, clean, dry containers. The various dried lots were composited to form a uniform CMA supply for experimental work.

The dry CMA was redissolved in water and analyzed for VFA and metals concentrations. Table 5.2 presents the results. Properties of the three lots differed very little. With the exception of iron content, they also differed little from the original aqueous material. There was no apparent tendency for drying either to concentrate or dilute metals.

5.7   PHYSICOCHEMICAL REACTIONS INVOLVING CMA IN SOILS
5.7.1 Test Soil Properties

Four soils were selected to be representative of roadsides. These soils were characterized by testing texture classification (ASTM D 422-63), organic content (by loss on ignition method), and cation exchange capacity and exchangeable cations (both with an ammonium acetate extraction procedure), and calcium carbonate content (by sodium hydroxide titration of hydrochloric acid extract). Table 5.3 summarizes soil properties.

5.7.2 Experimental Methods

A laboratory experiment was run to determine the extent of acetate and metals adsorption by soils and subsequent desorption by water flushing. This experiment was performed by applying duplicate 30 ml volumes of CMA solutions of various concentrations (0-5000 mg/l for metals analysis; 0-500 mg/l for acetate analysis) to 10 g of Lee Forest, Pack Forest, and Vantage soils in centrifuge tubes. A wider concentration range was applied for the metals study than for acetate because of the much greater cation than anion exchange capacity of soils. Each tube was stoppered with a freshly washed rubber cork and shaken every ten minutes for one hour. The tubes were then centrifuged at 5000 rpm for 10-to-15 minutes. The liquid in each tube was filtered using clean, 0.45 micron Millipore filter paper, and the filtrate was analyzed for acetate and metal concentrations. To determine desorption, 30 ml of distilled water was added to the solids. Shaking, centrifuging, filtration, and analyses were repeated.

Acetate and metal mobilities in the controlled field plot soils were investigated by sampling the soil water collected in suction lysimeters and the groundwater from well points. Organics were preserved in soil water samples by chloroform injected into the lysimeters, but there was no attempt at preserving organics in groundwater entering the well points.

5.7.3 Results

Table 5.4 summarizes the adsorption and desorption of metals measured in one soil in the laboratory experiment. Results for the other soils were very similar. The total quantity of metals applied in the 5000 mg CMA/l solution is less than 0.2 milliequivalents per gram of soil (meq/g), while the cation exchange capacities of the soils are approximately 1-2 meq/g. This observation suggests that the amount of metal adsorption should increase approximately linearly with the concentration of the CMA solution, since the metal concentrations in the CMA solutions are much lower than the maximum possible levels that can be adsorbed. Examining the computed adsorption data in the tables confirms this behavior for most metals studied. Exceptions to linear adsorption were observed for copper, zinc, and aluminum (Al) at all CMA concentrations in all soils. This result suggests that these ions were released from the soil rather than adsorbed. Apparently, they were replaced by calcium (Ca) and magnesium (Mg).

Table 5.5 summarizes the adsorption and desorption of acetate. The Vantage soil (lowest in organics) had the lowest anion exchange capacity, being able to adsorb less than 30 percent of the applied acetate. In other soils approximately 40-65 percent (mean 47.8) was adsorbed. The anion exchange capacity of soils is normally below 1 meq/100 g, which corresponds to 0.6 mg acetate/g soil. Acetate adsorption values in this experiment ranged up to 0.327 mg/g; therefore, maximum adsorption capacity was not reached with 500 mg CMA/l addition. It may be presumed that larger quantities could be adsorbed, at least in the more highly organic soils.

With the exception of two values, 30-55 percent (mean 38.9) of the adsorbed acetate was subsequently desorbed and flushed from the soil. Working with mean values, on an overall basis, then, approximately 70 percent of applied acetate could eventually be mobile, even in relatively organic soils. In a soil like Vantage, perhaps 85 percent or more could be transported in soil water.

Table 5.6 gives means of metal concentrations measured in lysimeters at 60 cm depth in controlled field plots over four periods before, during, and after treatments. There was no clear and consistent pattern of Ca, Mg, or K increase in the soil water at either depth during or after treatments. Control plot soil water was just as likely to experience increases in the major cations during the winter, probably because of increased infiltration. The same general statements apply to the trace metal ions; i.e., there was no explicit tendency overall despite some isolated readings that suggested trace metal elevation below CMA plots.

Figure 2 illustrates acetate variation in soil water over time. It generally took two-to-three weeks after the initial CMA treatment of the season for acetate to become detectable at depth. It then disappeared from the soil water within about the same period after the final treatment. During the first treatment season, little difference between spray and flood plots in soil water acetate was evident. In the second season, however, acetate did not rise as high in flood plot as in spray plot lysimeters.

Table 5.7 gives means of metal concentrations measured in shallow groundwater below controlled field plots over four periods before, during, and after treatments. Metals exhibited greater changes and some stronger trends in groundwater over the period of observation than they did in soil water. Still, the trends were not without exception. Ca increased notably below CMA Plots 3 and 8, but not 1 and 5, in the second winter. Elevation in Fe and Al was seen at three of the four CMA plots (excluding 5). Zn was substantially higher in the second winter at Plot 3 but not others. The remaining metals were relatively stable.

Figures 3 and 4 illustrate variations in specific conductivity and acetate in groundwater beneath CMA and water plots, respectively, over time. There was approximately a two-month lag between treatments and the appearance of acetate in the saturated zone, where it persisted until close to the next treatment period. It sometimes appeared in groundwater drawn from wells in control plots long after treatments. Obviously, the speed of groundwater movement was sufficient to cover the approximately 10 m between CMA and control plot wells in less than a year. It is notable that small amounts of acetate remained undegraded in the saturated zone after months.

5.7.4 Discussion

A result of potential environmental significance apparent in the laboratory data is the tendency for displacement of some trace metals from soils when contacted by CMA. This tendency was also noted in the CalTrans soils experiments (16), but others have found it to decrease as Ca and Mg raise soil base saturation (17). Displaced metals would be mobilized in the soil water and would have a tendency to be transported to surface water or groundwater. Al has implications for drinking water quality, and Al, Cu, and Zn can be toxic to aquatic life.

Field results provide an opportunity to check the existence and extent of the laboratory-based finding. Groundwater exhibited elevated Fe and Al, and occasionally Zn, after CMA treatments. However, lysimeters in the unsaturated zone did not exhibit this tendency. It remains a matter for further investigation, especially in roadside soils subjected to actual CMA applications.

Also of potential interest from the environmental impact standpoint is the indication from laboratory testing that the majority of applied acetate could be mobile in water rather than bound to soil. This finding means that during periods of high runoff and soil percolation at low temperatures, when biological activity is at a minimum, there may be a high potential to carry acetate to surface water or groundwater. There, decomposition would tend to reduce dissolved oxygen and diminish water quality.

The field measurements again offer an opportunity to check the realism of laboratory findings. When acetate was measured in soil water at 30 and 60 cm depths, it never exceeded 8 percent of the concentration applied on the surface, despite the fact that at least 90 percent of the water normally percolated. When seen in groundwater at 3 m, acetate was never higher than 3 percent of the surface concentration, except in one measurement where it rose to 10 percent. Clearly, mechanisms in natural soil columns assist the removal of acetate to a greater degree than predicted by laboratory-scale work. These mechanisms probably include biodegradation and chemical reactions in addition to adsorption working in concert.

This finding, to a large extent, relieves the concern about high acetate mobility in soils. Nevertheless, its demonstration in the laboratory shows the potential for extensive transport with high water flushing. Coarse soils with low exchange capacities would be most subject to passing acetate to a surface receiving water or groundwater.

5.8 CMA DECOMPOSITION IN SOILS

5.8.1 Experimental Methods

Laboratory batch experiments were performed using the four soils whose properties were summarized above. Twenty grams of air-dried, sieved soil were placed in 60 x 15 mm sterile, plastic, disposable Petri dishes. CMA was added in solution form, by pipet. The liquid volume that would moisten the soil to field capacity was used. A range of CMA concentrations (0, 125, 1250, 2500, and 5000 ppm) was applied to represent CMA in highway runoff under varying dilutions.

The Petri dishes were placed in sealed containers. Acetate decomposition was determined by sampling CO_2 in the container head space. Respiration experiments were done at 2 C and 10 C to simulate winter and spring soil temperatures, respectively.

Gas samples were obtained through a septum port glued into the lids of the containers. Vacutainer tubes (4 ml, BD6490) were used to collect gas samples. CO_2 was measured by isothermal (80 C) gas chromatography (Perkin-Elmer Sigma 300) using a thermal conductivity detector.

5.8.2 Results

Figure 5 illustrates CO_2 production over time at 2 C in the four test soils. There was no significant difference between controls and 125 ppm CMA treatments in any soil or on any day. Peak respiration rates tended to increase with increasing CMA concentration. A longer lag in reaching peak rate was generally evident in the more concentrated treatments. Development of the microbial community was probably hindered by the osmotic effect of high cation additions.

Figure 6 compares peak respiration rates versus CMA loading at the two incubation temperatures. The peak respiration rate increased with loading for each of the four soils, at both 2 and 10 C. No saturation of rate with loading was observed up to 3000 g CMA/g soil. There was no significant difference in respiration rates at 0 and 38 g CMA/g soil.

The maximum respiration rates measured in the four soils were ordered as follows: Lee > Puyallup > Pack Forest > Vantage. At each loading the peak respiration rate was directly proportional to the percent organic matter found in the soil sample.

Development of maximum respiration rates lagged by 13-29 days after the beginning of incubation at 2 C and by 7-9 days at 10 C. Acetate was measured at the end of the experiment, when respiration had returned to the baseline level. Measured acetate concentration was below the detection limit.

5.8.3 Discussion

The principal result of these experiments that has implications for the use and environmental impact assessment of CMA is that acetate decomposition in soils can be incomplete, even after a period of weeks. This observation is especially true at near-freezing temperatures. Very elevated CMA concentration can create a lag in respiration while the osmotic effect of high cation concentrations is mitigated or decomposers adapt. If undegraded acetate should be released to a receiving water by surface runoff or interflow, decomposition in water could diminish dissolved oxygen. Likewise, if acetate reaches groundwater and then decomposes, it could deteriorate its quality. If the groundwater is aerated, aerobic decay could reduce oxygen. If oxygen is already depleted, anaerobic decomposition could release indesirable by-products.

5.9   EFFECTS OF CMA ON TERRESTRIAL VEGETATION
5.9.1 Experimental Methods

The initial investigation of CMA effects on plants involved pot tests performed in a greenhouse. These tests were conducted in two soils: (1) the sandy loam from Puyallup, WA, and (2) the Pack Forest (Murphy's Ranch) soil mixed 2:1 with Indianola sand, also from Pack forest. The mixture was used to provide data from soils of substantially different texture. A portion of soil (1000 to 1200 g, depending on the density of the particular soil) was mixed with a suitable amount of NPK fertilizer to stimulate good growth. The soil was then further mixed with the desired amount of dry CMA. Then, the soil was poured into a standard 5-inch plastic pot, moistened with distilled water, and planted with one or more seedlings. CMA quantities were 0 (control), 0.5, 1.0, 2.0, and 4.0 g/kg soil. These amounts corresponded to soil solution CMA concentrations of 0, 1250, 2500, 5000, and 10,000 ppm, respectively, when the soil was moistened to field capacity.

In most instances the planting was done shortly after the moistening, but in some treatments planting was delayed for 48 hours to permit cation exchange reactions between the CMA and the soil to take place. Species used for these trials were Romaine lettuce, buckhorn plantain (Plantago lanceolata), sunflower (Helianthus annus), Douglas fir (Pseudotsuga menziesii), balsam fir (Abies balsamina), and red maple (Acer rubra). After a suitable growing period, the plants were harvested and dry weights of shoots and roots recorded.

Controlled field plots were established as described above. On July 9, 1986, following one winter's treatments, a biomass harvest was made by dividing the area into eight rectangles and then cutting all vegetation at 5 cm from the soil level. Prior to this total harvest, the individual buttercup and plantain plants were harvested (but central bud and developing young leaves were left intact). In all cases the material was dried and then weighed.

In September, 1986, a partial harvest was made for determination of relative cover of the grass, clover, and other plants. Four locations on each plot were randomly identified for harvest by blindly tossing a 10 cm ring. All vegetation rooted within the ring was cut at 5 cm from the soil level, returned to the greenhouse, dried, and separated. Collections were then weighed to determine relative representation of grass, clover, and other species in the biomass.

Tree seedlings in boxes and stands of larger trees were examined visually for symptoms of injury or stress.

5.9.2 Results

Table 5.8 summarizes shoot yields measured in the pot tests. The principal points to be noted are:

1. The addition of 0.5 g CMA stimulated growth in most cases, although this stimulation was large enough for statistical significance only in a few instances, mostly with the coarser soil (Pack/sand).
2. Most plants tolerated 1.0 g CMA, and growth was mostly equal to or a little greater than in the controls. However, Douglas fir was sharply retarded by this addition on the Puyallup soil and died in the Pack/sand. Sunflower died in the Pack/sand soil with this addition.
3. At 2.0 g addition, Douglas fir died in both soils, and sunflower died in the Pack/sand soil. Most of the species survived this addition, with yields depressed in plantain but increased in balsam fir and red maple.
4. All plants died in soil to which 4.0 g CMA was added, if planting occurred at once. However, all plants survived with this much CMA if planting was delayed 48 hours after the CMA addition.

At the harvest of July 1986, all controlled field plots showed a vigorous growth of fescue and clover, with a small admixture of volunteer weedy species. The growth of the grass and clover had become so tall that the plantain, buttercup, Douglas fir, and maple seedlings were severely shaded; their harvest in sufficient numbers for comparisons of treatments was impractical. The values for the biomass of major species in the plots are given in Table 5.9a. There were no statistically significant differences between these yields for the eight plots, although there was a trend toward higher yields with flood applications of either CMA or water. All plots were vegetated with grass and clover to the near exclusion of other species, although there was always a very small percentage of weedy invaders. Grass "cover" exceeded clover "cover" somewhat on five of the eight plots, four of which received spray applications. Table 5.9b summarizes the cover data. For grass, cover was significantly greater on water- than on CMA-treated plots, but mode of application made no difference. However, for clover, material (CMA or water) made no difference, while flooding with either material significantly increased cover over spraying. There was some spatial heterogeneity in cover of other plants, with the first three plots having more of these species than other plots.

On all plots there was intense rooting to a depth of 5 to 7 cm, and some roots penetrated to 20 to 25 cm. No differences could be distinguished between the plots.

Only qualitative assessments can be made of the results of spraying tree seedlings in soil boxes and larger tree stands. They were carefully examined on 24 April 1987, with no differences among treatments or adverse symptoms being apparent on controls or CMA sprayed specimens. Potential symptoms that were examined included chlorosis, necrosis, abscission, pigmentation, and morphological changes.

5.9.3 <u>Discussion</u>

A trend seen in most of the pot tests was for some improvement in growth and yield ι ith the smaller additions of CMA (0.5 to 1.0 g/kg soil). This effect is not too surprising, because even on a fertile but somewhat acid soil such as the Puyallup, additions of calcium and/or magnesium are often recommended for maximizing yields.

Adverse effects of larger CMA additions (2.0 and 4.0 g/kg soil) were evident in reduced growth or, often, death of the plants. Both of these additions are higher than ever expected to result from highway runoff or spray of CMA. These effects were almost certainly caused by osmotic water stress from concentrations of salts being too high in the soil solution in contact with the roots.

Delay of planting for two days after mixing of CMA into the soil and moistening dramatically improved plant survival. This result can be explained by the exchange reactions between the added CMA and the clay and organic mater of the soil, as well as the consumption of acetate by microorganisms. Both of these processes reduce the concentrations of ions in the soil solution and alleviate the osmotic stress.

There were no statistically significant effects of CMA applications on the yields of the controlled field plots. Rooting depth and intensity also were very similar in all of the plots. Further, there was no evidence of transitory injury such as leaf spotting, curling, or chlorosis from the treatments.

The means of treating was associated with a significant effect on plant cover. CMA versus water and spray versus flood appeared to shift the species composition somewhat, according to the statistical analysis more than would be expected on chance. However, this shift is not believed to have any implications for the use of CMA or assessment of its environmental impacts. Both grass and clover provide good, stabilizing roadside covers.

5.10 DEPLETION OF AQUATIC DISSOLVED OXYGEN RESOURCES
5.10.1   Experimental Methods

A biochemical oxygen demand (BOD) experiment was conducted following the basic procedure described by the American Public Health Association (19), with some modifications. The experiment was run for twenty days rather than the standard five. Concentrations of CMA solutions used in the experiment were 10, 100, 1000, and 5000 mg CMA/l, made up by dissolving dry CMA in the prescribed nutrient-enriched dilution water. In addition to these concentrations, two sets of blanks were also prepared, one with the nutrient water and bacterial seed that was added to the CMA solutions, and the other with nutrient water without the seed. The bacterial seed added to the bottles was obtained from the primary clarifier at a local sewage treatment plant.

To determine a rate of decomposition, duplicate samples from each of the four CMA concentrations and both blanks were analyzed for dissolved oxygen concentration on each day of the twenty-day experiment. Sample bottles were placed in the dark in a constant temperature room set at 20 C. Dissolved oxygen analysis was done using the azide-modified Winkler method (19). Rate constants were determined by the method of Thomas (20).

A multiple-temperature BOD experiment was set up in a similar manner to the first experiment, with the following exceptions. It was run using only one CMA concentration (10 mg CMA/l, the appropriate concentration determined from the first experiment), but with bottles incubated in triplicate at three different temperatures, 2, 10, and again 20 C. Dissolved oxygen readings were made with a YSI dissolved oxygen meter equipped with a BOD bottle probe, calibrated by the modified Winkler method.

Dissolved oxygen was measured in the field ponds approximately biweekly. Samples were collected near midday in standard 300 ml BOD bottles and analyzed by the modified Winkler method.

5.10.2   Results

Figure 7 illustrates the results of the first BOD experiment using the four different CMA concentrations. Oxygen was completely depleted in the 100, 1000, and 5000 mg CMA/l concentrations after only two days. Using the points from the 10 mg CMA/l curve, the calculated BOD rate constant of the decomposition reaction was k = 0.107/day, and the calculated ultimate BOD was 77 percent of the initial applied CMA concentration. Ultimate BOD is the theoretical maximum oxygen demand that would be exerted in an unlimited time. According to these results, approximately 69 percent of the ultimate BOD would be exerted in the first five days after entering a water body at 20 C, 83 percent in the first 10 days, 88 percent in 15 days, and 92 percent in 20 days.

Figure 8 shows the amount of oxygen consumed in bottles incubated at each of the three temperatures in the multiple-temperature experiment, calculated by subtracting the mean of the three CMA replicates from the mean of the three seeded blanks. A significant drop in DO was seen after a one-day lag in the 20 C samples, while the 10 C and 2 C samples had lag times before depletion of 3 and 9 days, respectively. Once oxygen depletion commenced at 10 C, it proceeded at approximately the same rate as at 20 C, reached a temporary plateau from days 5 to 8, and then rose to the level in the 20 C bottles by day 12. In addition to the much longer lag time at 2 C, the oxygen was depleted at a slower rate than at the other temperatures once depletion did start. Because of this slower rate, bottles were monitored after 30 and 40 days to determine if depletion would eventually reach the level of the other two temperatures, and indeed it did. Rate constants calculated for the 20, 10, and 2 C experiments were 0.130, 0.064, and 0.020/day, respectively.

Depletion of dissolved oxygen was very evident in the field ponds as a result of CMA input. Figure 9 shows dissolved oxygen from September, 1986, to March, 1987, in CMA ponds. Marks and values at the top of the CMA graphs indicate the times and volumes (liters) of direct CMA runoff that each pond received from treatments of the plots that drained into them.
None of the control ponds had DO's below 10 mg/l during or after the treatment period, while all CMA ponds had measurements below this level, except Pond 5. This exception is significant, because Pond 5 received only a small amount of direct runoff during the applications. Totaling the measured runoff amounts, Ponds 1, 3, 5, and 8 received 198, 71, 7, and 227 liters of direct CMA runoff, respectively, or an average of 1.4-45 liters per treatment. These numbers correlate very well with the observed oxygen depletions in each pond. Assuming complete mixing, 33 liters of runoff would give a pond CMA concentration of about 10 mg/l after a treatment.

5.10.3   Discussion

Using stoichiometry, it was estimated that the ultimate BOD should be approximately 62 percent of the initial applied CMA concentration. This concentration is considerably lower than the experimentally determined value of 77 percent from the first experiment. This result is unusual, in that research with pure glucose has shown that experimentally determined ultimate BOD values are consistently lower than theoretical values by about 10-20 percent. Apparently, glucose is not completely converted to carbon dioxide and water (21). The opposite result in the CMA experiments suggests that biodegradable materials besides acetate were present in the deicer. Butyrate was measured in the CMA, and lactate, sugars, and residual dead bacterial cells from the fermentation may also occur.

Using the stoichiometric value, the ultimate BOD would be exerted some time in the interval 30 to 40 days for both 10 C and 20 C temperatures, but not until approximately day 100 at 2 C, according to logarithmic or power regression equations derived from the data. The consequence of these results is that CMA runoff into cold surface waters would have an extended oxygen depletion effect compared to the situation at higher temperatures. On the other hand, the effect could occur quickly enough at higher temperatures to prevent sufficiently rapid reaeration from the atmosphere and, therefore, cause a greater effect on aquatic biota.

It is important to consider these results in management decisions concerning CMA application. Runoff from a spring snowstorm could reach a receiving water at a time when it is beginning to warm up. Increased water temperatures would cause more rapid decomposition. Also, increased algal production corresponding to this period of higher light and temperature could cause increased respiration at night. These factors together could result in critical oxygen levels being reached, unless CMA runoff is highly diluted.

5.11 AQUATIC MICROORGANISM GROWTH

5.11.1   Experimental Methods

A laboratory phytoplankton bioassay was performed according to the procedure of the American Public Health Association (19), using the green alga _Selenastrum capricornutum_. Triplicate flasks were set up for a control and seven CMA concentrations (1, 10, 50, 100, 500, 1000, and 4000 mg/l). As a control for osmotic effects of CMA, an identical set of flasks was set up using sodium chloride in the same concentrations. Test flasks were placed on a shaker table in a constant temperature room set at 24 C under continuous cool white fluorescent lighting at 3800 lux. Algal cell concentrations were determined indirectly using a Coulter Counter Model ZBI electronic particle counter, and cell volumes were calculated using a Coulter Channelyzer mean cell volume computer. Bacteria populations were also enumerated using the Coulter Counter.

Field pond water was grab sampled biweekly and analyzed for chlorophyll $\underline{a}$. From the start of field sampling in January 1986 through September of 1986, measurements were made using a Turner model 110 fluorometer calibrated with a spectrophotometer (22). At that point the fluorometer malfunctioned, and all subsequent measurements analyses were done directly on a spectrophotometer (23).

5.11.2   Results

In the laboratory bioassay, algal biomass reached its peak and then either leveled off or declined after day 7 of the experiment. Therefore, biomass levels from this day were used in statistical analyses of maximum standing crops. Figure 11 displays day 7 biomass for CMA and NaCl bioassays. Analysis of variance showed a statistical difference between means ($p < 0.001$) in the CMA bioassay. Using the Student-Newman-Keuls multiple comparison test with alpha = 0.05, biomass on day 7 was significantly greater in the control than in the 1 mg CMA/l, greater in the 1 mg/l than in the 10 mg CMA/l, and greater in the 10 mg CMA/l than in the 50 mg CMA/l treatment. Significant differences could not be detected between the 50 mg CMA/l concentration and all higher concentrations. In summary:

Control > 1 > 10 > 50 = 100 = 500 = 1000 = 4000

The parallel experiment using sodium chloride in the same concentrations as CMA exhibited considerable differences from the results of the CMA additions. There was no significant difference (alpha = 0.05) in biomass on day 7 between the control and all concentrations up through 1000 mg NaCl/l. Inhibition from NaCl was only evident in the 4000 mg NaCl treatment, which had significantly lower biomass than all other treatments. Summarizing:

Control = 1 = 10 = 50 = 500 = 1000 > 4000

Measurements of bacteria populations in the control and CMA treatments on five days during the bioassay showed highest cell numbers in the three highest CMA concentrations. Again using analysis of variance and the Student-Newman-Keuls test, cell counts on day 7 were significantly different between each of the three highest concentrations but not significantly different between the control, 1, 10, 50, and 100 mg CMA/l. In summary of bacteria counts:

4000 > 1000 > 500 > 100 = 50 = 10 = 1 = control

Measurements to detect responses of field pond microorganisms to CMA input produced far less obvious results than the laboratory experiments. Figure 11 displays chorophyll $\underline{a}$ concentrations in all ponds through the entire duration of field measurements. All ponds had an algal bloom during the months of June and July. Although two control ponds had the two highest peak concentrations, variability was too great to conclude that algal populations were suppressed in CMA ponds.

5.11.3   Discussion

Based on the results of these experiments, it is difficult to predict the effects of CMA on algae in natural systems. While the Selenastrum bioassay showed strong inhibition of this planktonic species by CMA, the field tests demonstrated no obvious effect at all. CalTrans bioassays exposing two algal species to CMA and NaCl demonstrated agreement with the bioassay reported here (6). CMA was inhibitory to both species at much lower concentrations than was NaCl.

It is not completely clear why CMA so strongly inhibited Selenastrum growth. Although it must be noted that equal concentrations of NaCl and CMA do not impose equivalent osmotic potentials as a result of their differences in molecular weight and valence, it is very clear in comparing NaCl to CMA treatments that CMA inhibition was something other than osmotic stress. What is more likely is toxic inhibition by contaminants in CMA such as heavy metals, or competition from bacteria whose growth is stimulated by acetate. With the relatively low trace metal content of the CMA and the demonstrated bacterial growth in the test vessels, the latter is the more likely explanation. The clearest result of this experiment is the intense stimulation of saprophytic, aerobic organisms in the presence of CMA.

5.12 EFFECTS OF CMA ON FISH
5.12.1   Experimental Methods

Acute bioassays were performed on rainbow trout (Salmo gairdneri) in a constant temperature room at 10 C. Clean, four-liter glass jars each holding two liters of solution were used. Following a preliminary range-finding bioassay, CMA concentrations used were 0, 10,000, 15,000, 20,000 and 25,000 mg CMA/l. Three replicate jars for each concentration were each stocked with 10 rainbow trout in the sac-fry stage. Jars were aerated using compressed air bubbled through aquarium airstones to insure survival of controls. Therefore, these experiments represent impacts of CMA attributed strictly to factors other than oxygen depletion. Fish were counted every 12 hours for the four-day duration of each test, and dead fish were removed immediately. Temperature, pH, and dissolved oxygen were measured daily. Values of LC_{50} were calculated as in Standard methods (19).

Chronic fish bioassays were also performed to assess the effects of CMA on hatching and subsequent survival of rainbow trout eggs. The experimental setup was very similar to the acute trout bioassay, except that eyed eggs were used rather than sac-fry, and the CMA concentrations were 0, 10, 100, 500, 100, and 5000 mg CMA/l. Also, the sets of 10 eggs in each jar were suspended in baskets. The baskets were constructed of 2-inch long sections of 2-inch diameter PVC pipe with one end covered with fine-mesh plastic screen secured to the pipe with silicone seal. These baskets were suspended in the solutions from above by fine fishing line. Together with aeration as in the acute fish bioassays, the set up provided a constant water flow around the eggs.

The chronic experiment was monitored daily to determine hatching of the eggs and survival of the fry. Dead fish were removed immediately. Daily measurements were also made of pH, temperature, and dissolved oxygen. The experiment was run for 35 days with aeration and an additional 10 days with the air supply turned off.

Ponds at the Pack Forest site were monitored regularly for survival of stocked bluegill sunfish. The ponds were also stocked with fathead minnows, which were not enumerated but were observed qualitatively.

5.12.2   Results

Preliminary range-finding tests using rainbow trout indicated that a concentration series extending well above levels expected in actual highway runoff would have to be run to determine a 96-hour LC_{50} value. Test specimens had greater than 90 percent survival in both 5000 and 10,000 mg/l treatments but greatly reduced survival at 15,000 mg/l. None survived 20,000 or 25,000 mg/l treatments through the 4-day experiment. The 96-hour LC_{50} was calculated to be 17,000 mg CMA/l.

Table 5.10 summarizes the results of the chronic rainbow trout experiment. With the exception of one egg in a 1000 mg/l replicate, every egg in all concentrations hatched by day 8. The control and all treatments up to 1000 mg CMA/l had at least 90 percent survival of the sac fry through the end of the regular portion of the experiments. In the 5000 mg CMA/l treatment, all died by day 21.

After the 35 days of the regular experiment, the aeration was shut off and the systems continued to be monitored for another ten days, in order to determine the extent of oxygen depletion and impacts on the fry. All those alive in the control, 10, and 100 mg/l treatments survived the remaining 10 days, and oxygen never dropped below 8.5 mg/l in any of these vessels. CMA was probably mostly decomposed by that time in these treatments. Dissolved oxygen fluctuated unpredictably between replicates of both the 500 and 1000 mg/l treatments. When it remained above 2 mg/l, all or most of the remaining fish survived; when dissolved oxygen dropped below that level, however, most fish died quickly.

As might be expected from the laboratory fish bioassays, which demonstrated great tolerance to very high CMA concentrations, none of the bluegills in any of the ponds died. There was some concern that, because the laboratory bioassays were artificial aerated, low dissolved oxygen concentrations created by CMA input could result in mortalities in the field not seen in the laboratory. Because of relatively low pond CMA concentrations, dissolved oxygen concentrations never dropped below 4 mg/l in any of the ponds, however, a level sufficient for bluegills to survive.

After a short period following stocking, numbers of fathead minnows in each pond were not counted. With their rapid reproduction and small size, netting the fish was too difficult. However, observations during pond sampling indicated that all ponds continued to maintain healthy populations of fathead minnows.

5.12.3   Discussion

CMA concentrations up to the maximum level expected in actual highway runoff had no deleterious effect on rainbow trout survival or hatching success. Rainbow trout fry exhibited nearly complete survival in acute bioassays at 5000 ppm CMA and even higher. In a chronic bioassay hatching was similar at 5000 ppm and lower concentrations, but the hatchlings exhibited lower survival at the higher concentrations. A concentration (1000 mg/l) well above the level routinely expected did not impede survival of the young. It must be noted that aerating the test chambers removed the effect of oxygen depletion that appears to pose the greatest threat to fish exposed to CMA. After a period of decomposition, CMA concentrations expected routinely (up to 100 mg/l) did not reduce dissolved oxygen or increase long-term mortality after aeration ceased. Even after an equal period of decomposition, higher concentrations, that could occur from time-to-time in practice, did result in greatly reduced oxygen and rapid, high mortality. However, this hazard should be much smaller in natural running waters.

In field ponds no conclusive differences were noted in bluegills or fathead minnows between control and treatment ponds. Fish survival was very high in all ponds, and minnows reproduced a number of generations. These observations lend support to the conclusions reached after laboratory bioassays; viz., if oxygen depletion is avoided, CMA concentrations expected to result in natural waters after mixing of even relatively concentrated highway runoff should not create short- or long-term impacts on aquatic animals. Excessive oxygen depletion, however, is a definite possibility. It could occur prior to mixing and dilution of runoff, or as a consequence of rapid washoff of a large CMA volume. The latter effect would be most likely if the receiving water had already been warmed well above freezing, when acetate decomposition and oxygen removal would be most rapid. In the field ponds, dissolved oxygen did drop as low as 4 mg/l. This concentration was sufficient for the requirements of the resident organisms, but could affect salmonid fish negatively if prevalent for a period of days or longer.

5.13 CONCLUSIONS

Overall, the results of this research indicated that the leading potential environmental impact of CMA use is dissolved oxygen depletion in surface water receiving highway runoff. While results were inconclusive, the potential for mobilization of soil metals was demonstrated in this and other research. This possibility must remain a matter of concern until research can be performed under actual highway operating conditions. Mobilization of acetate and retarded decomposition in low-temperature soils is of some, although lesser concern. Mobile acetate could travel to groundwater and reduce oxygen there. It was clear from the results that even the highest CMA concentrations and loadings anticipated in highway runoff pose no threat to vegetation survival and growth. Also, the danger to fish from these CMA levels is small, if acetate is removed before discharge or diluted sufficiently to avoid substantial oxygen depletion.

It was a general goal of this research program to draft interim guidelines for CMA use based on conclusions reached in interpreting results. These guidelines were intended to apply until a full-scale demonstration could be carried out and either verify their appropriateness or suggest modification.

The conclusions summarized above and elaborated upon earlier have suggested the following guidelines. In most cases the guidelines should be applied especially in certain circumstances that make a negative environmental impact more likely. These special circumstances are noted with each guideline, as appropriate. One or more reasons, rooted in the research results, exists for each guideline. These reasons are noted generally in parentheses following the guideline statement.

1. Do not apply CMA in catchments where highway runoff can directly reach receiving waters that will have less than 100:1 dilution available in the runoff season.

    Conditions warranting special protection:

    a. When receiving water is inhabited by protected aquatic species, especially salmonid fishes.

    b. When receiving water is a small, poorly flushed pond or lake and a large quantity of runoff would enter at once.

    c. Late-season snowstorms, when receiving water may already have warmed.

    (Protection against dissolved oxygen depletion.)

2.   Provide vegetated drainage courses between highways and receiving waters to the extent possible.

       (For removal of acetate and other constituents from runoff.)

3.   Avoid CMA use when runoff will pass over very coarse soil that overlies a sensitive aquifer or is adjacent to a receiving water that should be protected from dissolved oxygen depletion. Condition warranting special protection:

       a.   When soil infiltration could occur in large volumes over short time spans.

              (To avoid high acetate mobility in soil with low anion exchange capacity)

4.   Consider carefully the use of CMA along highway segments that may have relatively high heavy metal concentrations in roadside soils (from high traffic volumes or atmospheric sources), and that are above important aquifers or adjacent to surface water that has a protected beneficial use (e.g., drinking water source or protected aquatic life habitat).

       (To avoid possibility of heavy metal mobilization through cation exchange reactions, until that question can be considered in additional research.)

5.   Take care to avoid spills and runoff from CMA stock piles in the vicinity of receiving waters at all times.

       (To avoid acute oxygen depletion, osmotic, and nutrient-enrichment effects.)

These guidelines pertain almost entirely to settings with restricted direct receiving waters, coarse soils, or unusual combinations of circumstances. Thus, it is concluded that CMA could be used in the vast majority of cases without a threat to environmental quality.

ACKNOWLEDGEMENTS

       The research reported in this paper was sponsored by the National Cooperative Highway Research Program of the Transportation Research Board. The authors wish to acknowledge gratefully the assistance of the NCHRP project manager, Crawford F. Jencks.

REFERENCES

1   F.W. Holmes and J.H. Baker, "Salt Injury to Trees, II. Sodium and Chloride in Roadside Sugar Maples in Massachusetts," Phytopathology Vol. 56:6 (1966).

2   E.F. Button, "Ice Control Chlorides and Tree Damage," Public works Vol. 93:3 (1965).

3   G.P. Lumis, et al., "Salt Damage to Roadside Plants," Ontario Ministry of Agriculture and Food, Ottawa, Ont., Canada (1971).

4   R. Field, E.J. Struzeski, Jr., H.E. Masters, and A.N. Tafuri, "Water Pollution and Associated Efects from Street Salting," U.S. Environmental Protection Agency, National Environmental Research Center, Cincinnati, OH, EPA-R2-73-257 (1973).

5   R. Field, E.J. Struzeski, H.E. Masters, and A.N. Tafuri, "Water Pollution and Associated Effects from Street Salting," Journal of Environmental Engineering Vol. 100:459-477 (1974).

6   R.E. Hanes, et al., "Effects of Deicing Salts on Plant Biota and Soils--Experimental Phase," National Cooperative Highway Research Program, Transportation Research Board, Washington, D.C., NCHRP 170 (1976).

7   G.A. Prior and P.M. Berthoux, "A Study of Salt Pollution of Soil by Highway Salting," Highway Research Record Report No. 193, Highway Research Board, Washington, D.C. (1976).

8   F.E. Hutchinson and B.E. Olson, "The Relationship of Road Salt Applications to Sodium and Chloride Ion Levels in the Soil Bordering Major Highways," Highway Research Record Report No. 193, Highway Research Board, Washington, D.C. (1967).

9   G. Feick et al., "Release of Mercury from Contaminated Freshwater Sediments by the Runoff of Road Deicing Salt," Science Vol. 175:1142 (1972).

10   R.E. Hanes et al., "Effects of Deicing Salts on Water Quality and Biota - Literature Review and Recommended Research," National Cooperative Highway Research Program, Washington, D.C., Report 91 (1970).

11   E.E. Huling and T.C. Hollocher, "Groundwater Contamination by Road Salt--Steady State Concentrations in East Central Massachusetts," Science Vol. 176:288 (1972).

12   D.C. Murray and U.F.W. Ernst, "An Economic Analysis of the Environmental Impact of Highway Deicing," U.S. Environmental Protection Agency, Municipal Environmental Research Laboratory, Cincinnati, OH, EPA-60/2-76-033 (1976).

13   Commonwealth of Massachusetts, "Legislative research Council Report Relative to the Use and Effects of Highway Deicing Salts," Comonwealth of Massachusetts, Boston (1965).

14   S.A. Dunn and R.O. Schenk, "Alternates to Sodium Chloride for Highway Deicing," Transportation Research Record Vol. 776:12-15 (1980).

15   B.H. Chollar, "Federal Highway Administration Research on Calcium Magnesium Acetate - an Alternative Deicer," Public Roads Vol. 47:113-118 (1984).

16   G.R. Winters, J. Gidley, and H. Hunt, "Environmental Evaluation of Calcium Magesium Acetate (CMA)," California Department of Transportation, Sacramento, CA, FHWA/CA/TL-84/03 (1984).

17   H.A. Elliott and J.H. Linn, "Effect of Calcium Magnesium Acetate on Heavy Metal Mobility in Soils," Journal of Environmental Quality Vol. 16:222-226 (1987).

18   R. Horner, "Environmental Monitoring and Evaluation of Calcium Magnesium Acetate (CMA)," Transportation Research Board, Washington, D.C. (1988).

19   American Public Health Association, Standard Methods for the Examination of Water and Wastewater, 16th Ed., American Public Health Association, Washington, D.C. (1985).

20   Metcalf and Eddy, Inc., <u>Wastewater Engineering</u>, McGraw-Hill, Inc., New York, N.Y., pp. 248-249 (1972).

21   C. Sawyer and P. McCarty, <u>Chemistry for Environmental Engineering</u>, McGraw-Hill Book Company, New York (1978).

22   J.P.H. Strickland and T.R. Parsons, "A Practical Handbook on Seawater Analysis," Bulletin of the Fisheries Research Board of Canada 167 (1972).

23   C.J. Lorenzen, "Determination of Chlorophyll and Phaeopigments: Spectrophotometric Equations," <u>Limnology and Oceanography</u> Vol. 12:343-346 (1967).

Table 1. Laboratory Analysis Specifications

Constituent	Method	Method Number (19)	Equipment
pH	Potentiometric	423	Cole-Parmer Model 5985-80 meter
Specific Conductivity	Wheatstone bridge-type meter	205	Barnstead Meter
Dissolved Oxygen	Azide modification of Winkler Method	421B	--
Total Alkalinity	Titrimetric	403	--
Total Suspended Solids (TSS)	Gravimetric	209C	Mettler Type H15 analytical balance
Biochemical Oxygen Demand (BOD)	Azide modification of Winkler Method	507	--
Volatile fatty acid salts (VFA)	Gas chromatographic	504A	Hewlett-Packard Model 5840A gas chromatograph
Metals (Al, Ca, Cu, Fe, K, Mg, Pb, Se, Zn)	Inductively coupled plasma (ICP)	305	Jarrell-Ash ICP
Total Phosphorus	Ascorbic acid following persulfate digestion	424F	Perkin-Elmer Lambda 3 spectrophotometer Barnstead autoclave

Table 2. Comparison of Properties of Redissolved Dry CMA and Original Drum Contents[a]

Sample	Acetate	Ca	Mg	K	Al	Se	Fe	Cu	Pb	Zn
Dried Lot 1	21180	2480	2920	1290	0.46	1.04	1.38	0.04	0.23	0.16
Dried Lot 2	21150	2500	3030	1260	0.49	1.08	0.88	0.00	0.25	0.30
Dried Lot 3	20620	2600	3060	1300	0.52	1.10	0.98	0.01	0.28	0.19
Mean of Dried Lots	20980	2530	3000	1280	0.49	1.07	1.08	0.02	0.25	0.22
Standard Deviation of Dried Lots	315	64	74	21	0.03	0.03	0.26	0.02	0.03	0.07
Mean of Original Drum Contents	19506	2790	3101	1330	0.58	1.49	9.03	0.02	0.68	0.13

[a]All values are in parts per million.

Table 3. General Properties of Test Soils

Soil	Textural Class	% Organics	Cation Exchange Capacity (meq/100 g)	Exchangeable Cations (ppm) Ca	Mg	K
Pack Forest	Loam	6.0	12.6	754	125	217
Lee Forest	Silt loam	12.3	26.5	1030	94	148
Puyallup	Sandy loam	5.2	8.8	899	65	155
Vantage[a]	Sandy loam	2.3	12.8	1320	365	337

[a]Calcareous soil with 1.25% calcium carbonate.

Table 4. Summary of Adsorption and Desorption of Metals in the Pack Forest Soil

CMA (mg/l)	Ca	Mg	K	Fe	Cu	Zn	Se	Pb	Al
Adsorption in Soil (μg/g of soil)									
0	-19	-5	-16	-2	-0.040	-0.174	-0.100	-0.082	-5.430
10	-16	0	-14	-2	-0.036	-0.199	-0.099	-0.110	-4.134
100	8	38	-4	-1	-0.030	-0.200	-0.071	-0.092	-3.306
500	106	193	50	0	-0.025	-0.310	0.002	-0.042	-1.539
1000	250	378	129	1	-0.024	-0.250	0.104	-0.010	-1.467
5000	1267	1603	687	7	-0.035	-0.409	0.881	0.326	-1.811
Desorption from Soil (μg/g of soil)									
0	8	2	11	3	0.015	0.133	0.075	0.053	6.315
10	6	2	9	2	0.014	0.107	0.056	0.045	2.665
100	9	3	11	1	0.015	0.174	0.091	0.060	3.551
500	14	6	14	1	0.015	0.133	0.094	0.100	2.106
1000	24	14	22	2	0.015	0.126	0.086	0.069	4.198
5000	90	85	73	0	0.017	0.152	0.104	0.076	0.971
Fraction Desorbed									
0	-0.41	-0.44	-0.66	-1.32	-0.37	-0.76	-0.74	-0.64	-1.16
10	-0.39	-4.04	-0.65	-1.02	-0.39	-0.54	-0.57	-0.41	-0.64
100	1.18	0.08	-2.65	-1.52	-0.52	-0.87	-1.27	-0.66	-1.07
500	0.13	0.03	0.28	2.80	-0.61	-0.43	49.85	-2.41	-1.37
1000	0.10	0.04	0.17	2.00	-0.64	-0.50	0.83	-7.30	-2.86
5000	0.07	0.05	0.11	0.05	-0.49	-0.37	0.12	0.23	-0.54

Table 5. Acetate Adsorption and Desorption for Three Soils

CMA (mg/l)	Acetate (mg/l)	Equilibrium Acetate (mg/l)	Acetate Adsorbed (mg/g soil)	% Acetate Adsorbed	% Acetate Desorbed
		Pack Forest Soil			
10	5.8	2.0	0.011	65.5	54.5
20	11.6	5.5	0.018	52.6	44.4
40	23.2	8.5	0.044	63.4	34.1
80	46.4	23.0	0.070	50.4	31.4
100	58.0	29.5	0.086	49.1	12.8
160	92.8	54.5	0.115	41.3	41.7
500	290.0	197.5	0.278	31.9	34.5
		Lee Forest Soil			
100	58.0	35.5	0.068	38.8	30.9
500	290.0	181.0	0.327	37.6	40.4
		Vantage Soil			
100	58.0	41.0	0.051	29.3	5.9
500	290.0	250.5	0.118	13.6	38.1

Table 6. Means of Metal Concentrations in Soil Water at 60 cm Depth Over Four Periods During Field Plot Experiment

Treatment	Period[a]	Concentration (mg/l)								
		Ca	Mg	K	Fe	Cu	Zn	Se	Pb	Al
CMA-Spray	P	4.8	1.3	1.0	0.28	0.01	0.08	0.02	0.00	0.33
	T1	9.1	4.9	2.1	0.15	0.01	0.05	0.02	0.00	0.17
	I	2.5	1.1	0.8	0.04	0.00	0.03	0.00	0.00	0.08
	T2	10.1	6.3	2.1	0.05	0.00	0.09	0.02	0.01	0.07
Water spray	P	6.0	1.6	1.3	0.16	0.01	0.16	0.02	0.00	0.22
	T1	16.7	1.4	1.7	0.15	0.02	1.65	0.02	0.02	0.20
	I	1.8	0.6	1.0	0.22	0.01	0.04	0.01	0.02	0.13
	T2	1.9	0.6	0.9	0.11	0.00	0.03	0.02	0.03	0.13
CMA-Flood	P	11.0	2.2	1.8	0.12	0.01	0.30	0.02	0.00	0.12
	T1	12.4	5.6	2.5	0.05	0.01	0.22	0.02	0.05	0.07
	I	5.5	1.8	1.7	0.10	0.00	0.06	0.02	0.03	0.08
	T2	3.9	2.3	1.7	0.08	0.00	0.07	0.02	0.04	0.10
Water-Flood	P	10.6	2.9	3.4	0.96	0.01	0.31	0.03	0.10	0.65
	T1	3.8	1.3	2.8	0.96	0.01	0.05	0.02	0.00	0.52
	I	3.2	1.0	1.7	0.25	0.01	0.11	0.00	0.00	0.25
	T2	4.0	1.1	1.9	0.16	0.01	0.11	0.02	0.02	0.25

[a] P -- Prior to first treatment (1/86-2/86)
T1 -- First treatment period and one month afterward (3/4/86-4/25/86)
I -- Intermediate period between T1 and T2 (4/26/86-12/17/86)
T2 -- Second treatment period and one month afterward (12/18/86-3/17/87)

Table 7. Means of Metal Concentrations in Groundwater at 3 m Depth Over Four Periods During Field Plot Experiments

Treatment	Period[a]	Concentration (mg/l)								
		Ca	Mg	K	Fe	Cu	Zn	Se	Pb	Al
CMA-Spray	P	5.4	2.6	1.2	5.9	0.03	0.22	0.05	0.04	6.1
	T1	5.6	1.6	1.1	13.2	0.05	0.07	0.08	0.05	10.5
	I	6.4	2.6	1.5	22.0	0.05	0.10	0.09	0.08	13.8
	T2	11.5	1.7	1.7	15.1	0.05	0.68	0.06	0.05	7.4
Water-spray	P	15.3	9.0	1.3	1.1	0.05	0.14	0.01	0.02	0.0
	T1	7.8	4.1	1.0	0.8	0.01	0.05	0.03	0.02	0.4
	I	6.5	3.2	1.4	5.7	0.02	0.04	0.03	0.05	3.2
	T2	6.1	4.1	1.9	5.7	0.01	0.02	0.03	0.03	2.1
CMA-Flood	P	14.0	9.2	0.5	0.2	0.01	0.01	0.02	0.03	0.1
	T1	6.9	3.7	0.8	4.8	0.02	0.10	0.04	0.04	2.8
	I	5.5	1.5	1.2	5.4	0.03	0.08	0.05	0.06	5.2
	T2	23.5	1.1	1.6	0.5	0.02	0.16	0.02	0.02	1.5
Water-Flood	P	23.2	3.3	0.8	3.3	0.01	0.59	0.03	0.03	2.4
	T1	15.8	3.5	0.7	3.9	0.02	0.12	0.03	0.03	1.8
	I	4.0	1.8	0.5	2.7	0.02	0.05	0.03	0.03	1.2
	T2	2.9	1.1	0.4	1.2	0.02	0.04	0.04	0.03	0.7

[a] P -- Prior to first treatment (1/86-2/86)
T1 -- First treatment period and two months afterward (3/4/86-5/25/86)
I -- Intermediate period between T1 and T2 (5/26/86-12/17/86)
T2 -- Second treatment period and one month afterward (12/18/86-3/17/87)

Table 8. Summary of Shoot Yields in CMA Pot Tests

Note: All values are in percent of control, except last column, which gives control yields in g/pot. Values in parentheses are for plants that replaced the original ones.

Plant Species	Soil[1]	CMA Addition (g per kg soil)						Control Yield (g/pot)
		0(Control)	0.5	1.0	2.0	4.0	4.0*	
Romaine lettuce	Puyallup	100^a	106^a	103^a	85^b	$(68)^{3b}$	--	5.84
	Pack/Sand	100^{ac}	122^{ac}	$(128)^a$	$(160)^b$	$(181)^c$	95	2.84
Plantain (Plantago lanceolota)	Puyallup	100^a	104^a	102^a	65^a	Died	91^a	2.47
	Pack/Sand	100^a	162^a	137^a	55^a	Died	183^a	1.22
Sunflower (Helianthus annuus)	Puyallup	100^a	104^a	94^a	87^a	Died	87^a	6.46
	Pack/Sand	100^a	181^b	Died	Died	Died	227^b	2.01
Douglas Fir (Pseudotsuga menziesii)	Puyallup	100^a	104^a	18^a	Died	Died	111^a	2.18
	Pack/Sand	100^a	177^a	Died	Died	Died	65^a	1.67
Balsam Fir (Abies balsamina)	Pack/Sand	100^a	122^a	104^a	191^a	Died	135^a	0.23
Red Maple (Acer rubra)	Pack/Sand	100^a	141^a	149^a	234^a	Died	63^a	2.01

* Plants were placed in the pots two days after the addition of CMA and moistening.

a,b,c Treatment means that have the same superscript letter are not significantly different statistically, but are significantly different from those with a different letter. Significance was determined at $P < 0.05$ by anova, followed where appropriate by the Student -Newman-Keuls multiple range comparison. Use of an underline with the superscript indicates that only a few plants survived.

Table 9. Summary of Plot Biomass Yields and Cover

a) Biomass Yields (all values in g dry weight/m^2)

Plot	1	2	3	4	5	6	7	8
Material Application	CMA Spray	Water Spray	CMA Spray	Water Spray	CMA Flood	Water Flood	Water Flood	CMA Flood
Mean Biomass	748^a_w	754^a_w	828^a_w	868^a_w	881^a_x	1026^a_x	944^a_x	1107^a_x
Standard Error	71	74	67	52	82	120	84	83

b) Cover (all values in g dry weight)

	1	2	3	4	5	6	7	8
Grass Mean	2.99^a_w	5.30^b_w	2.30^a_w	3.17^b_w	1.99^a_w	2.69^b_w	4.38^b_w	2.22^a_w
Standard Error	0.66	0.78	0.31	0.61	0.50	0.42	0.87	0.27
Clover Mean	2.57^a_w	1.56^a_w	1.80^a_w	2.57^a_w	3.05^a_x	3.22^a_x	2.06^a_x	3.94^a_x
Standard Error	1.00	0.31	0.28	0.41	0.55	1.14	0.48	0.83
Other Mean	0.50	0.32	0.48	0.16	0.10	0.11	0.07	0.07
Standard Error	0.28	0.11	0.24	0.05	0.05	0.04	0.03	0.01

[a] Superscript series signifies statistical significance of differences based on Material (CMA or water), and subscript series signifies statistical significance of differences based on Application in a two-way anova. Treatment means that have the same letter are not significantly different, but are significantly different at P < 0.05 from those with a different letter. Statistical procedures were as designated in Table 8.

Table 10. Chronic Rainbow Trout Eyed-Egg Bioassay Summary

Treatment	% hatched by Day:[a]			% Alive by Day[a]					
	6	7	8	12	14	18	21	24	28
Control	43	97	100	97	97	97	97	97	97
10 mg CMA/l	40	93	100	100	100	100	97	97	90
100 mg CMA/l	33	97	100	100	100	100	97	97	97
500 mg CMA/l	43	97	100	97	97	97	90	90	90
1,000 mg CMA/l	70	97	97	97	97	97	97	97	97
5,000 mg CMA/l	83	97	100	87	67	40	0	0	0

[a]Mean of three replicates.

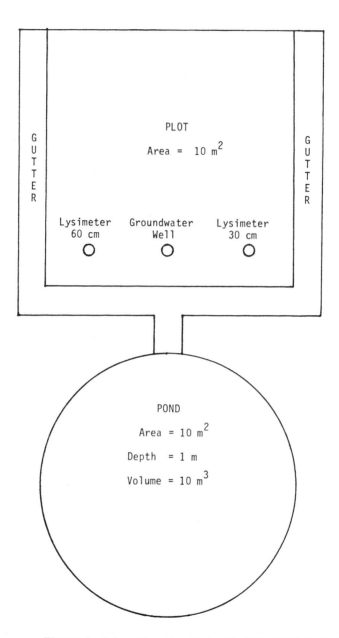

Figure 1. Schematic of a Controlled Field Plot System

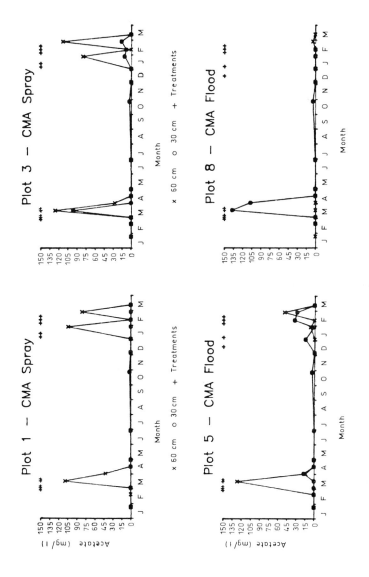

Fig. 2. Specific Conductivity and Acetate in CMA Plot Soil Water.

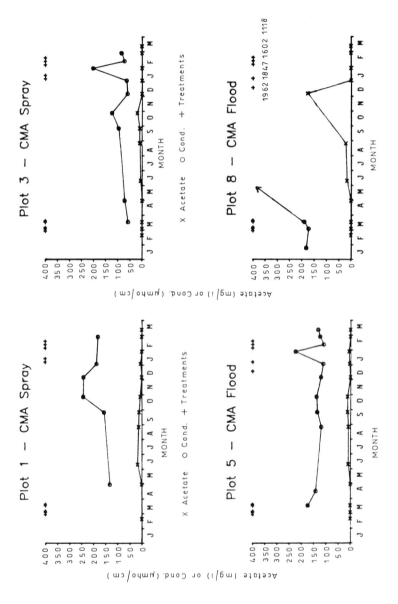

Figure 3. Specific Conductivity and Acetate in CMA Plot Groundwater

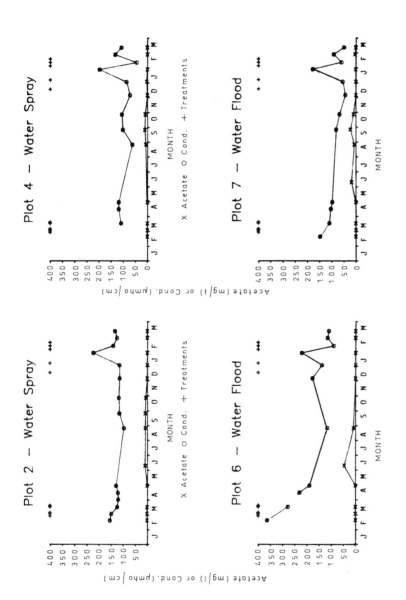

Figure 4. Specific Conductivity and Acetate in Water Plot Groundwater

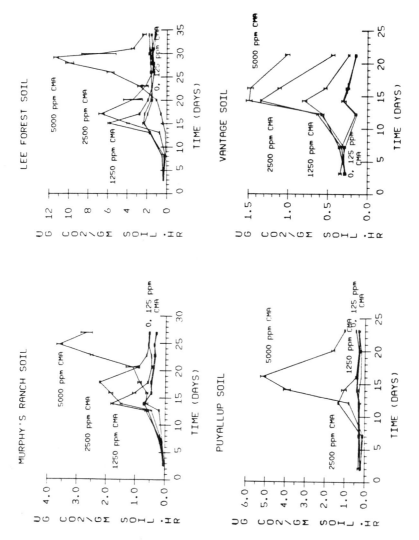

Figure 5. CO_2 Production Over Time at 2 C in Four Test Soils

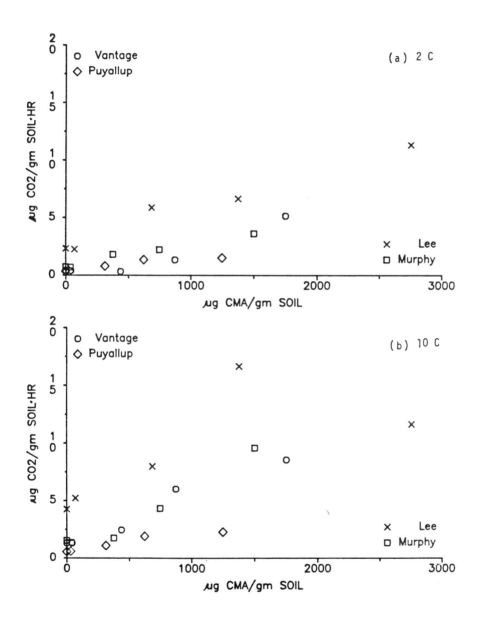

Figure 6. Peak Respiration Rates Versus CMA Loadings

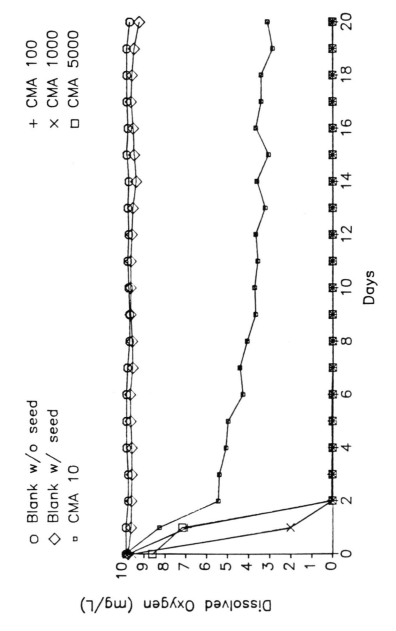

Figure 7. Results of 20 C BOD Experiment with Several CMA Concentrations

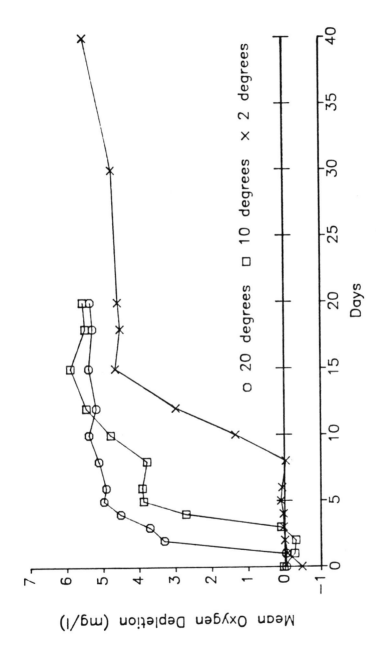

Figure 8. Results of Multiple–Temperature BOD Experiment

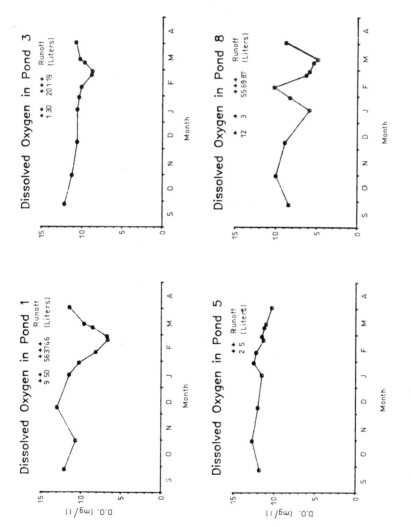

Figure 9. Dissolved Oxygen in CMA Ponds (September, 1986, to April, 1987)

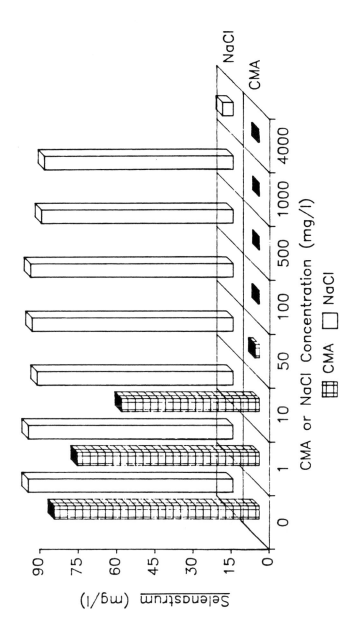

Figure 10. Algal Biomass on Day 7 in CMA and NaCl Bioassays

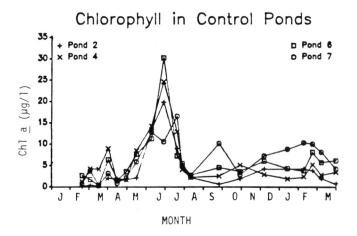

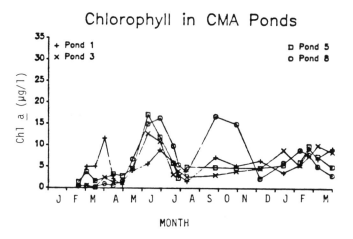

Figure 11. Field Pond Chlorophyll Over Period of Field Experiments

Chapter 6

A SUMMARY OF CMA USE AND RESEARCH DEVELOPMENT, SPECIFICALLY IN CALIFORNIA

Jeffrey L. Gidley
Associate Environmental Planner, State of California, Department of Transportation, Division of New Technology and Research

6.1   INTRODUCTION

Currently, according to the Salt Institute, approximately ten million tons of sodium chloride (rock salt) are used annually to provide snow and ice control on the Nation's highways (1). California has averaged over 17,000 tons annually for the last ten years and near 30,000 tons for the last two years (2) with the majority being used on the major trans-Sierra routes. Nationwide, the extensive use of salt for deicing is a recent development. For example, in 1950, less than one million tons of salt were used for deicing. The very rapid growth of salt use occurred between 1965 and 1971 when salt use doubled from about five million to over ten million tons. Since then, annual salt use has fluctuated depending on weather conditions but averages near ten million tons (3).

The fact that salt damages the highway system, the environment, the public infrastructure and private vehicles is now accepted by highway agencies. Even the Salt Institute estimates that the real cost of deicing salt to the public is in excess of $300 per ton (1). Since the average purchase price of salt is estimated at an amount less than $30 per ton, the purchase price of salt represents less than 10% of the estimated real cost. Other agencies, such as the New York State Energy Research and Development Authority (NYSERDA), have estimated the real cost of salt use significantly higher. NYSERDA, in this case, estimates real cost of salt use to be near $0.80 per pound or $1600.00 per ton (4). In California, Caltrans Structural Maintenance estimated in 1985 that the cost to rehabilitate bridge structures damaged by salt use would eventually cost about $300. In 1976, the US EPA estimated that the cost of deicing to the motoring public in terms of vehicle damage was two billion dollars annually (5).

The public is also very aware that deicing operations cause damage. In New York, salt damage to the Williamsburg Bridge forced closure of the structure. This closure was followed just two weeks later by an additional example, when New York had to severely restrict subway speeds on the Manhattan Bridge, again due to salt induced corrosion. Other examples are available from across the snow-belt. In California, motorists have been delayed over Interstate 80 while extensive repairs have been made to bridges corroded by deicing salt. In Massachusetts, water wells had to be closed and new water systems have been developed by the Massachusetts Department of Transportation because of sodium contamination from deicing operations. Additionally, throughout the snow-belt, vegetation has been severely damaged by salt. As an example, in 1973, the U.S. Forest Service estimated that approximately 3000 trees were killed by deicing operations during the previous winter in the Lake Tahoe Basin (6).

Because of the requirement to provide a safe and efficient highway system during the winter and the known, deleterious effects of salt, there has been an ongoing, nationwide search for an alternative product which could be an effective deicer and yet be environmentally safe and non-destructive when used. Calcium Magnesium Acetate (CMA) has been developed to meet these goals. In extensive testing, it has been found to be an effective deicer, but with different operational characteristics than road salt. Cost, however, remains CMA's principle draw-back and an impediment to its' increased use.

This report presents a summary of the research and product trials conducted on CMA. Included in the report is information from the major corrosion, environmental, manufacturing and product development reports completed and the conclusions from the current Caltrans implementation study on CMA.

6.2   CMA RESEARCH

In the mid-seventies, the Federal Highway Administration (FHWA) began a search for an alternative deicer capable of providing similar deicing qualities as salt, but with not environmental or corrosion impacts. In 1979, Bjorksten Laboratories, under contract to FHWA, identified CMA as a potential alternative deicing material (7) and CMA became the focus of an intensive research program covering the environmental, toxicological, corrosion, manufacturing and performance characteristics of the product. This program has included work performed under contract to FHWA, by the National Cooperative Highway Research Program (NCHRP), individual States and private industry. A great deal of research data is available now from these various research projects and, summarizing the results presented below, the FHWA has been remarkably successful in developing a material that meets most of the stated goals for the research project. However, CMA, as it currently is available is till not the total answer to all deicing problems.

6.2.1 Corrosion

In the area of corrosion, several different investigators have compared the corrosion potential of CMA and road salt for many materials found in the highway environment. Michigan DOT (8) and Locke and Kennelley, FHWA/RD-86/064, (9) both compared corrosion caused by salt and CMA on bridge and highway construction metals including A-36, A-588 weather, stainless, prestressed, galvanized and

reinforcing steels plus grey cast iron and rebar in concrete. In all cases salt
was more corrosive than CMA. The Michigan DOT reports:

"CMA does appear to be considerably less corrosive than NaCl. For most
materials tested the average corrosion losses in CMA were only one-third
to one-tenth of those occurring in NaCl. Initial results also suggest
that materials exposed to CMA probably experience more uniform pitting
and, consequently smaller reductions in the original strength. NaCl also
appears to be more likely to create stress concentrations that will
adversely affect fatigue performance."

Locke also found significant reductions in corrosion losses with CMA. Table 6.1
gives weight loss data for various materials comparing NaCl and CMA during full
immersion, alternating immersion and spray fog tests conducted by various
investigators.

Some controversy still exists concerning the effect of CMA on corrosion
rates of steel rebar in concrete slabs. Locke conducted potentiometric tests on
rebar in concrete slabs that had been contaminated with NaCl to simulate long
term contamination from deicing operations. In the tests, upper and lower rebar
mats, imbedded in a contaminated concrete slab, were connected electrically and
the potential was measured between mats in the slab when each slab was ponded in
a CMA solution. After 15 months, the top mat potential was -677 mV. From this
Locke concluded that steel embedded in salt contaminated concrete may have
significant corrosion activity. Chollar and Virmani (10) point out, however,
that the exhibited potential is "not surprising since Cl^- ions are known to cause
corrosion of reinforcing steel." Since there was no control where water alone
was ponded on the contaminated slabs, it cannot be known whether CMA increased,
decreased or had any effect on corrosion in this situation.

Continuing, comparing concrete reinforced slabs during a four year FHWA
study, Chollar and Virmani (10) concluded that CMA had no effect on the slabs
while NaCl caused significant impacts. Following the study period, the CMA slab
had no signs of surface deterioration, while the NaCl slabs had significant
cracks and rust stains. In potentiometric test, the CMA slab had potentials near
zero, while the NaCl slab had potentials around -600 mV and these results were
consistent with the observed conditions of the slabs. Chollar and Virmani (10)
concluded:

"Our results show that the potential of the black steel rebars in slabs
ponded with salt solutions started increasing numerically within the first
3 months of exposure, while that of rebars in slabs ponded with CMA
solution did not increase at all during that time period. The CMA
solutions did not cause any significant potential shift or corrosion after
4 year on/off ponding in an outdoor environment. This contradicts the

Oklahoma results {Locke and Kennelley (9)} on this subject, where it was concluded that CMA solutions have a potential to corrode uncoated rebars embedded in portland cement concrete to a somewhat similar extent as salt. This conclusion was based on changes in the electrical potentials for the black steel rebars exposed to simulated pore solutions, embedded in mortar cylinders and concrete slabs. The Oklahoma study did not confirm the corrosion of the rebars for CMA or salt-ponded slabs or cylinders by breaking them and observing the rebars."

Whatever the true nature of CMA potential to affect corrosion in contaminated concrete, it is clear from Chollar and Virmani (10) that CMA can be used on new steel reinforced concrete with no resultant increase in corrosion or slab deterioration.

In addition to metals, other materials, common in the transportation industry, have also been tested comparing the effects of NaCl and CMA. Slick (11), testing paints, coatings, rubbers, plastics, concrete, brake hydraulic systems and linings, concluded that CMA's effect on these materials was significantly less than for NaCl or there was no effect at all. Figure 1 shows the extent of material testing that has been completed to determine the impact of CMA on transportation related materials and to contrast its' effects with salt.

6.2.2 Environment and Toxicology

The first comprehensive environmental test of CMA conducted for the FHWA was performed by The California Department of Transportation, Transportation Laboratory (12). This study was conducted to determine the environmental impacts of CMA on surface water, ground water and air quality as well as on the aquatic and terrestrial ecosystems. The potential impacts were evaluated using a literature study, laboratory bioassays of aquatic organisms, pot tests of vegetation common to snow-belt states and a soil leaching study. Since no commercial source of CMA was available, an analytical grade CMA was produced using a mix of calcium acetate and magnesium acetate purchased from a chemical supply house.

The study showed that CMA would probably be environmentally safe if used for road deicing. Specific results from the study showed that CMA was less toxic to fish than NaCl, but that at 5000 ppm CMA will slightly retard rainbow trout egg hatching. Additionally, in bioassays on Daphnia (water fleas), CMA had a lower toxicity concentration level than salt and CMA-Daphnia bioassays were affected by bacteria (probably through oxygen depletion further decreasing the total number of survivors. These toxicity levels, however, were in excess of expected runoff and concentration levels in roadside aquatic environments. Salt, however, inhibited Daphnia reproduction at levels expected to occur in runoff. Algae growth was inhibited by CMA and the study results indicated that a continuous concentration of less than 50 mg/liter would be necessary to eliminate any deleterious effects of CMA on algae. In soil studies, leaching with very high concentrations of CMA resulted in some removal of metals and nutrients from test soils. Completing the study, the most striking environmental difference between salt and CMA was the effect on plants. CMA, overall, was significantly less deleterious to the plants tested In conclusion, the study showed that: "At the CMA concentrations likely to be generated by the use of snow and ice control, CMA may be less environmentally damaging than NaCl"(12).

Following the Caltrans study, a second FHWA environmental study was conducted by Horner, University of Washington (13). The purpose of this study was to examine the environmental effects of CMA using a "commercial grade" CMA, in this case, using an acetate produced by fermentation of corn. This study included both laboratory studies and large-scale field testing. Conclusions from this study show that:

1. CMA in field trials had no deleterious effect on aquatic organisms. In laboratory tests, CMA in high concentrations affected organisms by osmotic stress and oxygen depletion. Fish were more resistant than aquatic invertebrates. Invertebrates survived and reproduced optimally at 500 ppm CMA, while fish showed good survival and hatching at levels in excess of 1000 ppm.

2. Plants tested during field applications by spray and flood up to 3000 ppm CMA showed no affect in yield, cover, vigor or rooting. No external damage symptoms occurred in any plants subjected to the treatments.

3. CMA depleted oxygen levels of aquatic ecosystems in laboratory and field trials. The decrease was due to the decomposition of acetate and could be deleterious to aquatic life.

4. Biodegradation of CMA is temperature dependent. At temperatures near freezing, CMA will be present in soil or water for weeks, but at higher temperatures (20°C) degradation occurs within one to two weeks. Therefore, at lower temperatures, oxygen demand would be moderated.

5. CMA does not affect soil properties except for permeability, which may be increased. A significant portion of the acetate was captured on the soil surface; less than 10 percent of applied acetate appeared in soil water or groundwater in field plots. Calcium and magnesium are immobilized in soils, but laboratory studies indicated a potential to release trace metals pre-existing in soils.

The principal environmental concerns expressed in Horner's report are the potential for oxygen depletion in the aquatic environment due to biological oxygen demand and the potential to mobilize trace metals. However, in a study of the effects of CMA on small roadside ponds in Alaska, Rea and LaPerriere (13) did not find a significant reduction in oxygen levels in two of three paired ponds treated with CMA. The study was conducted during the Summer, when according to results from Horner, the effect of biological oxygen demand should have been greatest. Further, in a study conducted at Pennsylvania State University, Elliot and Linn (14) found that CMA would not cause a significant increase in metal mobilization from soils.

Fig. 6.1. List of materials tested, comparing CMA and salt for corrosion and deterioration.

- A-36 Steel: Plain, Galvanized, Painted and Stressed
- A-588 Weathering Steel: Weathered, Unweathered and Painted Grey Cast Iron
- A-325 Stressed Bolts
- Aluminum: 6061-T6, 2024, 5052 and 6063-T6
- Reinforcing Steel: Plain, Epoxy coated, In concrete, stainless and stainless clad
- Prestressing Strand: ASTM Grade 270
- Automotive Components: Steels, Stainless steels, Aluminum alloys, Tires and rubber, Plastics and elastomers, Adhesives, Brake linings, Paints, Primers and Undercoatings
- Highway Components: Asphalt, Portland cement concrete, Pavement markers and adhesives, Sign reflective sheeting and paint, Joint sealant, Plastics, Bridge bearing materials and Drainage pipes.

The effect of CMA on public health has been investigated by Chevron Chemical Company. The tests conducted included studies of the effects on air quality, skin and eye sensitivity and ingestion poisoning and followed procedures in accordance with guidelines from the Federal Inter-agency Regulatory Liaison Group. The results (15) from the test are as follows:

1. Sub-chronic oral toxicity: no effect at 1000 mg/kg/day.
2. Acute inhalation $Lc_{50} > 5000$ mg/m^3.
3. Acute dermal $Ld_{50} > 5000$ mg/kg.
4. Acute dermal irritation: non-irritant.
5. Acute ocular irritation: mild-to-moderate eye irritation.
6. Acute oral LD_{50} approximately 3150 mg/kg.
7. Dermal sensitization: negative.

In all studies, CMA was found to have very low mammalian toxicity and to have similar, but less sever, effects as salt. Figure 2 is the current material data safety sheet from Chevron.

Based on environmental studies conducted on sodium chloride and CMA, CMA is superior from an environmental aspect. There appears to be few environmental effects from CMA that are creditable under conditions found during deicing operations. Table 6.2 gives a summary of the environmental effects of CMA compared to road salt. Additionally, from toxicological studies conducted, CMA appears to be as safe as salt for use in highway deicing.

6.2.3 Manufacturing

CMA is currently manufactured in the United States in commercial quantities only by Chevron Chemical Company. During the winter of 1987-88, CMA was purchased from Chevron for $563.00 per ton delivered to Crestview Maintenance Station from Kennewick, WA. The delivered price increased during the winter of 1988-89 to $657.50 per ton due to an increase in the cost of acetic acid. As currently produced, the possibility of a significant decrease in price per ton seems remote, however, there are ongoing experiments to produce CMA at a significantly reduced cost.

The most common approach taken has been to produce acetic acid by bacterial fermentation using sugars derived from a vegetation feed stock. Under FHWA sponsorship, the first such study was conducted by SRI International (16). The study identified a potential bacteria capable of producing acetic acid, developed a procedure and a cost estimate. However, the amount of acetic acid produced during this research was only three percent. This level of production was not sufficient to be commercially acceptable. Therefore, a second FHWA study was conducted by the University of Georgia (17), whose purpose was to develop a strain of bacteria capable of producing higher quantities of acetic acid from feed stocks. Several strains were investigated and promising strains were developed using mutagenic techniques. The strains produced acetate, but at low concentration level. An additional study currently being conducted to produce CMA by bacterial fermentation has been undertaken by NYSERDA and several co-sponsors (4). They hope to derive acceptable levels of acetic acid from low grade biomass sources, such as cheese whey, distressed corn, or hydrolyzates of hemicellulose from wood or corn cobs.

These fermentation studies have only been marginally successful. Currently the only acetic acid available in sufficient quantity and quality is produced from natural gas. If these studies are successful, the cost of CMA could be substantially reduced. Hudson (4) estimated that the cost of CMA using acetic acid produced by bacterial fermentation could be reduced form the current level of about $0.25 per pound to about $0.13 per pound. Thus, the cost per ton for CMA, reduced to about $260.00, would be less than the real cost of road salt as estimated by the Salt Institute (1).

6.2.4 Performance Testing

Initial performance tests of CMA were conducted for FHWA by Michigan DOT (18) and Washington DOT (19) using 200 tons of CMA produced by SRI International beginning in 1983-84. This CMA was a fine powered material and there were some significant handling and storage problems. None the less, CMA was judged to be an effective deicer; however, not, as convenient as salt nor as rapid acting as rock salt. Michigan concluded that CMA could be effective for snow and ice control especially on structures where control of chlorides was critical.

Implementation research on CMA in California began in 1985. The program included a small scale study utilizing one maintenance station at Caples Lake on Highway 88. The CMA used was purchased from Verdugt, Inc., a Dutch company and imported for use during the Winter of 1985-86. The CMA was a powdered material with a Ca:mg ratio of about 1:1. It was mixed with cinders, loaded in trucks and spread on a test section. Various problems occurred, but basically, this initial study was encouraging. The report (20) concluded that CMA was as effective as rock salt in deicing operations and worked well when combined with cinders, but that it had handling difficulties including mixing and application problems when combined with wet sand.

Following further research sponsored by the FHWA to improve the ice melting characteristics of CMA (21) and manufacturing development by Chevron Chemical Company to produce a pelletized form of CMA, additional performance testing was conducted during the winter of 1986-1987 by three states (Wisconsin, Massachusetts and California) and the province of Ontario, Canada. The CMA used in these studies was either a pelletized pure CMA with a Ca:Mg ratio of 3:7 or the same CMA coated on sand at about 25-30 percent CMA to sand by weight. In findings, summarized by Chollar (22), CMA was judged as effective as salt at deicing roads, but required a longer time to react. Additionally, application rates for CMA were higher than for salt (1.4-1.6), but fewer applications were required overall bringing the total weight ratio of CMA to salt to about 1.1 to 1 for most storms. Handling still proved to be a problem, but much less than when using previous CMA products. Dust and clumping were the most significant handling concerns, but these problems could be easily overcome. CMA showed a tendency to maintain a deicing effect from storm to storm, decreasing the need for early applications in later storms and CMA exhibited the ability to prevent pack formation allowing easier snow removal by snow plows. The results from CMA-sand mixtures showed varying degrees of success and, at least in California, due to handling problems was judged as not successful.

In California, CMA produced by the Chevron Chemical Company has been utilized on the three main trans-Sierra routes, Interstate 80, Highway 50 and Highway 88 and in Mono County on Highway 395 and secondary routes. Initial testing was done on the trans-Sierra routes in 1986. The Interstate 80 and Highway 88 sites received pure CMA, bagged, with a Ca:Mg ratio of about 3:7. The Highway 50 site received CMA coated on sand. Each site had an approximate three mile test area with a comparable control area where salt was used for deicing. During storms, pure CMA, mixed with sand or cinders, or CMA coated on sand was spread in the test area and the results obtained were compared with results obtained using salt in the control area.

Results from the initial testing were very mixed. On Interstate 80 and Highway 50, maintenance personnel concluded that CMA was less effective than salt for road deicing. Although some attributes of CMA were judged superior, such as the tendency to control pack formation and the faster removal of pack after the storm, overall CMA was considered significantly inferior. Maintenance workers also expressed health concerns due to dust, smell and possible skin irritation. There was a negative public reaction, principally due to health concerns, but also due to perceived poor results from Interstate 80 that were reported in local newspapers. Conversely, on Highway 88 CMA was judged as superior to salt. Maintenance personnel liked it because it kept pack from forming, worked in areas where salt would not and it had a residual effect. Unlike the other sites, on Highway 88, there were no reported health concerns.

Following the Winter of 1986-1987, because of these ambiguous results, a single maintenance station was chosen to convert to CMA use entirely. This plan was developed because most problems with CMA use observed during 1986-87 were thought to be due to lack of acceptance by maintenance personnel and poor public understanding leading to a public relations problem. the site chosen was the Crestview Maintenance Station on Highways 395 in Mono County. This site was chosen because the highway receives significant snowfall, but does not have the high traffic levels which preclude experimentation with different maintenance procedures. The area also has a small population that has a high environmental awareness. During the Winter of 1987-88 about 30 miles of highway were deiced using CMA. The CMA used was pelletized, pure CMA which was received in bulk.

Following this successful use, the implementation area was expanded to include all of Mono County for the Winter of 1988-89, a total test length of about 150 miles of highway. Almost 900 tons of CMA were purchased and CMA was successfully used to provide snow and ice control within the County. A maintenance operational report written by Highway District 09 detailing procedures and findings during the Mono County tests is at Appendix A.

Based on results from the ongoing California implementation program, the following general conclusions can be drawn:

1. CMA can be an effective deicer comparable to salt. It is, however, slower working than salt and is not effective at very low temperatures in low humidity areas for removing ice-pack.

2. CMA has operational characteristics different from salt and, therefore, a thorough training program is essential to maintenance personnel acceptance and good product performance. Except for warm storms, CMA does not melt pack like salt, rather CMA produces an effect where snow does not form a pack allowing snow plows to effectively remove snow down to the pavement.

3. CMA requires different maintenance practices from salt, including early application during storms, higher application rates but lower application intervals and a longer working time prior to plowing.

4. CMA has had no observable deleterious environmental consequences to roadside vegetation nor an observable corrosion effect to maintenance vehicles. It does, however, form a film on vehicles which must be scrubbed to remove.

5. CMA acceptance by the public requires a successful public relations program to demonstrate the advantages of CMA while alleviating public concern for health effects. Once accepted, for an environmentally aware public, CMA becomes the deicer of choice as opposed to salt.

6. CMA (Chevron) does not dissolve as rapidly as salt, and in areas, such as Mono County, where there is low humidity and low temperatures, may not go into solution as rapidly as is desirable. During low temperature conditions (below 20° F) mixing with small amounts of salt, about 10-20 percent, increases the mix effectiveness.

7. Use and storage in areas with low as opposed to high humidity areas requires different techniques. Over summer storage in low humidity areas has no effect on CMA structure, whereas long term storage in more humid areas can cause an increase in moisture content. Additionally, CMA may form a solid mass if it becomes wet. In areas with wet storms, CMA trucks must be covered to prevent caking. In other areas this precaution is not necessary.

6.3   CONCLUSIONS

CMA has undergone the most extensive testing program of any alternative deicing product now available. It has been found to be environmentally safe and non-corrosive. It has an effective road deicer, but it has different operational characteristics than road salt. Consequently, maintenance personnel need to receive training and experience before optimum results can be obtained. There are still handling difficulties with CMA, such as dust and storage, but these can be easily overcome. The most significant draw-back to CMA is the cost, a problem that will be difficult to overcome. However, when comparing CMA and salt costs for deicing use, the long term life-cycle repair and replacement costs to the highway system, environmental degradation of the aquatic and terrestrial ecosystem and vehicle deterioration experienced by the motoring public should be considered.

ACKNOWLEDGEMENT

The California Department of Transportation (Caltrans) formed the Caltrans Snow Removal Policy Task Force because of concern for environmental damage from increased use of salt as a deicing chemical. As a part of the Task Force's overall policy review, the Chairman requested that a summary report be prepared on Calcium Magnesium Acetate (CMA), an alternative deicing chemical, currently being tested by Caltrans. CMA has been shown to be an environmentally safe deicing chemical and has been undergoing implementation testing for several years, but the research report has not been completed.

The purpose of this chapter has been to summarize research conducted on CMA throughout the Nation to apprise Task Force members and Caltrans policy makers on completed and ongoing research efforts and results, and to present conclusions from the current Caltrans implementation program.

114

REFERENCES

1   "The Great Salt Debate." Better Roads. June, 1987. pp. 30-35.
2   Personal Communication, Roy Kelly, California Department of
     Transportation, July, 1989.
3   F.O. Wood. 1981. "Survey of Salt Usage for Deicing Purposes." in
     Automotive Corrosion by Deicing Salts. National Association of Corrosion
     Engineers, Houston, TX. pp. 25-34.
4   L.R. Hudson. 1987. "Calcium Magnesium Acetate (CMA) From Low-Grade
     Biomass." New York State Energy Research and Development Authority.
     Paper presented at IGT Conference, Energy From Biomass and Wastes XI,
     March 18, 1987, Orlando, FL.
5   D. Murray, and U. Ernst. 1976. An Economic Analysis of the Environmental
     Impact of Highway Deicing. U.S. EPA. EPA-600/2-76-105.
6   R.F. Scharpf and M. Srago. 1974. Conifer Damage and Death Associated
     with the Use of Highway Deicing Salt in the Lake Tahoe Basin of California
     and Nevada. U.S.D.A., U.S. Forest Service, Forest Pest control-Technical
     Report 1. 16 pp.
7   S.A. Dunn, and R.U. Schenk. 1980. Alternative Highway Deicing Chemicals.
     Federal Highway Administration. FHWA-RD-79-108. 158 pp.
8   R.L. McCrum, J.W. Reincke, and J.W. Lay. Evaluation of Calcium Magnesium
     Acetate (CMA) as a Deicing Agent: Corrosion Phase-A Comparative
     Evaluation of the Effects of CMA vs. Salt (NaCl) on Highway Metals (3
     Month Exposure). Michigan Transportation Commission. Research Report No.
     R-1258, March, 1985. 37 pp.
9   C.E. Locke and K.J. Kennelley. 1986. Corrosion of Highway and Bridge
     Structural Metals by CMA. Federal Highway Administration Report No.
     FHWA/RD-86/064. June, 1986. 292 pp.
10   B.H. Chollar and Y.P. Virmani. 1988. "Effects of Calcium Magnesium
     Acetate on Reinforced Steel Concrete." Public Roads. Vol. 51, No. 4.
     pp. 113-115.
11   D.S. Slick. 1987. Effects of Calcium Magnesium Acetate (CMA) on
     Pavements and Motor Vehicles. Federal Highway Administration. Report No.
     FHWA/RD-87/037. April, 1987. 187 pp.
12   G.R. Winters, J. Gidley, and H. Hunt. 1985. Environmental Evaluation of
     Calcium Magnesium Acetate (CMA). Federal highways Administration. Report
     No. FHWA/RD-84/094. California Department of Transportation. June, 1985.
     74 pp.
13   C.L. Rea and J.D. LaPerriere. 1985. Effects of Calcium Magnesium
     Acetate, A Road Deicer, On the Lentic Environment in Interior Alaska.
     Institute of Water Resources/Engineering Experiment Station, University of
     Alaska-Fairbanks. Alaska Department of Transportation and Public
     Facilities, Report No. AK-RD-86-02, 48 pp.
14   H.A. Elliot and J.H. Linn. 1987. "Effect of Calcium Magnesium Acetate on
     Heavy Metal Mobility in Soils." J. Environmental Quality. Vol. 16, No.
     3, 1987: pp. 222-226.
15   Chevron Chemical Company. "Technical Bulletin - Toxicity." July, 1987.
16   C.W. Marynowski, J.L. Jones, R.L. Boughton, D. Tuse, J.H. Cortopassi and
     J.E. Gwinn. 1983. Process Development for Production of Calcium
     Magnesium Acetate (CMA). SRI International, Report No. FHWA/RD-82/145,
     Federal Highway Administration, March, 1983.
17   L.G. Ljungdahl, L.H. Carreira, R.J. Garrison, N.E. Rabek, L.F. Gunter and
     J. Wiegel. 1986. CMA Manufacture (II): Improved Bacterial Strain for
     Acetate Production. University of Georgia, Research foundation. Report
     No. FHWA/RD-86/117, Federal Highway Administration, June, 1986.
18   DeFoe, J.H. 1984. Evaluation of Calcium Magnesium Acetate (CMA) as an
     Ice Control Agent. Testing and Research Division, Michigan Department of
     Transportation, Report No. R-1248. 28 pp.
19   G. Demich and T. Weiman. 1984. "CMA Research Project." Washington
     Department of Transportation.

20   J. Gidley. 1986. Preliminary Evaluation of Calcium Magnesium Acetate for use as a Highway Deicer in California. California Department of Transportation, Project No. E86TL63, September, 1986. 15 pp.

21   R.U. Schenk. 1985. Ice-Melting Characteristics of Calcium Magnesium Acetate. Bjorksten Research Laboratories. Federal Highway Administration, February, 1985. 63 pp.

22   B.H. Chollar. 1988. "Field Evaluation of Calcium Magnesium Acetate During the Winter of 1986-87." Public Roads. Vol 52, No. 1. June, 1988. pp. 13-18.

APPENDIX A - District 09 Operational Report on CMA Use - Prepared by George Nash, Deputy District Director, Maintenance and Operations, Caltrans District 09, 87/88 and 88/89 F.Y.

Background

In 1987, District 9 was selected to continue the operational testing of the road deicing material, CMA (Calcium Magnesium Acetate) that was started on a small scale in Districts 3 and 10 in previous winters. District 9 was selected because the BCP funds available would approximately provide enough CMA to match previous salt use and allow for total salt replacement in Mono County areas.

In the first year, 87/88, $200,000 was available which provided enough CMA to replace salt use in the Crestview Maintenance Station area, and in 88/89 $600,000 was available which allowed CMA substitution for salt throughout the Bridgeport Maintenance Territory, which is basically all of Mono County.

The District readily accepted the opportunity of testing CMA because we had been receiving considerable opposition to use of salt. Even though the amount of salt being used in Mono County was fairly small (2 ±/tons/lane mile), many residents were concerned about the corrosion it could cause to their vehicles, and others were concerned for the possible effect on roadside vegetation and water quality. Due to these concerns, the district received excellent support from residents, local politicians, the press, and other local agencies with the CMA project.

CMA Use in 87/88 FY

The BCP for the 87/88 FY provided $220,000 for purchase of CMA. The contract price for the CMA - Chevron ICE-B-GON - was $563.00, and we received 354 tons. We used 323 tons of this amount through the winter on Routes 395, 203 and 158 in southern Mono County out of the Crestview Maintenance Station. Approximately 31 tons were remaining in inventory at year end. Attached are the post mile locations where this material was used. Also attached is the District's salt use record for the 86/87 FY for comparison.

In 87/88 CMA was used on Route 203 through the downtown business area of Mammoth, even though this is outside of the Crestview Maintenance Station area of responsibility. This was done for two reasons: Primarily it was an opportunity to substitute for salt use in this area because the Town Council was concerned about the effect of possible salt accumulation on the bluegrass landscaping in that area. Second, it provided an opportunity to use CMA in an urban situation.

CMA Use in 88/89 FY

In 88/89, $600,000 was available for purchase of CMA. Again, Chevron was awarded the contract at a price of $675.50. Through the winter the District took delivery of 887 tons of ICE-B-GON CMA. A total of 686 tons were used and 224 tons were left in inventory at year end. The material was used throughout Mono County on Routes 395, 203, 158, 182, 108, and 270. (The attached summary shows amounts used on each route). It essentially totally replaced salt, with only 36 tons of salt used for the entire season; and most of this salt was used for CMA/salt mixture tests.

In 88/89, 12 tons of the bulk CMA were also used on Routes in northern Inyo County. In addition, we purchased 23 tons of the material at $746/ton delivered in 1600-pound sacks as a trial for this method, since no bulk storage facilities are available in this area. This sacked material was stored at the Independence and Olancha satellite stations and was used on Route 395 from Dunmovin to Big Pine (Iny-395-20/100). This CMA was used as a supplement to the salt use in this area of Inyo County.

CMA Effectiveness

Generally, the District's operational tests of CMA have shown that it can be used in lieu of salt for Class B route level of service needs.

CMA, as with previous years' salt use, was spread mainly to prevent formation of thick ice pack during storms. Usually R-1 chain controls were still needed (as before with salt use) during actual storm periods. However, by using the CMA, chain controls generally could be lifted shortly after snowfall ceased, since pack had not been allowed to build up.

We found that CMA spread rates of between 250 to 400 lb/lane mile prevented significant pack formation for up to six hours. This is accomplished by spreading the CMA as soon as pavement becomes white and pack begins to form. We found that once spread, it was necessary to not plow for at least 20 minutes to allow time for the CMA to work through and mix with the thin initial pack; after that, normal plowing would produce near black pavement for up to six hours, depending on snowfall intensity. During periods of heavy snowfall accumulations, the CMA, if spread before initial heavy pack buildup, changes the snow consistency to "oatmeal" slush which plows off readily when a plow can get to it, leaving little, if any, pack.

Generally, we found that CMA was as good as salt in removing black ice, except in cases where the temperature is so low that no free moisture is available to start the CMA working. On the negative side, we found, however, that if hard pack is allowed to form, CMA is very slow and not nearly as effective as salt in eating through and releasing pack from the pavement. The

CMA pellets will slowly drill through a pack, depending on temperature, but it appears that by the time they reach the bottom, little is left on the pellet, and no brine is formed to release the pack. In some cases where the temperature is near 20 degrees F or below, the CMA pellets are extremely slow to even drill through the pack, apparently because there is insufficient free moisture to activate the CMA reaction. Some tests with CMA/salt mixtures were done with the idea that the salt would produce enough free water to activate the CMA. This seemed to improve the situation, but additional tests are needed to reach any sound conclusion.

The overall quantity of CMA used in Mono County in the winter of 88/89 -651 tons - is 1.7 times the amount of salt used over the same area through the winter of 86/87. This 1.7 factor is partly due to the fact that more CMA is needed than salt to melt a given quantity of ice, and is partly due to increased use providing a better level of service. This was particularly true on Routes 203 and 158 where CMA was freely used as opposed to the previous policy of limiting salt use. Overall differences in winter conditions from year to year also account for part of the difference. Generally, our observations indicated that on a storm-for-storm basis, approximately the same quantity or slightly more CMA by weight is needed as salt to produce the same results.

In the first winter of use, 87/88, we used as much as five tons/lane mile of CMA as opposed to the approximate salt use level of about two tons/lane mile. This was because we were still learning how to use it, and typically, it was not being spread soon enough to prevent pack formation. Consequently, large amounts were then used trying to remove the pack. As discussed earlier, we learned that if the CMA is spread on a timely basis at the beginning of the storm, pack is prevented, and much less CMA is needed.

Due to its relatively low density - 44#/C.F., storage for sufficient quantities of CMA was a concern. Also, due to the round pellet form of the ICE-B-GON material, the angle of repose in a storage pile is quite flat, further adding to the storage problem. Because of this, the District had to have frequent small CMA deliveries. If we begin using CMA on a long-term basis, larger storage facilities will be needed.

The operational test use of sacked CMA (1600-pound sacks) in Inyo County proved to be satisfactory. If funds became available to allow expanded use of CMA in Inyo County, the sacked storage method would be adequate until bulk storage facilities could be constructed.

Maintenance Crew Reaction

Prior to actual use of CMA, District, TransLab, and Chevron ICE-B-GON personnel met with Maintenance crews and supervisors to familiarize them with the product. Methods of use, reasons for the operational testing, and the material data safety sheet were discussed.

No negative reactions by crew members to using the material have been encountered. We did, however, note some reluctance on the part of several supervisors to using the material in sufficient quantities due to its high cost.

CONCLUSIONS

The District's operational use of CMA shows that CMA can be used to replace most salt used on highways such as Route 395 through Mono County with level of service B. Some modest salt use would still be needed for extreme conditions.

The CMA, if used timely, prevents pack buildup and allows bare pavement - normal conditions - to be achieved shortly after snowfall ceases. It, however, as with our previous level of salt use, does not eliminate the necessity of chain controls during the actual snowfall periods. Additional testing of application rates and timing are still needed to determine most economical use for Class B level of service. Also, more tests with CMA and small quantities of salt mixtures are needed to see if that would be an effective way to at least significantly reduce salt use.

Also, further tests are needed to determine if more frequent CMA applications, or CMA mixed with small quantities of salt, and more frequent plow passes would produce bare pavement - no chain control conditions - throughout storm periods. Pending further tests of this nature, the District is reluctant to use CMA on, for example, Route 58 freeway through the Tehachapi Pass area where it is imperative that a bare pavement condition be maintained. (Even R-1 conditions on Route 58 are essentially the same as closing the highway).

ATTACHMENT

87/88 CMA USE			
Mono-395-26.0/40.5	42	210	5.0
Mono-203-4.0/8.6	17.2	53	3.1
Mono-158-0.0/5.9	11.8	60	5.1

State of California   DISTRICT 09   WINTER
Department of Transportation   SUPT./C.C. 620   1988/1989
SALT USAGE SUMMARY   DATE 5/10/89
CMA   SHEET 2 OF 3

1	2 FROM		3 TO		4 LANE	5
ROUTE	COUNTY	POST MILE	COUNTY	POST MILE	MILES (TOTAL)	TONS SPREA
395	(48) INY	20	(48) INY	73	118	23
06 120 168 266	(47) MNO & (48) INY	VAR	(47) MNO & (48) INY	VAR	200 ±	12*

NOTE: CMA
* 12 tons various from Sherwin Shed - 630 inventory

SUPT. LEVEL

6 Territory Salt CMA Inventory / / 0 Tons	7 Total Salt CMA Received 35 Tons	8 Disbursed Winter of 88/89 35 Tons	9 Territory Inventory Balance / / 0 Tons	10 Physical Inventory Balance / / 0 Tons	11 Difference + ; - 0 Tons

DISTRICT LEVEL

/ / Tons	Tons	/ Tons	/ / Tons	/ / Tons	Tons

State of California			DISTRICT 09	WINTER	
Department of Transportation			SUPT./C.C. 630	1988/1989	
SALT USAGE SUMMARY			DATE 5/23/89		
CMA				SHEET 3 OF 3	

1	2 FROM		3 TO		4 LANE	5
ROUTE	COUNTY	POST MILE	COUNTY	POST MILE	MILES (TOTAL)	TONS SPREA
						TONS
395	MONO	0.0	MONO	120.49	329	563 1.7
203	MONO	1.0	MONO	8.6	25.40	24 0.9
158	MONO	0.0	MONO	6.0	12	32 2.7
108	MONO	10.0	MONO	15.0	10	6 0.6
270	MONO	0.0	MONO	9.8	19	10 0.5
182	MONO	0.0	MONO	12.65	25	16 0.6

SUPT. LEVEL

6 Territory Salt CMA Inventory / /	7 Total Salt CMA Received	8 Disbursed Winter of /	9 Territory Inventory Balance / /	10 Physical Inventory Balance / /	11 Difference + ; -
31 Tons	852 Tons	651 Tons	232 Tons	224 Tons	-7 Tons
DISTRICT LEVEL					
/ / 31 Tons	887 Tons	/ 686 Tons	/ / 232 Tons	/ / 224 Tons	-7 Tons

State of California   Business   and   Transportation
Agency

MEMORANDUM

E.B. Thomas, Chief   Date: 07/27/87
Office of Roadbed Maintenance   File's

Department of Transportation
District 9

Salt Usage 1986-1987

Salt used in District 9 during the 86-87 season was as follows:

Route	P.M.	Lane Miles	Tons Salt	Tons/L.M.
Mno-395	0.0.27.2	109	15	0.14
Mno-395	27.0-56.0	66	113	1.71
Mno-395	56.0.77.0	50	15	0.30
Mno-395	77.0-120.4	90	190	2.11
Mno-203	1.5-8.6	11	10	0.91
Mno-158	0.0-6.0	12	12	1.0
Mno-108. & 89		20±	25	1.25
Ker-58	80.0-102.0	44	41	0.93
Ker-178	62.5-82.5	12	10	0.83
(Various Locations)				
Ker-202	1.2-7.4	12.6	4	0.31

GEORGE NASH
Deputy District Director
Maintenance & Operations

TABLE 6.1

Corrosion Comparison Between Salt and CMA - Published Weight Loss Data*

FHWA (Locke)	FULL IMMERSION				ALT IMMERSION-DIP				SPRAY FOG TEST			
	H_2O	NaCl	CMA	RATIO	H_2O	NaCl	CMA	RATIO	H_2O	NaCl	CMA	RATIO
A-36 STEEL	4.80	10.1	1.7	5.9	6.0	10.8	3.4	3.2	2.80	6.9	3.8	1.8
		10.2	3.5	2.9		16.3	3.4	4.8		10.7	3.6	3.0
		7.6	3.3	2.3		14.0	2.7	5.2		25.3	6.8	3.7
A-36 WELDED	4.70	9.2	2.1	4.4	4.8	11.7	3.0	3.9	2.10	6.4	3.2	2.0
		8.8	2.1	4.2		13.5	3.4	4.0		10.2	3.0	3.4
		5.8	3.1	1.9		14.6	4.4	3.3		18.9	5.1	3.7
A-36 CREVICE	3.8	8.8	1.9	4.6	7.6	9.6	3.0	3.2	3.20	6.7	2.7	2.5
		9.2	3.0	3.1		13.4	2.7	5.0		5.7	3.5	1.6
		7.9	3.0	2.6		17.7	3.7	4.8		10.9	4.5	2.4
A-588 WEATHER	0.90	7.3	1.5	4.9	1.5	9.7	4.1	2.4	1.40	10.9	3.3	3.3
		11.8	1.7	6.9		13.6	2.3	5.9		22.6	4.4	5.1
		18.3	2.7	6.8		16.9	0.9	18.8		30.6	5.3	5.8
6061-T6 AI	0.95	1.28	0.00		0.1	0.37	0.04	9.2	0.12	3.25	0.01	3.25
		0.22	0.06	3.7		0.16				2.03	0.39	5.2
		1.29	0.05	25.8		0.18	0.01	18.0		1.69	0.02	84.5

MICHIGAN

A-36						44.2	12.2	3.6				
A588 WEATHER						84.3	8.8	9.6				
GALV GUARDRAIL						14.4	4.9	2.9				
GALV BRIDGERAIL						13.6	2.4	5.6				
REBAR						52.0	14.6	3.6				
TENSION STRAND						22.2	4.4	5.0				

ICE-B-GON[R]

A-36		0.79	0.00			7.91	0.52	15.2		49.0	10.0	4.9
6061 AI		0.07	0.03	2.3		0.03	0.01	2.7				

FHWA data from FHWA/RD-86/064 by Locke, et alia
  NaCl solutions at 0.3, 0.6, 1.2 wt%, by row
  CMA solutions at 0.5, 1.0, 2.0 wt%, by row
  H_2O was tap water for comparison

MICHIGAN data from Research Report No. R-1258, Mich D.O.T. 1985
  NaCl solutions at 3.5 wt%
  CMA solutions at 6.1 wt%

ICE-B-GON[R] data from internal report by Southwest Research Institute, 1988
  (Available from Chevron Chemical Company)
  NaCl solutions at 3.5 wt%
  Ice-B-Gon solutions at 3.5 wt%

*Data compilation from Hoenke, Karl A. "Materials Compatibility with Calcium Magnesium Acetate". Chevron Chemical Co., July, 1988.

FIGURE 6.2

Material Data Safety Sheet (MSDS) for CMA - Chevron Chemical Company

MATERIAL SAFETY DATA SHEET   **CHEVRON**
Prepared According to the OSHA Hazard Communication Standard (29 CFR 1910.1200).
(Formerly Called MATERIAL INFORMATION BULLETIN

CHEVRON ICE-B-GON™ Deicer   **CAUTION:**   **MAY BE HARMFUL IF SWALLOWED**

Typical Composition
         Calcium magnesium acetate
         (calcium to magnesium molar ratio 3:7) 91% Min.
         Water (free and hydration)   5% Max.
         Water-insoluble material   4% Max.

EXPOSURE STANDARD
         No Federal OSHA exposure standard or ACGIH TLV has been established for this material. Based on information reviewed to date, we recommend an exposure standard of 10 mg/m^3 total dust for a daily 8-hour exposure to this product.

PHYSIOLOGICAL & HEALTH EFFECTS

EYES

Expected to cause no more minor eye irritation. Application into the eyes of rabbits produced slight membrane irritation without corneal injury. All eyes were normal by 72 hours after application.

SKIN

Not expected to be irritating to the skin. Application onto the skin of rabbits produced no observable irritation. The Draize skin irritation score for a 4-hour exposure was 0.1 out of possible 8. Not expected to be acutely
toxic by skin absorption. The acute dermal LD_{50} (rabbit) was greater than 5000 mg/kg. See Additional Health Data.

INHALATION

Not expected to be acutely toxic by inhalation. The inhalation LC_{50} (rat) for a four-hour exposure was greater than 4600 mg/m^3. Breathing the dust may be irritating to the respiratory tract.

INGESTION

Expected to have slight acute toxicity by ingestion. The acute oral LD_{50} (rat) was 3145 mg/kg. See Additional Health Data.

Chevron Environmental Health Center, Inc.
PO Box 4054, Richmond, CA 94804-0054
Emergency Phone Number (415)233-3737
Rev. 6 08/03/87 - No. 3081

EMERGENCY & FIRST AID PROCEDURES

EYES

Flush eyes immediately with fresh water for at least 15 minutes while holding the eyelids open. If irritation persists, see a doctor.

SKIN

Wash skin thoroughly with soap and water. Launder contaminated clothing.

INHALATION

If respiratory discomfort or irritation occurs, move the person to fresh air. See a doctor if discomfort or irritation continues.

INGESTION

If swallowed, give water or milk to drink and telephone for medical advice. Consult medical personnel before inducing vomiting. If medical advice cannot be obtained, then take the person and product container to the nearest medical emergency treatment center or hospital.

FIGURE 6.2 - Material Data Safety Sheet-(page 2)

MATERIAL SAFETY DATA SHEET

ADDITIONAL HEALTH DATA

Calcium magnesium acetate (3:7) did not cause skin sensitization in the Modified Buehler test in guinea pigs.

Daily oral dosing of male and female rats at 1000 mg/kg/day for 28 days resulted in no signs of toxicity.

SPECIAL PROTECTIVE INFORMATION

Eye Protection: Do not get in eyes. Eye contact can be avoided by wearing chemical safety goggles.

Skin Protection: No special skin protection is necessary.

Respiratory Protection: No special respiratory protection is normally required. However, if operating conditions create high airborne concentrations, the use of an approved respirator is recommended.

Ventilation: No special ventilation is usually necessary. however, if operating conditions create high airborne concentrations of this material, special ventilation may be needed.

FIRE PROTECTION

Flash Point: n/a
Autoignition Temp.: n/a
Flammability Limits: n/a
Extinguishing Media: Material does not burn.

SPECIAL PRECAUTIONS

Read and observe all precautions on product label.

ENVIRONMENTAL PROTECTION

Environmental Impact: This material is not toxic to fish. The acute LC_{50} for rainbow trout and fathead minnow was 18,700 and 21,000 mg/liter respectively.

Precautions if Material is Released or Spilled: Clean up large spills, observing precautions in Special Protective Information.

Waste Disposal Methods: Place contaminated materials in disposable containers and dispose of in a manner consistent with applicable regulations. Contact local environmental or health authorities for approved disposal of this material.

REACTIVITY DATA

Stability (Thermal, Light, etc.): Stable
Incompatibility (Materials to Avoid): None

Hazardous Decomposition Products: None known.

Hazardous Polymerization: Will not occur.

PHYSICAL PROPERTIES

Solubility: Partially soluble in water. Appearance (Color, Odor, etc.): Yellow or white hard pellets.
Boiling Point: n/a
Melting Point: n/a
Specific Gravity: 1.2 Min.
Vapor Pressure: NDA
Vapor Density (Air=1): n/a
Percent Volatile (Volume%): n/a
Evaporation: n/a
pH: 8 to 10 in a 10% solution.

n/a = Not Applicable
NDA = No Data Available

The above information is based on data of which we are aware and is believed to be correct as of the date hereof. since the information contained herein may be applied under conditions beyond our control and with which we may be unfamiliar and since date made available subsequent to the date hereof may suggest modifications of the information, we do not assume any responsibility for the results of its use. This information is furnished upon the condition that the person receiving it shall make his own determination of the suitability of the material for his particular purpose.

Rev. 6 08/03/87
No. 3081

TABLE 6.2

A Partial List of Environmental Effects from Road Deicing: CMA and Salt

Environmental Impact	Calcium Magnesium Acetate (CMA)	Salt (NaCl)
Soils	- Biodegradable in soil. - Ion exchange of Ca, Mg displaces other cations (metals). - No adverse effect on soil compaction and Increases soil permeability.	- May accumulate in soil. - Ion exchange of Na displaces other cations. - Causes soil compaction which decreases permeability.
Vegetation	- Little or no adverse effect.	- Osmotic stress and soil compaction harm root systems. - Spray causes foliage dehydration damage.
Groundwater	- Poor mobility in soil, unlikely to reach groundwater. - Ca, Mg increase water hardness.	- Mobile Na and Cl ions readily reach groundwater. - Increases Na and Cl concentrations in well water as well as alkalinity and hardness.
Surface Water	- Potential for oxygen depletion through biological oxygen demand (BOD) at conc. greater than 100 ppm in closed systems. - Decomposes in 5 days at 20°C, 10 days at 10°C, 100 days at 2°C.	- Causes density stratification in ponds and lakes which can deplete oxygen. - Direct runoff or salty groundwater can raise Na and Cl concentrations in streams and lakes.
Aquatic Life	- Less toxic to trout than salt. - Minimal effect on trout eggs up to 5 times expected max runoff concentration of 1000 ppm. - No effect on food chain (zooplankton, daphnia, bluegill and fathead minnows) up to 1000 ppm.	- Monovalent Na, Cl ions stress osmotic balance. - Toxic levels: Na 500 ppm stickleback, Cl 400 ppm trout.

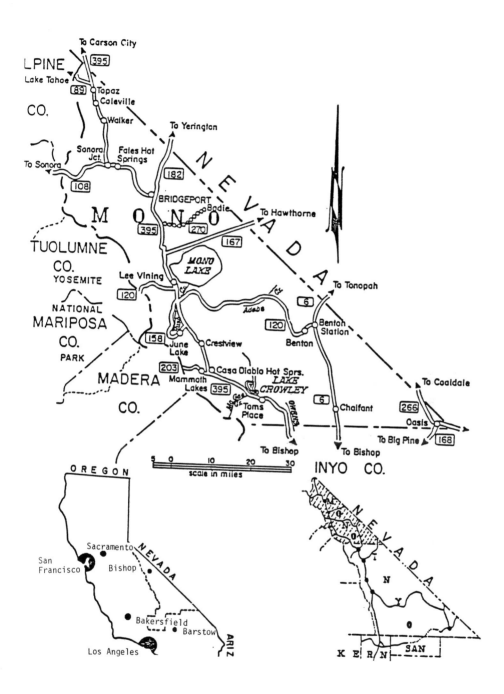

To Carson City
LPINE
395
Lake Tahoe
CO.
89 Topaz
Coleville
Walker
To Yerington
Sonora   Fales Hot
Jct.   Springs
To Sonora
182
108
BRIDGEPORT Bodie
M O N O   To Hawthorne
395 270
167
TUOLUMNE   MONO
CO.   LAKE
YOSEMITE   Lee Vining
120   To Tonopah
NATIONAL   Adobe   6
MARIPOSA   120   Benton
CO.   Station
PARK   158 June   Crestview Benton
           Lake
MADERA   203   Casa Diablo Hot Sprs.   LAKE
          Mammoth   CROWLEY   To Coaldale
CO.   Lakes 395   6   Chalfant 266
                  Toms   Oasis
                  Place   To Big Pine 168
                         To Bishop   To Bishop
                                      INYO CO.
OREGON   5 0   10   20   30
                    scale in miles

N E V A D A

Sacramento
San   NEVADA
Francisco Bishop

Bakersfield
          Barstow   ARIZ

Los Angeles

N E V A D A

K E R N   SAN

Chapter 7

THE CORROSIVITY AND ELECTROCHEMICAL BEHAVIOR OF A NEW DEICER, CALCIUM MAGNESIUM ACETATE[1]

Kevin J. Kennelley[2], Carl E. Locke[3]
[1]Work conducted while at the University of Oklahoma.
[2]Exxon Production Research Company, P.O. Box 2189, Houston, TX   77001
[3]University of Kansas, 4010 Learned Hall, Lawrence, KS   66045

7.1   INTRODUCTION

Deicing chemicals are widely used in the United States because of the bare-roads year-round policy existing in all the states. The public has grown accustomed to this policy and now demands that the roads, and especially bridges, be free of ice and snow at all times. As a result, the use of deicers has grown steadily since 1946. The most commonly used chemical for deicing is sodium chloride (NaCl), and 12 x 106 tons are placed on the highways and bridges in the United States annually [1].   This statistic can be stated in other ways.   In Wisconsin, approximately 450 pounds of salt are used for each lane-mile of highway each winter season [2]. Calcium chloride is also used as a deicer, but is used less because of its higher cost.

These chloride-containing salts are corrosive. Automobile corrosion is a widely recognized and costly problem in all developed countries faced with high use of deicing salts. It is estimated that 50 percent of the automobiles in the United States and Canada are located in areas where NaCl is used extensively [2]. In addition to contributing to automobile corrosion, the use of chloride deicers leads to corrosion of bridge structure materials, for example, the corrosion of the reinforcing steel embedded in concrete.

Bridge decks designed to last 50 years are being replaced in 10 to 20 years [3]. The accumulated damage on bridges located on the Federal-aid system has been estimated to have a cost of repair of $26.7 billion. Of this, $8.8 billion is needed for bridge deck restoration.   Annual cost for all bridge deck restoration is estimated to be $400 million [3].

Several different approaches are used to reduce the corrosion damage caused by the chloride-containing deicing salts: (a) automobiles are now manufactured with special primers and coatings to combat the corrosion process; (b) bridge structural materials are painted, and several methods are used for corrosion control of the concrete reinforcing steel; (c) epoxy-coated steel is the most commonly used corrosion preventive system for new construction; and (d) cathodic protection of reinforcing steel in existing construction seems to be an accepted method of halting corrosion in bridge decks.

The Federal Highway Administration (FHWA) has investigated the possibility of using a noncorrosive deicing material as another approach to reducing the cost of corrosion caused by the chloride-containing chemicals. A study commissioned by the FHWA in 1980 identified calcium magnesium acetate (CMA) as a possible alternate deicing compound [4]. This initial project included an exploratory corrosion experiment that indicated that CMA might be a noncorrosive deicer. Based on this project, the FHWA funded a relatively extensive corrosion project to determine the corrosivity of CMA to a range of bridge structural metals. The results of that study [5] are summarized in this chapter.

Calcium magnesium acetate is a name used to describe the reaction product of acetic acid and dolomitic limestone. This reaction product is a mixture of calcium acetate and magnesium acetate. Bjorksten Laboratories found that CMA had deicing properties close to those of sodium chloride and that the manufacturing costs were substantially greater [4]. It is thought that if CMA proves to be noncorrosive, the real cost of this material, with reduced damage caused by corrosion, would make it attractive for use as a deicer.

7.2   EXPERIMENTAL PRODEDURES

The corrosion behavior of a number of different metals in CMA was studied in different environments. The expansion joints, gutters, railings, and support beams on bridges can be constructed from steel, cast iron, weathering steel, galvanized steel, painted steel, and aluminum. Hence, these metals were immersed in solutions of CMA or exposed to the vapor space above the solutions utilizing static, dip, and spray tests. For comparison purposes, the metals were also exposed to solutions of sodium chloride, calcium chloride, and tap water. The effect of CMA on the corrosion of reinforcing steel in portland cement concrete was investigated. The corrosion electrochemistry of the metal solution systems was studied using several different experimental techniques. A brief description of the experimental details follows. Complete details of experimental procedures can be found in a dissertation by Kennelley [6] and a thesis by Boren [7].

7.2.1 Weight-Loss Studies

Three types of weight-loss tests were conducted at 25#C: static immersion tests, dip tests, and spray tests. The metals exposed to the corrosive solutions are listed in Table 7.1. Two duplicate samples were used for each test condition. These metals are representative of the materials used in the construction of bridge and highway structures.

The metals used met ASTM specifications for the type of metal specified. Galvanizing was done by a commercial galvanizer to a specification of 0.1 gm/cm^2

(3.3 oz./ft^2) over all surfaces of the coupons. Weathering of the galvanized and weathering steel samples was accomplished in Norman, Oklahoma by exposing the rectangular coupons outside during a 6-month period. Some of the steel coupons, after surface cleaning as described in National Association of Corrosion Engineers (NACE) TM-01-69 [9], were painted with a 3-layer linseed-oil-based system consisting of a red-lead primer, a red-lead intermediate coat, and an aluminum-pigmented top coat. This paint system is one used by the Oklahoma Department of Transportation for bridge structures. All samples other than the painted, weathered, and galvanized metals were surface-cleaned before and after exposure using procedures specified in ASTM G1-81 [8] and NACE TM-01-69 [9]. The weathering steel samples were cleaned in the same manner as other steel samples after exposure and before weighing. Weight-loss determinations were made by weighing before and after exposure for all but the painted samples, the U-bend samples of A-36, and the A-325 stressed bolts. These three groups of samples were visually inspected for relative damage of the painted samples and any evidence of stress cracking in the stressed samples.

The samples were exposed to 13 different solutions, listed in Table 7.2, and tap water. The reagent grade CMA (CMA(R)) was prepared from equimolar amounts of reagent grade calcium acetate and reagent grade magnesium acetate. The commercial grade CMA (CMA(C)) was furnished by the FHWA from a batch of CMA prepared from dolomitic lime and acetic acid. The commercial batch of CMA was supplied in pelletized form and contained approximately four to eight percent insoluble product with a biodegradable dye added.

The solution concentrations were chosen from a series of electrochemical tests conducted by Luster [10]. These tests indicated that one percent by weight solutions of CMA could be expected to have the highest corrosion rates of any of the compositions. One concentration (0.5 percent by weight) below this and one concentration (2 percent by weight) above were also chosen. The concentrations of sodium chloride used are equiosmolal (same number of ions) to the CMA concentrations. The calcium chloride concentrations were based on the same concentration of chloride ions as those in the sodium chloride solutions. Tap water was also used for purposes of comparison.

Each type of metal sample was exposed to each of the solutions, fully immersed, half immersed, and in the vapor space (see Fig. 1). The exposure time was 17 months, and the solutions were changed after eight months of exposure. Water lost by evaporation due to the bubbled air was replaced monthly. A set of metal samples was exposed to sprays of each of the 13 solutions. Some difficulty was encountered with the initial spray apparatus based on an aspirator; the arrangement shown in Figure 2 was then developed. Test time was 14 months, which included the exposure to the original setup and the arrangement shown in Figure 2. The equipment shown in Figure 2 was used for approximately 75 percent of the time. A dip apparatus was constructed to immerse each metal sample in the test solutions periodically and then remove it. The guidelines for a dip test specified for stress corrosion cracking in ASTM G-44 [8] were used in the selection of the fraction of time the samples were immersed. The metals were immersed for 10 minutes of each hour and suspended above the solutions for the remaining 50 minutes of each hour. A motorized chain-driven system alternately immersed and removed the coupons from the test solutions. The average duration of the test was 16 months.

7.2.2 Electrochemical Tests

The procedures and apparatus used for the polarization tests were similar to those described in ASTM G-5 [8]. Anodic polarization with reverse scans, cathodic polarization scans, and linear polarization scans were performed on A-36 steel in commercial and reagent grade CMA concentrations of 0.5, 1, 2, 5, 10, and 20 weight percent. These tests were performed to establish the electrochemical behavior of steel in CMA solutions and to calculate corrosion rates.

A Princeton Applied Research Potentiostat (Model 173), Model 175 Programmer, and Model 376 Logarithmic Current Converter were used for the electrochemical experiments. The apparatus that was used to hold the solution and electrodes was similar to that described in ASTM G-5 [8]. A saturated calomel reference electrode was used for all polarization experiments.

Air was vigorously bubbled through the CMA solutions for about one hour, and then the air flow was stopped. It was found that steel would temporarily passivate because of the oxygen saturation after the air bubbling, but the steel potential would shift to the active range (-680 to -700 mv versus a saturated calomel reference electrode) if sufficient waiting time was used. In some systems, 24 hours were required for the potential to shift to an active condition. The potential sweeps were conducted at 0.278 mv/sec and were started after the air was bubbled through the solutions.

Anodic polarization experiments were started at the active corrosion potential. The potential was shifted to about +1,100 mv (with respect to saturated calomel reference electrode), and the scan direction was then automatically reversed. The potential was then scanned back to the original corrosion potential. Cathodic polarization curves were also begun at the active corrosion potential, and the potential was shifted to about -1500 mv. The potential was scanned only in the active direction, with no reverse scan for the cathodic polarization. A plot of potential versus log current was generated for each experiment.

Linear polarization experiments were conducted to determine corrosion rates by scanning the potential +30 mv from the corrosion potential. The scan direction was then automatically reversed, and the potential was scanned to a value -30 mv from the corrosion potential. Results of these tests were then used to calculate corrosion rates. The linear polarization scans were conducted at both active and passive potentials.

7.2.3 Steel in Corcrete

Reinforced concrete slabs were constructed to determine the effect of CMA solutions on the electrochemical potential of the reinforcing steel. Each slab contained two rebar mats that were not electrically connected in the slab, as shown in Figure 3. Slabs were made with bare steel bars in both mats, epoxy-coated bars in both mats, and epoxy-coated bars in the top mat and bare bars in the bottom mat. A standard-mix design of portland cement concrete was used for the blocks. In some of the blocks, sodium chloride was added to the mix water for the portion of the concrete poured around the top mat to provide a 0.2-percent chloride ion concentration in the concrete, based on cement content. Solutions of CMA, sodium chloride, and calcium chloride were ponded on top of these slabs after a curing time of 28 days under wet curing mats. The potentials of the top and bottom mats of copper/copper-sulfate reference electrodes were recorded periodically over an 18-month period. Current flow between the top and bottom mats was taken each time after the potential measurements. The mats were not connected electrically during the time of the measurements.

The potentials of the reinforcing steel were measured in simulated pore solutions and mortar cyclinders soaked in CMA solutions. The simulated pore solutions' compositions were based on work done by Farzammehr [11]. CMA was added to these pore solutions to obtain 0.5, 2.0, and 10 percent by weight with chloride ion compositions up to 1.0 percent by weight.

7.3   RESULTS
7.3.1 Weight-Loss Tests

Corrosion rates of the metals were calculated from the weight-loss tests and are reported in mils per year (mpy). The results for the steel samples (A-36, A-588, A-325), the cast iron, and the aluminum samples are reported for each of the test conditions in Tables 7.3-7.7. The results of the rectangular, the angle iron, the welded, and the crevice samples of A-36 have been averaged and reported as such. The tabulation of the complete results is available in a report on which this paper is based [5].

The corrosion rates from all the test conditions have a similar pattern. The sodium chloride solutions are the most corrosive. Calcium chloride solutions are less corrosive than sodium chloride solutions. The CMA solutions are less corrosive than either of the chloride-containing solutions, with corrosion rates that are at least two to five times less than those found for the sodium chloride solutions.

Solutions of calcium chloride had the same chloride concentration as the NaCl solutions, yet the weight-loss results showed that the $CaCl_2$ solutions were less corrosive than the NaCl solutions. Since the solutions of sodium chloride and calcium chloride contained the same chloride concentration, the differences in corrosion rates must be associated with the cations. The Ca^{2+} ion has a higher charge/ionic radius ratio than the Na^+ ion. The water molecules hydrating the calcium ions will be more acidic than the water molecules hydrating the sodium ions. This suggests that the calcium chloride solutions should be more corrosive than the sodium chloride solutions because of the more acidic environment of the $CaCl_2$ solutions. The experimental weight-loss results show just the opposite: calcium chloride solutions are less corrosive than sodium chloride solutions. This difference is due to the fact that the calcium chloride solutions precipitate calcium carbonate out of solution onto the metal coupon surface. The insoluble carbonate tends to protect the steel and reduce the corrosion rate [12] The presence of calcium carbonate in the corrosion product layer of steel coupons tested in calcium chloride solutions was confirmed by X-ray diffraction analysis.

The dependence of the weight-loss results on the concentration of the NaCl and $CaCl_2$ solutions shown in Tables 7.3-7.7 varies somewhat with solution composition and test condition. The corrosion rates of the metals exposed to the CMA (reage.t grade) (CMA(R)) and CMA (commercial grade) (CMA(C)) solutions did not vary widely with respect to solution concentration. The variation is so slight in most of the test conditions that it is hard to make any such determination.

Exposure conditions greatly affect the corrosion rates. The dip test and the half-immersed test conditions resulted in the highest corrosion rates. This was due to water droplet impingement in the spray test or wave action for the half-immersed coupons. The vapor-space exposure resulted in the lowest corrosion rates of all the test conditions used. Corrosion of the coupons in the vapor space was lowest because the corrosive environment was mainly the humidity in the vapor space of the buckets.

7.3.2 Stressed Samples

The A-325 stressed bolts were tested in the fully immersed and vapor-space immersion tests. None of the bolts broke, and no cracks were observed. The threaded regions of the bolts fully immersed in NaCl and $CaCl_2$ solutions corroded to a greater extent than the bolts in CMA(R), CMA(C), and tap water, as noted visually.

None of the A-36 stressed U-bend steel coupons broke, nor were any cracks found after exposure in the dip, spray, or immersion test. Severe thinning of the U-bend sample in the stressed region of the coupons was visually observed on the samples exposed to the NaCl, $CaCl_2$, and tap water solutions. Coupons exposed to CMA(R) and CMA(C) were not corroded as much as those in the other solutions. A visual comparision of the corrosion experienced for A-36 U-bend samples exposed to 1.2 weight
percent NaCl versus one weight percent commercial grade CMA is shown in Figure 4.

7.3.3 Painted Samples

The effect of the solutions on the painted samples was determined visually for all samples. The paint coating on the A-36 steel samples exposed to NaCl solutions was blistered in nearly all of the test conditions (Fig. 5). The painted samples exposed in the vapor space above the NaCl solutions were in good condition. All other solutions ($CaCl_2$, CMA, and tap water) had little or no effect on the paint coatings.

The effect of the solutions on the painted weathering steel samples was somewhat more severe. The NaCl solutions were the most damaging of all the solutions to the painted coatings. Most were blistered, and in some instances, the paint flaked off the metal. The CMA solutions did not harm the paint except in two of the tests using CMA(C). Some flakes of paint were removed from the samples during exposure to 0.5 percent CMA(C), and a few blisters were found in the paint on samples exposed to 1.0 percent CMA(C). All other samples were in good condition after exposure, which was for approximately 12 months.

7.3.4 Galvanized Samples

It was not possible to make quantitative determination of the corrosion rates of the galvanized coupons because of problems in removing the corrosion product without also removing the galvanized layer. Several different acidic solutions were used in an attempt to accomplish the removal. The cleaning solutions either did not adequately remove the corrosion product layer, resulting in an apparent weight gain, or did not remove all of the galvanized layer totally. Only visual observations, therefore, are reported for these samples.

Sodium chloride solutions appeared to be the most corrosive of all the solutions. Tap water was the least corrosive of all solution to the galvanized layer.

The same difficulty was encountered in determining the effect of the solutions on corrosion of the galvanic couples. For all the samples immersed in the corrosive solutions, the steel coupons were cathodically protected by the galvanized coupon. Corrosion rates for steel were reduced in all solutions to less than 1 mpy. The galvanic coupling had little effect on the samples exposed to the vapor space, dip test, and spray test. This was a result of the lack of a continuous electrolyte to conduct current between the metals in those tests.

7.3.5 Electrochemical Tests

Results from polarization experiments conducted by Kennelley [6] and Luster [10] indicated that steel in CMA solutions behaved as an active-passive metal. Figure 6 illustrates the anodic polarization behavior of A-36 steel in a 1.0-percent CMA(R) solution, and Figure 7 illustrates comparable behavior in a 1.0-percent CMA(C) solution. The directions of the arrows indicate the directions of the potential scan. The curves are typical of active-passive behavior. The A-36 steel was at a potential of about -700mv in relation to the calomel reference electrode before the scan, which is an active condition. Luster [10] found that steel in most of the CMA solutions was at a potential of about -150 to -200 mv in relation to the calomel, which is in the passive range displayed in the polarization curves. Kennelley [6] found that air must be bubbled through the solutions at a very high rate to maintain the steel at the passive potentials. He found that with a low rate of bubbled air or in stagnant condition, the steel would initially be at a passive potential and then with time (2 to 20 hours) would spontaneously shift to an active value.

The polarization curves contained several steps during the anodic portion of the scan. These are thought to be due to changes in the nature of the passive layer on the steel. Similar behavior was observed during the forward anodic polarization scans for both the commercial and reagent grade solutions at the same concentrations.

The reverse scans in Figures 6 and 7 are typical for an active-passive metal susceptable to localized corrosion. The hysteresis loop observed in these reverse scans is typical of systems known to display pitting corrosion. Localized corrosion can propagate at potentials more noble than the the point at which the return loop intersects the forward scan. This point is discussed in ASTM Standard G-61 [8].

Reverse scans during anodic polarization showed a difference in behavior between commercial grade and reagent grade CMA solutions. As the concentration of reagent grade CMA increased in solution, the hysteresis loop on the reverse anodic scan decreased and was not present at concentrations of five weight percent or higher. At all concentrations of commercial grade CMA, a large hysteresis loop was present. This difference in electrochemical behavior must be associated with the dye present in the commercial grade CMA or with trace elements in the dolomitic limestone used to synthesize commercial grade CMA.

For all commercial grade CMA solutions and reagent grade CMA solutions of less than five weight percent, the reverse scan intersects the forward scan at potentials in close proximity to the active-passive transition. It is probable that whenever the steel is in the passive region, pitting may occur. For steel at passive potentials in reagent grade CMA solutions at five weight percent and higher, the steel should not be susceptible to pitting.

Reverse scans for steel in reagent and commercial grade CMA solutions of two weight percent and higher indicated a very unusual behavior (Fig. 8). At potentials of about -200mv, the direction of current flow to the metal shifted from anodic to cathodic, which peaked, then decreased to zero at about -500 to -600 mv. At that potential, the current again changed direction and once again was anodic. The anodic currents also peaked and decreased to zero with a subsequent shift to the cathodic direction. The potential of the steel was being shifted at a uniform rate of 0.278 mv/sec in the cathodic direction during this interesting current behavior.

This unique double-loop behavior consists of a cathodic loop followed by an anodic loop. The cathodic loop is due to the current required to break down the passive film. During the initial stages of the cathodic loop, it is likely that calcium carbonate and/or magnesium hydroxide precipitates at cathodic sites on the surface of the steel electrode. With a decrease in the number of cathodic sites on the steel surface, the exchange current for the cathodic reaction is lowered. This can result in a shift of the corrosion potential to more cathodic potentials, causing net anodic current flow at the steel electrode. The polarity of current at the steel surface would change from cathodic current to anodic current and would result in the double loops [6].

The active-passive behavior exhibited by steel in the CMA solutions can lead to widely different corrosion rates dependent on the potential of the metal. Linear polarization experiments for steel in CMA solutions exhibit this type of behavior (see Table 7.8). The corrosion rates for steel at passive potentials were about two orders of magnitude less than the corrosion rates of the steel at an active potential in the same solution. The corrosion rates of the steel at active potentials, as determined by linear polarization, were higher than the rates determined by weight loss, but were of the same order of magnitude.

The differences between corrosion rates calculated from electrochemical tests and those measured from weight-loss tests are probably due to different surface conditions as a consequence of exposure times. The weight-loss samples were exposed to the solutions for 17 months, and substantial amounts of protective corrosion products accumulated on the surface. The samples used in linear polarization experiments were exposed with a clean surface, and, therefore, initial corrosion rates were measured. The initial rates can be expected to be higher than those measured with protective corrosion product layers in place.

7.3.6 Steel in Concrete

Simulated pore solutions were prepared in order to study the effect of CMA on the chemistry of those fluids. The electrochemical potentials and polarization behavior of steel were measured. Potentials of steel samples made from reinforcing steel immersed in simulated pore solutions are given in Table 7.9. The potentials of the solutions containing no CMA are in the passive region except for the ones containing chloride ions in an amount equivalent to 0.6 and 1.0 percent by weight in a mortar sample. The accepted value of the active-passive difference for steel in concrete is -275 mv with the saturated calomel electrode (-350 mv with the copper-copper sulfate). All samples containing CMA were at an active condition. The potentials shown in Table 7.9 were taken after several months of exposure in the solutions.

Table 7.10 gives the potential of reinforcing steel in mortar samples that had been soaked in a CMA solution for several months. All potentials were in the same range as the potentials of the pore solutions containing the 10 percent CMA, which indicated that the CMA had reached the rebar surface. All potentials were in the active range. The addition of CMA changed the pH of the pore solutions because of the precipitation of calcium hydroxide and magnesium hydroxide [5]. These components were identified by X-ray diffraction. The potential of the steel shifted to more active values when the lower-pH solutions were encountered. It is thought that the potential of the steel in concrete and mortar samples shifted to the active values because the pore-solution chemistry was changed by the addition of the CMA to the concrete.

Table 7.11 contains the potential and corrosion current data taken from the concrete slabs on which different solutions were ponded. These data are quite scattered and varied. They do indicate that the amount of corrosion current is dependent upon two main factors: (a) the potential difference between the top and bottom mats of steel and (b) the presence or absence of the epoxy-coated bars. In all slabs on which CMA was ponded except three, it appears that the CMA had not reached the top mat of steel. In one slab with both mats bare and salt in the top portion of the concrete, the potential of the top mat was in the same range as that found for the mortar samples. Similar results were found in both slabs with the top mat of epoxy-coated bars and salt in the top portion of the slab. In several of the slabs on which chloride-containing solutions were ponded, similar shifts were also found.

In a few slabs, the bottom mat potential was shifted into the active range, with the top mat remaining in the passive range. This can be explained by channels that were left in these slabs during preparation. The solutions could then reach the bottom mats in preference to the top mat, which would produce the observed results.

7.4   CONCLUSIONS

Steel exposed to CMA solutions behaves like an active-passive metal. If it were possible to maintain the steel in a passive condition, CMA would be an inhibitor. Results of the weight-loss study indicate that for the conditions under which metals would be exposed to the CMA as a deicer, it is likely that steel would be in an active condition and not passive. The active corrosion rates for the CMA solutions tested are about two to four times less corrosive than the corrosion rates of steel in sodium chloride. The corrosion rates of steel in calcium chloride and tap water were found to be both higher and lower than those in CMA solutions depending upon the exposure conditions. Cast iron in the CMA solutions corroded at a rate about half that in the sodium chloride solutions. Aluminum in most of the CMA solutions did not have a measurable corrosion rate, but was pitted by all concentrations of CMA. Aluminum was also pitted by the chloride-containing solutions.

Therefore, CMA used as a deicer should result in a lower corrosion of the exposed metals in bridge structures. Expansion joints, bridge railings, metal beams, and metal piers should last longer if CMA is substituted for sodium chloride. Automobile bodies may also be corroded less by CMA than by the sodium chloride deicer.

The effect of CMA on the corrosion of reinforcing steel would possibly not be substantially different from that of sodium chloride. The most damaging factor in the effect of sodium chloride is the corrosion currents forced as a result of potential differences between the steel in the decks. Large potential differences have been found across the surface of the top mat caused by uneven diffusion of the chloride ions to the top mat surface, and potential differences exist between top and bottom mats. The resulting macro cells are thought to be the reason the corrosion rates of the reinforcing steel in bridge decks have been high enough to cause costly damage in a relatively short time period.

The potential of steel in portland cement mortar and in simulated pore solutions shifts to active values greater than -600mv by CMA. This is a result of changes in the pore solution composition. This shift in rebar potential caused by the presence of CMA may lead to the same sort of potential difference for reinforcing steel in bridge decks as that found with NaCl as a deicer. However, the diffusion rates of CMA compared to NaCl are not known. Other studies are necessary, therefore, to determine if the possibilities found in this study will be a serious problem.

This study was completed in late 1985. Other investigators have continued research aimed at the possibility of using CMA as an alternate deicer. Both Salcedo et. al. [13] and McCrum [14] have verified that CMA is considerably less corrosive to bridge structural metals than sodium chloride. However, these investigations did not examine the effect of CMA on steel in concrete. In 1988, Chollar and Virmani [15] presented results of a limited corrosion study that investigated the effect of CMA on steel in concrete. Their ponding of CMA concrete slabs containing steel rebar indicated that CMA solutions did not cause significant potential shift or corrosion of the steel in concrete. However, their study was limited and did not investigate the effect of CMA on rebar corrosion in salt-contaminated concrete.

In conclusion, CMA will corrode exposed metal less than will sodium chloride, but there may be little difference in the corrosion of reinforcing steel when CMA is used. Large-scale ponding studies similar to those presented in this study should be conducted in an outdoor environment to clarify the effect of CMA on the corrosion of steel in concrete.

ACKNOWLEDGEMENT

The authors would like to acknowledge the support of the Federal Highway Administration for this project.

REFERENCES

1   F. O. Wood, Survey of Salt Usage for Deicing Purposes (R. Baboian, ed.), Automotive Corrosion by Deicing Salts, National Association of Corrosion Engineers, Houston, Texas, 1979.
2   J. Hale, Proceedings of the Northstar Workshop on Noncorrosive Winter Maintenance, Report No. FHWA/MN/RD-84/03, Minnesota Department of Transportation, St. Paul, 1983.
3   America's Highways: Accelerating the Search for Innovation, in Special Report 202: America's Highways: Accelerating the Search for Innovation, TRB, National Research Council, Washington, D.C., 1984.
4   S.A. Dunn and R.U. Schenk, Alternative Highway Deicing Materials, Bjorksten Research Laboratory, Inc., National Information Service, Springfield, Virginia, 1980.
5   C.E. Locke and K.J. Kennelley, Corrosion of Bridge Structural Metals by CMA, Final Report, FHWA Contract DTFH 61-83-0045, Washington, D.C., June 1986.
6   K.J. Kennelley, Corrosion Electrochemistry of Bridge Structural Metals in Calcium Magnesium Acetate, Ph.D. dissertation, University of Oklahoma, Norman, Oklahoma, 1986.
7   M.D. Boren III, The Effect of Calcium Magnesium Acetate on the Corrosion of Reinforcing Steel in Concrete, M.S. thesis, University of Oklahoma, Norman, Oklahoma, 1986.
8   1979 Annual Book of ASTM Standards, Part 10, Metals--Physical, Mechanical, Corrosion Testing (R.P. Lukens et al., eds), American Society for Testing and Materials, Philadelphia, Pennsylvania, 1979.
9   Test Method Tm-01-69, National Association of Corrosion Engineers, Houston, Texas, 1969.
10 V.G. Luster, Corrosion Electrochemistry of Steel and Aluminum in Various Deicing Salts, M.S. thesis, University of Oklahoma, Norman, Oklahoma, 1985.
11 H. Farzammehr, Pore Solution Analysis of Sodium Chloride and Calcium Chloride Containing Cement Pastes, M.S. thesis, University of Oklahoma, 1985.
12 Corrosion Basics, National Association of Corrosion Engineers, Houston, Texas, 1984.
13 R.N. Salcedo and W.N. Jensen, Corrosivity Tests Pit Road Salt vs. CMA, Public Works, Vol. 118, No. 11, November 1987, pp. 58-61, 90-91.
14 R.L. McCrum, Corrosion Evaluation of Calcium Magnesium Acetate (CMA), Salt (NaCl), and CMA/Salt Solutions, National Association of Corrosion Engineers Corrosion/89, Paper No. 127, New Orleans, Louisiana, April 17-21, 1989.
15 B.H. Chollar and Y.P. Virmani, Effects of Calcium Magnesium Acetate on Reinforced Steel Concrete, Public Roads, Vol. 51, No.4, March 1988, pp. 113-115.

TABLE 7.1

Metals Tested in Weight-Loss Experiments

METALS	SHAPE	SURFACE
A-36	Rectangular Angle Iron Crevice U-bend Rectangular Welded	Cleaned Cleaned Cleaned Cleaned Cleaned
A-36 Steel Galvanized	Rectangular Rectangular	Cleaned Weathered 6 months
Grey Cast Iron	Rectangular	Cleaned
A-325 Steel	Circular Stressed Bolt	Cleaned Cleaned
A-588 Weathering Steel	Rectangular Rectangular	Weathered 6 months Weathered 6 months then painted
6061-T6 Aluminum	Rectangular	Cleaned
Galvanic Couples: A-36/Galvanized	Rectangular	Cleaned
A-588/Galvanized	Rectangular	Cleaned

TABLE 7.2

Solutions Used in the Weight-Loss Tests (wt. %)

REAGENT GRADE CMA	COMMERCIAL GRADE CMA	NaCl	$CaCl_2$
0.50	0.50	0.30	0.29
1.00	1.00	0.60	0.57
2.00	2.00	1.20	1.14
Note: Solutions were made using tap water as the solvent.			

TABLE 7.3

Corrosion Rates of Metals Fully Immersed in Corrosive Solutions
(mils/year)

Concentration (wt. %)	Solution	Steel			Cast Iron	Aluminum
		A-36	A-588	A-325		
0.3	NaCl	2.9	7.3	5.3	6.2	1.3
0.6	NaCl	8.6	11.8	6.6	4.3	0.2
1.2	NaCl	6.6	18.3	3.9	4.1	1.3
0.29	CaCl$_2$	3.0	7.1	1.9	2.2	0.0
0.57	CaCl$_2$	2.3	5.2	2.5	1.6	0.0
1.14	CaCl$_2$	1.8	13.6	2.7	1.8	0.1
0.5	CMA(R)	1.6	1.3	2.2	3.3	0.0
1.0	CMA(R)	1.2	1.5	2.4	1.6	0.0
2.0	CMA(R)	0.9	1.6	2.2	1.9	0.0
0.5	CMA(C)	1.8	1.5	3.0	2.2	0.0
1.0	CMA(C)	2.7	1.7	3.8	1.6	0.1
2.0	CMA(C)	2.7	2.7	2.5	2.0	0.1
Tap Water		4.1	0.9	5.7	4.3	0.9

TABLE 7.4

Corrosion Rates of Metals Half-Immersed in Corrosive Solutions
(mils/year)

Concentration (wt. %)	Solution	Steel			Cast Iron	Aluminum
		A-36	A-588	A-325		
0.3	NaCl	10.2	10.6	9.3	5.5	0.6
0.6	NaCl	13.1	17.9	9.7	5.5	0.2
1.2	NaCl	13.9	18.4	12.7	6.5	0.8
0.29	CaCl$_2$	4.1	7.9	4.9	3.5	0.0
0.57	CaCl$_2$	3.9	6.7	6.7	3.3	0.1
1.14	CaCl$_2$	3.1	10.4	7.9	3.1	0.0
0.5	CMA(R)	2.5	1.9	1.9	2.9	0.0
1.0	CMA(R)	1.9	2.0	4.0	1.7	0.0
2.0	CMA(R)	1.4	2.6	2.2	1.8	0.0
0.5	CMA(C)	2.0	2.8	4.0	2.9	0.0
1.0	CMA(C)	2.2	3.1	6.4	3.4	0.0
2.0	CMA(C)	2.4	2.7	2.8	1.6	0.5
Tap Water		3.7	3.1	5.4	3.5	0.4

144

TABLE 7.5
Corrosion Rates of Metals in Vapor Space Above Corrosive Solutions
(mils/year)

Concentration (wt. %)	Solution	Steel			Cast Iron	Aluminum
		A-36	A-588	A-325		
0.3	NaCl	3.5	6.1	0.5	0.2	0.1
0.6	NaCl	5.7	7.2	8.4	0.3	0.3
1.2	NaCl	3.8	7.3	8.3	0.3	0.3
0.29	CaCl$_2$	1.4	1.3	5.3	0.9	0.0
0.57	CaCl$_2$	5.1	1.2	7.7	0.7	0.1
1.14	CaCl$_2$	2.6	2.0	8.9	1.0	0.2
0.5	CMA(R)	2.8	1.3	0.1	0.5	0.0
1.0	CMA(R)	2.1	1.3	0.4	1.1	0.0
2.0	CMA(R)	1.7	1.6	3.1	0.5	0.0
0.5	CMA(C)	1.4	1.6	4.6	1.2	0.0
1.0	CMA(C)	1.5	1.7	4.0	2.8	0.0
2.0	CMA(C)	1.2	2.4	2.7	1.6	0.0
Tap Water		1.2	3.9	1.1	0.5	0.2

TABLE 7.6

Corrosion Rates of Metals Exposed to Corrosive Solutions by Dipping
(mils/year)

Concentration (wt. %)	Solution	Steel			Cast Iron	Aluminum
		A-36	A-588	A-325		
0.3	NaCl	10.7	9.7	13.6	7.6	3.2
0.6	NaCl	14.5	13.6	13.4	8.9	2.0
1.2	NaCl	14.1	16.9	12.8	8.1	1.7
0.29	CaCl$_2$	7.3	7.1	5.7	5.8	0.1
0.57	CaCl$_2$	8.9	8.2	5.6	6.8	0.1
1.14	CaCl$_2$	12.8	9.3	6.8	8.0	0.1
0.5	CMA(R)	2.7	1.3	2.6	4.0	0.0
1.0	CMA(R)	1.3	2.3	2.5	3.5	0.0
2.0	CMA(R)	1.4	4.8	2.4	3.5	0.0
0.5	CMA(C)	2.4	4.1	3.1	3.7	0.0
1.0	CMA(C)	2.4	2.3	2.7	4.1	0.4
2.0	CMA(C)	2.7	0.9	3.1	4.5	0.0
Tap Water		4.7	1.5	4.0	6.5	0.1

TABLE 7.7

Corrosion Rates of Metals Exposed to Sprays of Corrosive Solutions
(mils/year)

Concentration (wt. %)	Solution	Steel			Cast Iron	Aluminum
		A-36	A-588	A-325		
0.3	NaCl	6.4	10.9	7.6	6.3	3.2
0.6	NaCl	8.9	22.6	11.3	11.6	2.0
1.2	NaCl	16.6	30.6	17.2	15.2	1.7
0.29	$CaCl_2$	6.8	7.1	7.0	5.7	0.1
0.57	$CaCl_2$	7.7	7.5	7.5	6.9	0.1
1.14	$CaCl_2$	10.2	7.8	8.1	11.2	0.1
0.5	CMA(R)	3.2	2.9	3.3	5.2	0.0
1.0	CMA(R)	2.8	3.8	4.5	6.7	0.0
2.0	CMA(R)	1.9	4.0	5.4	7.8	0.0
0.5	CMA(C)	3.2	3.3	5.0	5.5	0.0
1.0	CMA(C)	3.3	4.4	4.2	8.0	0.0
2.0	CMA(C)	5.5	5.3	6.1	7.5	0.0
Tap Water		2.6	1.3	2.5	2.6	0.1

TABLE 7.8

Corrosion Rates from Linear Polarization Experiments
A-36 Steel in CMA Solutions

Solution	Concentration (%)	Active Corrosion Potential		Passive Corrosion Potential	
		Potential (mv)	Corrosion Rate (mils/yr)	Potential (mv)	Corrision Rate (mils/yr)
CMA(R)	0.5	-698	9.2*	-171	0.04
CMA(C)		-708	5.4*	-183	0.04
CMA(R)	1.0	-700	7.7*	-181	0.03
CMA(C)		-690	3.5*	-198	0.10
CMA(R)	2.0	-682	5.8*	-233	0.08
CMA(C)		-647	1.7*	-200	0.11
CMA(R)	5.0	-725	4.2	-143	0.01
CMA(C)		-729	2.9	-322	0.05
CMA(R)	10.0	-721	4.6	-182	0.20
CMA(C)		-748	2.2	-379	0.16
CMA(R)	20.0	-711	2.6	-246	0.01
CMA(C)		-659	0.5	-345	0.06

*Data from Luster [9]; other data from Kennelley [5].

TABLE 7.9

Potentials of Rebar in Simulated Pore Solutions
as a Function of Cl- and CMA Concentration

Cl⁻ Content in Mortar Sample (%)	CMA Concentration, Weight(%)			
	0.0 (mv)	0.5 (mv)	2.0 (mv)	10.0 (mv)
0.0	-178	-312	-292	-601
0.036	-179	-300	-340	-671
0.18	-249	-301	-319	-667
0.4	-235	-393	-336	-648
0.6	-328	-406	-480	-619
1.0	-557	-450	-468	-651

Note: Potentials versus saturated calomel electrode (SCE).

TABLE 7.10

Potentials of Rebar in Mortar Cylinders
Soaked in a CMA Solution

Cl⁻ Content in Mortar Cylinder (%)	Potential (mv) Versus Cu/CuSO₄
0.0	-701 / -697
0.036	-693 / -680
0.18	-669 / -704
0.4	-672 / -697
0.6	-694 / -671
1.0	-760 / -636

TABLE 7.11

Final Potential and Corrosion-Current Data from The Concrete Blocks

Potential vs. CuCuSO$_4$	Ponding Solution	Top Mat (mV)	Bottom Mat (mV)	Potential Difference (mV)	Corrosion Current (mA)
Top Epoxy	CMA	-254	-21	233	0.007
Bottom Epoxy		-13	-12	1	0.0002
Top Bare	CMA	-136	-17	119	0.033
Bottom Bare		-208	-151	57	0.027
Top Epoxy	CMA	-27	-24	3	0.0002
Bottom Bare		-19	-20	-1	-0.0003
Top Epoxy/Salt Top	CMA	-28	-25	3	0.0003
Bottom Epoxy		-51	-23	28	0.0008
Top Bare/Salt Top	CMA	-51	-105	-54	-0.0065
Bottom Bare		-677	-46	631	1.50
Top Epoxy/Salt Top	CMA	-740	-96	644	0.70
Bottom Bare		-656	-24	632	0.97
Top Epoxy	NaCl	-547	-344	213	0.008
Bottom Epoxy		-505	-158	347	0.083
Top Bare	NaCl	-185	-327	-142	-0.238
Bottom Bare		-401	-102	299	0.325
Top Epoxy	NaCl	-266	-80	186	0.036
Bottom Bare		-514	-147	367	0.70
Top Epoxy	CaCl$_2$	-488	-394	94	0.072
Bottom Epoxy		-107	-459	-352	-0.072
Top Bare	CaCl$_2$	-38	-36	2	0.0009
Bottom Bare		-118	-85	33	0.008
Top Epoxy	CaCl$_2$	-89	-100	-11	-0.0025
Bottom Bare		-294	-85	209	0.087

148

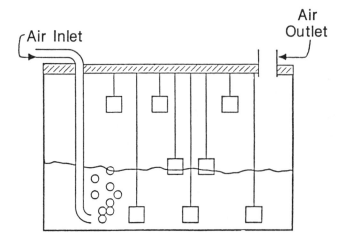

Figure 1   Experimental setup for static immersion tests.

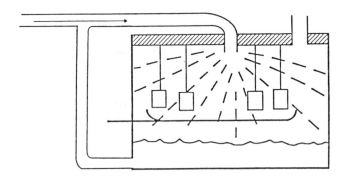

Figure 2   Atomized spray test setup.

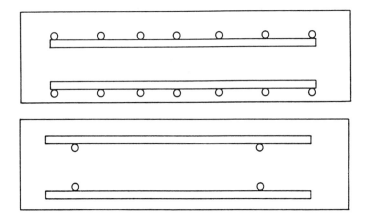

Figure 3   Arrangement of reinforcing in concrete slabs used in ponding tests.

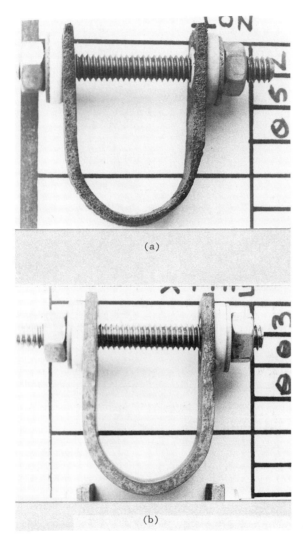

(a)

(b)

Figure 4   A-36 U-bend stressed coupons that had been in the fully immersed
condition of the static weight-loss test (17-month exposure):
(a) 1.2 wt.% NaCl solution exposure; (b) 1 wt.% commercial grade
CMA solution exposure.

150

Figure 5   A-36 painted steel exposed to 1.2 wt.% NaCl in the spray test
for 14 months.

Potential, mv

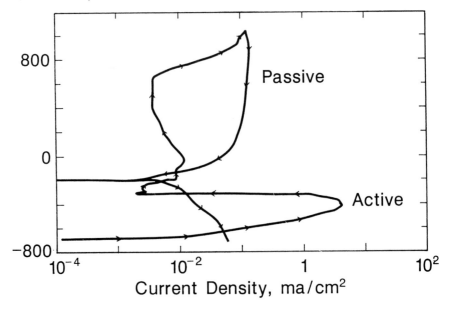

Figure 6   Anodic polarization with reverse scan for A-36 steel in 1.0%
reagent grade CMA.

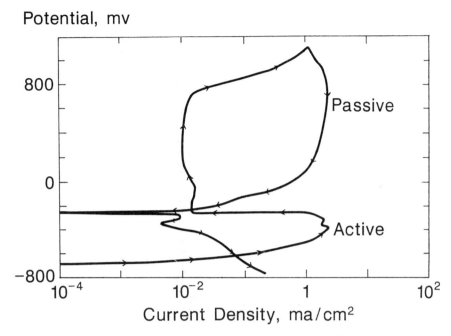

Figure 7   Anodic polarization with reverse scan for A-36 steel in 1.0%
           commercial grade CMA.

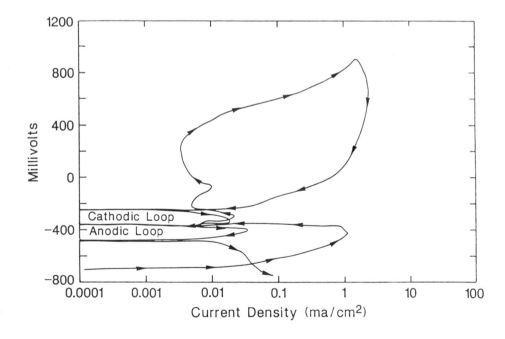

Figure 8   Anodic polarization with reverse scan for A-36 steel in 2.0%
           commercial grade CMA (arrows indicate scan direction).

Chapter 8

ENVIRONMENTAL EVALUATION OF CMA

Jeffrey L. Gidley and Harold Hunt
State of California Department of Transportation, Division of New Technology
and Research

8.1 EXECUTIVE SUMMARY
8.1.1 Purpose of this Study

The purpose of this study was to evaluate, by means of a literature survey and a limited laboratory study, how calcium magnesium acetate (CMA) interacts with the environment, and to identify CMA's beneficial or detrimental environmental impacts.

8.1.2 Introduction

In recent years, transportation agencies have endeavored to improve the safety and convenience of winter travel by attempting to keep roadways free of ice and snow. Therefore, over the last twenty years, the use of sodium chloride (NaCl) for snow and ice removal has risen dramatically. Large scale use of NaCl has significant negative economic and environmental impacts. NaCl corrodes metal and degrades pavement, thereby damaging bridges, road surfaces, and vehicles. Heavy NaCl use damages or kills roadside vegetation, degrades aquatic ecosystems, and pollutes domestic water supplies (1,2). The damage done by NaCl is ultimately paid for by the public.

Because of the negative aspects of NaCl use, research is underway to identify agents which are effective deicers, but which are less deleterious than NaCl. Recent research by Bjorksten Research Inc.(3), conducted for the Federal Highway Administration, identified CMA as a potential alternative to NaCl as a deicing agent. The Bjorksten research showed that CMA is effective in removing snow and ice from pavement. Bjorksten's limited environmental analysis of CMA found no significant negative environmental impacts for CMA. However, a more thorough environmental analysis of CMA was necessary to make sure that CMA would have no deleterious environmental impacts, and to identify any positive impacts that this chemical might have.

This report is a more comprehensive examination of the impacts of CMA on surface water quality, groundwater quality, air quality, aquatic ecology, and soils. These areas were studied in varying degrees of detail. A literature search was performed for each subject, and in some areas a limited laboratory analysis was performed. While more extensive than the earlier environmental work on CMA, this study does not address the long term impacts of CMA on the environment. A 1:1 equivalent weight mixture of analytical grade calcium acetate and magnesium acetate was used to manufacture CMA for this study. This was essentially a pure material which contained no byproducts. The by-products which may be present in industrially produced CMA are unknown. The by-products resulting from various manufacturing processes can sometimes pose environmental hazards even when the pure compound is innocuous. Consequently, the environmental impacts of industrially produced CMA were not determined in this study.

8.1.3 Environmental Evaluation

(i) Surface Water Quality, Groundwater Quality, and Air Quality

The impacts of CMA on surface water quality, groundwater quality, and air quality were determined by means of a literature survey. The literature data bases surveyed included: Aqualine, Aquatic Sciences and Fisheries Abstracts, Air Pollution Technical Information Center, Biological Abstracts (Biosis Previews), CAB Abstracts, Enviroline, Environmental Bibliography, Life Sciences Collection, Instructional Resources Information System (IRIS), National Technical Information Service (NTIS), and Pollution Abstracts (TRISNET).

No source specifically identifying the impacts of calcium acetate, magnesium acetate, or CMA on surface water quality, groundwater quality or air quality was found during the literature search. Up to the present time, it seems that not enough of these materials are being used to generate much environmental information about them. Some transportation agencies have reported dust problems while using CMA. Consequently, the impacts of CMA on these resources is unknown. Some estimates of the impact of CMA on surface water quality are discussed below.

The addition of calcium and magnesium to a water body could result in an increase of water hardness. However, this would not be expected to be significant unless a relatively large amount of CMA entered a water body. There would probably be no other significant effect from slightly elevated calcium and magnesium levels.

While the acetate ion is mildly toxic to some fish, it is readily decomposed and would probably not reach toxic levels. The decomposition of acetate could result in localized oxygen depletion in water bodies, however, the acetate concentrations necessary for this would probably not be reached during deicing operations.

(ii)   Aquatic Ecology

The potential impacts of CMA on aquatic ecosystems were investigated by a literature review and by laboratory bioassays. A bioassay is a toxicity procedure in which living organisms are exposed to known concentrations of a potentially toxic substance for a standardized period of time. The following organisms were used in the bioassays conducted for this study: Rainbow Trout (Salmo gairdneri), Fathead Minnow (Pimephales promelas), Waterflea (Daphnia magna), Green Algae (Selenastrum capricornutum), and Blue Green Algae (Anabaena flos-aquae). These organisms are commonly used in bioassays, and represent different groups of organisms in the aquatic food web.

Fish occupy the highest trophic level in the waterbodies most likely to receive CMA runoff. Rainbow Trout (Salmo gairdneri) and Fathead Minnows (Pimephales promelas) were used to represent this trophic level. Short term static bioassays and long term renewal bioassays were performed.

The short term static bioassays were used to determine the acute toxicities of calcium acetate, magnesium acetate, and CMA to Rainbow Trout and Fathead Minnow fingerlings. The results are expressed as 96 hour LC_{50}'s. The 96 hour LC_{50} is that concentration of a substance which kills 50% of the exposed test organisms in a period of 96 hours. The methods used in the static fish bioassays were the standard methods developed by the California Department of Fish and Game. The results of these bioassays are summarized below:

Short Term Fish Bioassay LC_{50} (mg/liter)				
	NaCl	CMA	Ca Acetate	Mg Acetate
Rainbow Trout	12,200	18,700	16,200	4,300
Fathead Minnow	11,400	21,000	14,300	9,000

The long term renewal bioassays were used to determine the impacts of CMA on the development of Rainbow Trout eggs. The long term bioassays developed problems which made analyzing the results difficult. However, the long term bioassays indicated that continuously maintained CMA concentrations >5,000 mg/liter could reduce Rainbow Trout hatching success.

Zooplankton are small aquatic animals which occupy an intermediate position in the food web between aquatic plants and fish. To address the effects of CMA on the zooplankton component of the food web acute short-term static and long-term static bioassays were conducted on waterfleas (Daphnia magna). The ability of Waterfleas to tolerate low dissolved oxygen levels,

limited genetic variability among parthenogenetic offspring, high
susceptibility to pollution, and ease of culture makes them ideal for
bioassays.

Short-term static bioassays were used to determine the acute median
lethal concentrations (96 hr LC_{50}'s) for CMA, calcium acetate, magnesium
acetate, and NaCl (kiln-dried, road salt). The methods used in the Waterflea
short-term static bioassays were adapted from Greenberg et al (4). The
results are summarized below:

Short Term Daphnia magna Bioassays 96 hr LC_{50}'s (mg/liter)

CMA when bacteria are present	<384
CMA when bacteria are absent	1421
Ca acetate when bacteria are present	482
Mg acetate when bacteria are absent	127
NaCl	4500

Long-term chronic bioassays were used to determine the No Observable
Effects Concentration (NOEC). Reproductive success was used as the measure of
the chronic effects of the tested chemicals on waterfleas. The assumption
inherent in this testing is that the toxicant level which causes no
reproductive impairment will have no chronic effects on individuals or
populations in natural environments. The results of the long-term Waterflea
bioassays are shown below:

Long Term Daphnia magna Bioassays No Observed Effects Concentration (mg/liter)

CMA	125
Ca acetate	100
Mg acetate	100
NaCl	500

Green Algae (Selenastrum capricornutum) and Blue Green Algae (Anabaena
flos-aquae) were used to determine the impacts of CMA, calcium acetate,
magnesium acetate, and NaCl (kiln-dried, road salt) on primary producers. The
method used for the algae tests was a modification of the EPA bottled algal
assay test. Both types of algae were grown in natural water, and in
artificial algae growth medium. Algal growth was determined by measuring the
fluorescence of samples by using a Turner Model III fluorometer. A
conservative estimate of the maximum concentration at which little effect from
CMA, calcium acetate, or magnesium acetate would probably be less than 50
mg/L; for NaCl S. capricornutum was not affected by 1000 mg/L concentrations,
similarly, A. flos-aquae was not affected by any concentration of NaCl tested.

(iii) Terrestrial Ecology

The impacts of CMA on terrestrial ecology were examined by limited laboratory studies on vegetation and soil. The limited time spent on this project and the lack of industrially produced CMA precluded an in depth field evaluation on vegetation or soils. The vegetation study for this investigation measured the impacts of CMA on selected plants when applied in the irrigation water, or in foliar sprays. The soils study measured the nutrients and metals that leached out of selected soils when 1 N solution of CMA was passed through the soil.

There have been many studies measuring the impacts of NaCl on roadside vegetation. There have been no similar studies on CMA. During this study, it was assumed that CMA applied to highway would leave the highway either as runoff or as traffic generated aerosols, both of which would impact vegetation. Therefore, selected plant species were irrigated with a CMA solution. The species selected for this study are listed in Table 8.1. NaCl (kiln dried, road salt) was also applied in irrigation water and as a foliar spray so that a comparison could be made between CMA and NaCl. The irrigation water contained 5, 10, 50 or 100 milliequivalents of CMA or NaCl. The foliar sprays were 0.1, 0.5, 1.0, 2.0 N solutions of CMA or NaCl. These levels were selected because they bracket the concentrations of NaCl and CMA expected within 25 feet of a highway. The results of the vegetation study are summarized on p. 194.

Whenever there are long term additions of chemicals to the soil, there is always a question of whether or not the addition of the chemicals will reduce soil fertility. Various acetates, ammonium, sodium, and calcium are used for extractive purposes in soil chemistry procedures. Therefore, it is possible that long term CMA use might result in a disruption of the soil chemistry and a loss of soil fertility. To investigate the potential effects of CMA laden water might have on soil chemistry, a 1N solution was passed through seven selected soil samples. The liquid that leached through the soils was chemically analyzed to determine if significant amounts of metals or plant nutrients were removed from the soil. As a control, deionized water was passed through other samples of the same soils. The liquid that leached through these samples was chemically analyzed to determine if significant amounts of heavy metals or plant nutrients were removed from the soil. The results of the samples of the control series were compared with the results of the CMA series. It appears that 1N CMA has the potential to remove significant amounts of iron, aluminum, sodium potassium and hydrolyzable orthophosphate from soil.

CMA/NaCl Damage to Plant Species

	Calcium Magnesium Acetate (CMA)		Sodium Chloride (NaCl)	
	Soil	Spray	Soil	Spray
Abies concolor (White-fir)	Low	Low	High	High
Acer saccharum (Surgar maple)	Low	Low	Moderate	Moderate
Amelanchier canadensis (June berry)	Low	Low	High	High
Arctostaphylow patula (G. manzanita)	None	None	High	High
Betula papyrifera (Paperbark birch)	Very low	Very low	Moderate-high	Moderate-high
Calodecrusdecurrens (Incense cedar)	Low	Low	Moderate	Moderate
Cornus florida (Flowering dogwood)	Moderate	None	High	High
Elaeagnus angustifolia (Russian olive)	High	Moderate	Low-moderate	Moderate
Fraxinus pennsylvania (White ash)	Low	Low	Moderate	Moderate
Malix 'Hopa' (Flowering crab)	Moderate	Low	Moderate	High
Pinus jefferyi (Jeffery pine)	Low	None	High	High
Pinus lambertiana (Sugar pine)	Low	Low	High	High
Quercus alba (White oak)	Low-moderate	Low-moderate	Moderate-high	Moderate-high
Quercus rubra (Red oak)	Low-moderate	Low-moderate	High	High
Salix sp. (Willow)	None	None	None	None
Thuja occidentalis (American arborvitac)	Low	Low	Moderate	Moderate
Viburnum lantana (Wayfaring tree)	Low	Low	High	High

Low = 0% - 25% treatment related damage
Moderate = 26% - 75% treatment related damage
High = 76% - 100% treatment related damage

8.1.4 Conclusions

1. CMA is less toxic to Rainbow Trout (Salmo gairdneri) and Fathead Minnows
    (Pimephales promelas) than NaCl, calcium acetate, or magnesium acetate.
    Magnesium acetate is slightly more toxic to Rainbow Trout and Fathead
    Minnows than NaCl. Calcium acetate is less toxic to these two fish
    species than NaCl. A continuously maintained concentration of 5000
    mg/liter of CMA, slightly delays the hatching of Rainbow Trout, but does
    not influence the number of eggs that hatch.

2. Waterflea (Daphnia magna) bioassays indicate that the 96 hr LC_{50} CMA is
    304 mg/liter when the CMA is associated with bacteria, and 1421 mg/liter
    when the CMA is bacteria free. The 96 hr LC_{50} is that concentration of
    a chemical which kills 50% of the test animals within 96 hours. The
    long term chronic bioassays indicated that waterflea reproduction was
    significantly inhibited at 250 mg/liter of CMA. The 96 hour LC_{50} was
    for NaCl determined to be 4500 mg/liter. Waterflea reproduction was
    significantly inhibited at 125 mg/liter of NaCl.

3. The algae bioassays indicate that CMA, calcium acetate, and magnesium
    acetate are more toxic to algae than NaCl. It is estimated that a
    concentration of less than 50 mg/liter of CMA is necessary to eliminate
    any deleterious effects of CMA on algae.

4. CMA leached through soil resulted in some removal of iron, aluminum, and
    selected nutrients from the tested soils.

5. In general, NaCl is more injurious to plants that CMA. Only one species
    of plant, the Russian olive (Elaeagnus Angustifolia) was damaged more by
    CMA tan by NaCl.

6. At the present time CMA's impacts on the public health and safety
    aspects of surface water quality, ground water quality and air quality
    and air quality are unknown. An extensive literature search found no
    information relating to these areas.

7. At the CMA concentrations likely to be generated by the use of CMA in
    snow and ice control, CMA may be less environmentally damaging than
    NaCl.

8. Workers exposed to CMA dust should wear dust masks.

8.1.5 Recommendations

    CMA appears to be less deleterious to aquatic and terrestrial ecosystems
than NaCl. However, these results are based on a literature search and
limited laboratory study, and as such suffer the limitations inherent in such
evaluations.

Based on the results of this study, it is recommended that:

1. Additional laboratory studies be conducted on the way and the rate by which bacteria degrades CMA under various oil and temperature conditions.

2. Controlled field testing should be conducted to determine the fate of CMA in the soil and vegetation and its impacts on ground water quality, aquatic ecosystems (particularly the lower trophic levels), and soil chemistry and physics over an extended period of time. These studies should be performed in climate zones where CMA is likely to be used.

3. Research should be conducted into how to reduce the dust problem associated with CMA.

LITERATURE CITED IN EXECUTIVE SUMMARY

1   D.M. Murray and V.F.W. Ernest. 1976. An Economic Analysis of the Environmental Impact of Highway Deicing. EPA Series, EPA 600/2-76-105.

2   F.S. Adam. 1973. Highway Salt: Social and Environmental Concerns. Highway Research Record No. 425, Highway Research Board, Washington, D.C.

3   S.A. Dunn and R.V. Schenk. 1971. Alternative Highway Deicing (CMA). Bjorksten Research Laboratories.

4   A. Greenberg et al editors. 1980. Standard Methods for the Examination of Water and Wastewater. 15th ed. A.P.H.A. Washington, D.C.

8.2 INTRODUCTION

The use of Sodium Chloride (NaCl) for snow and ice removal in the United States has risen dramatically during the last 20 years. Currently, between nine and ten million tons of NaCl are used annually (1). In some areas as much as 25 tons of NaCl per lane mile are applied each winter. Many highway departments have adopted a bare pavement policy in snowy areas and this results in high NaCl use. Bare, i.e., snow and ice free, pavement is safer and allows higher speeds than snow covered pavements.

Highway departments rely on NaCl rather than abrasives for snow and ice control because the deicing salt melts the snow and ice, thereby breaking the bond between the ice and the pavement. Abrasives do not melt the snow and ice on the road, rather they coat the snow and ice covered road. Abrasives are more readily blown from the roadway, take longer to apply, require more cleanup, and are more expensive than NaCl. An abrasive coated ice pack is a less fuel efficient road surface than bare pavement, and traffic speeds are reduced.

The use of NaCl for highway snow and ice control has serious drawbacks. NaCl damages vehicles, pavement, structures, vegetation, wildlife, and domestic water supplies (2,3). Therefore, the Federal Highway Administration is researching alternatives to NaCl.

The recent study "Alternative Highway Deicing Chemicals," by Bjorksten Research Inc., has identified calcium magnesium acetate (CMA) as a potential alternative to NaCl as a deicing agent. This research showed that CMA is effective in melting ice and snow, is easily stored, can be dispersed with existing equipment, and unlike NaCl is noncorrosive and environmentally acceptable.

Bjorksten's research included a limited environmental evaluation of CMA, consequently a more thorough analysis of CMA's interactions with the environment was necessary. The study reported here is a literature and limited laboratory study which addresses the potential impacts of CMA n surface water quality, ground water quality, aquatic ecology, terrestrial ecology and soils. These areas were studies in varying degrees of detail.

No commercial source of CMA existed when this research was conducted. The CMA used in this study was prepared by mixing analytical grade calcium acetate and magnesium acetate in a 1:1 equivalent weight ratio. NaCl used for comparison with CMA during testing was kiln-dried, road salt.

The impacts of CMA on surface water quality, ground water quality and air quality were studied via a literature search. The impacts of CMA on aquatic ecosystems were evaluated via a literature search and via a series of bioassays. A bioassay is a toxicity test in which living organisms are exposed to known concentrations of a potentially toxic substance for a standardized period of time.

Standard static fish bioassay procedures were used to determine the effects of CMA on Rainbow Trout (Salmo gairdneri) and Fathead Minnows (Pimephales promelas). Both short term acute and long term chronic bioassays were conducted.

CMA's impacts on aquatic producers were studied by means of algae bioassays. The method used for the algae was a modification of the U.S. Environmental Protection Agency's (EPA) bottled algal bioassay test (4). The test determines the effects of a substance on the maximum growth rate and the maximum standing crop of selected algae species. Single species tests were performed with the green algae Selenastrum capricornutum, and the blue green algae Anabaena flos-aquae. Natural water and an artificial medium were both used.

CMA's impacts on zooplankton were studies by means of waterflea (Daphnia magna) bioassays. Both short term acute and long term chronic D. magna bioassays were performed. Both NaCl and CMA were tested. Plants were subjected to either irrigation water containing one of the tested chemicals, or to a foliar spray containing one of the tested chemicals. The plants subjected to the chemicals were compared to controls to determine if the plants were damaged by the chemicals.

Whenever there are long term additions of chemicals to the soil, there is always a question of whether or not the addition of the chemicals will reduce soil fertility. Various acetates of ammonium, sodium, and calcium are used for extractive purposes in soil chemistry procedures. Therefore, it is possible that long term CMA use might result in a disruption of the soil chemistry and a loss of soil fertility. To investigate the potential effects that CMA laden water might have on soil chemistry, a 1 N solution was passed through seven selected soil samples. The liquid that leached through the soils was chemically analyzed to determine if significant amounts of metals or plant nutrients were removed from the soil. As a control, deionized water was passed through other samples of the same soils. The liquid that leached through these samples was chemically analyzed to determine if significant amounts of metals or plant nutrients were removed from the soil. The results of the samples of the control series were compared with the results of the CMA series.

8.3   CONCLUSIONS AND RECOMMENDATIONS

8.3.1 Conclusions

1.   CMA is less toxic to Rainbow Trout (Salmo gairdneri) and Fathead Minnows

(Pimephales promelas) than NaCl, calcium acetate, or magnesium acetate. Magnesium acetate is slightly more toxic to Rainbow Trout and Fathead Minnows than NaCl. Calcium acetate is less toxic to these two fish species than NaCl. A continuously maintained concentration of 5000 mg/liter of CMA slightly delays the hatching of Rainbow Trout, but does not influence the number of eggs that hatch.

2. Waterflea (Daphnia magna) bioassays indicate that the 96 hr LC_{50} CMA is 304 mg/liter when the CMA is associated with bacteria, and 1421 mg/liter when the CMA is bacteria free. The 96 hr LC_{50} is that concentration of a chemical which kills 50% of the test animals within 96 hrs. The long term chronic bioassays indicated that waterflea reproduction was significantly inhibited at 250 mg/liter of CMA. The 96 hr LC_{50} was for NaCl determined to be 4500 mg/liter. Waterflea reproduction was significantly inhibited at 125 mg/liter of NaCl.

3. The algae bioassays indicate that CMA, calcium acetate, and magnesium acetate are more toxic to algae than NaCl. It is estimated a concentration of less than 50 mg/liter of CMA is necessary to eliminate any deleterious effects of CMA on algae.

4. CMA leached through soil resulted in some removal of iron, aluminum, and selected nutrients from the tested soils.

5. In general, NaCl is more injurious to plants than CMA. Only one species of plant, the Russian Olive (Elaeagnus angustifolia) was damaged more by CMA than by NaCl.

6. At the present time, CMA's impacts on the public health and safety aspects of surface water quality, groundwater quality and air quality are unknown. An extensive literature search found no information relating to these areas.

7. At the CMA concentrations likely to be generated by the use of CMA in snow and ice control, CMA may be less environmentally damaging than NaCl.

8. Workers exposed to CMA dust should wear dust masks.

8.3.2 Recommendations

CMA appears to be less deleterious to aquatic and terrestrial ecosystems than NaCl. However, these results are based on a literature search and limited laboratory study, and as such suffer the limitations inherent in such evaluations.

Based on the results of this study, it is recommended that:

1. Additional laboratory studies be conducted on the way and the rate by which bacteria degrades CMA under various soil and temperature

conditions.

2.   Controlled field testing should be conducted to determine the fate of
CMA in the soil and vegetation and its impacts on groundwater quality,
aquatic ecosystems (particularly the lower trophic levels), and soil
chemistry and physics over an extended period of time. These studies
should be performed in climate zones where CMA is likely to be used.

3.   Research should be conducted into how to reduce the dust problem
associated with CMA.

8.4   ENVIRONMENTAL EVALUATION

8.4.1 Surface Water Quality, Groundwater Quality and Air Quality

This study addressed the impacts of CMA on surface water quality and
groundwater quality, air quality, aquatic ecology, terrestrial ecology and
soils in varying depths.

The surface water quality, groundwater quality and air quality
investigations were limited to literature surveys. Databases searched
included: Aqualine, Aquatic Sciences and Fisheries Abstracts, Air Pollution
Technical Information Center, Biological Abstracts (Biosis Preview), CAB
Abstracts, Claims/vs. Patent Abstracts, Enviroline, Environmental
Bibliography, Life Sciences Collection, Instructional Resources Information
System (IRIS), National Technical Information Service (NTIS) and Pollution
Abstracts.

No information on calcium acetate's, magnesium acetate's or CMA's
impacts on surface water quality, or groundwater quality was found during the
literature search.

During field tests conducted by Iowa and Michigan significant amounts of
fine CMA dust were generated during deicing operations (5,6). The dust may be
due to the small CMA particle size used during the testing. No information
concerning the medical implications of a CMA dust inhalation was discovered
during our literature search. However, inhalation of any dust can potentially
damage lung tissue. Further research should be conducted into how the CMA
dust problem can be resolved. Until effective dust control can be achieved,
workers using CMA should wear dust masks.

An increased input of calcium and magnesium in a waterbody could lead to
an increase in water hardness. However, this would not be expected unless
relatively large amounts of CMA enter a small waterbody. With the exception
of an increase in water hardness, it is difficult to envision significant
deleterious effects from slightly elevated calcium and magnesium levels.

The acetate ion, while mildly toxic to some fish, will probably be
decomposed by bacteria before it accumulates to troublesome levels. The

decomposition of acetate could result in localized dissolved oxygen depletion within water bodies, however, it is anticipated the levels necessary for this to occur probably will not be reached during deicing procedures.

8.4.2 Aquatic Ecology

(i) Fish Bioassays

CMA, when used as a roadway deicer, will find its way in runoff to streams and lakes. Because fish occupy the highest tropic level in the water bodies most likely to receive CMA laden runoff, two species of fish were tested to observe the effects of CMA, calcium acetate, and magnesium acetate. The two species chosen were a cold water species, the Rainbow Trout (Salmo gairdneri), and warm water species, the Fathead Minnow (Pimephales promelas). The effect of CMA on eyed eggs of Rainbow Trout was also tested.

Short-term, static bioassays and long-term, renewal bioassays were performed. The static bioassays were performed on both Fathead Minnow fingerlings and Rainbow Trout fingerlings to determine the acute toxicity of CMA, calcium acetate and magnesium acetate. The chronic bioassays were performed on eyed eggs of Rainbow Trout to determine the effect of CMA on egg development, hatching, and larval development. The bioassays and interpretation of results were performed by personnel from the California Department of Fish and Game, Water Pollution Control Laboratory.

a. Materials and Methods

(1) Short-term Static Bioassays - Standardized methods developed by the California Department of Fish and Game were used (7). The results were expressed as 96 hr LC_{50}'s. The 96 hour LC_{50} is that concentration of a toxicant which kills 50% of the test organisms during a 96 hour period.

The Rainbow Trout were acquired from the Nimbus Fish Hatchery in Rancho Cordova, CA. The Fathead Minnows were acquired from the Chico Fish Farm in Chico, CA.

The water used for testing was sand filtered water from the American River in Rancho Cordova, CA. Prior to testing, all fish were acclimated for seven days in American River water.

Preliminary range finding bioassays were conducted to establish the chemical concentration ranges which would later be tested in the definitive bioassays. In the range finding bioassays, 1 gallon wide mouthed jars containing 21 liters of solution were used. Two to four fish were placed in each jar for 24 hours.

Definitive bioassays were conducted using 20 liter aquariums containing 10 liters of solution. For each concentration tested, there were three replicates. Ten fish were added to each replicate, so that a total of 30 fish were exposed to each concentration of toxicant.

CMA was tested on Rainbow Trout, while calcium acetate and magnesium acetate were tested on both Rainbow Trout and Fathead Minnow.

Rainbow Trout were tested at $15°C \pm 1°C$ while Fathead Minnow were tested at $20°C \pm 1°$. The aquariums were checked daily for dead fish which were immediately removed. The dissolved oxygen level, pH, and water temperature were also measured. All results were recorded daily. At the end of 96 hours, the total number of fish that died at each concentration was determined. The LC_{50} values were calculated by linear interpolation of mortality vs. concentration data plotted on probability paper.

(2)   Long-term Renewal Bioassay - A long-term bioassay was used to test the impacts of CMA on rainbow trout eggs and larva Eyed Rainbow Trout eggs were obtained from the American River Fish Hatchery in Rancho Cordova, CA.

Thirty eggs were placed in each of the specially constructed egg cups for part of the bioassay. The egg cups were constructed from 38mm diameter acrylic tubing. The tubing was cut into segments that were 80mm long. Nylon netting was secured to one end of each tube with silicone sealant. The egg cups were attached by nylon monofiliment fishing line and fishing tackle swivel snaps to oscillating arms driven by an electric motor (7). The whole apparatus was placed in 20 liter glass aquariums which contained 10 liters of test solution. The egg cups were adjusted so that the top of the egg cup did not become submerged, while the eggs remained submerged. Air was bubbled through the solution through 1 ml. milk pipettes.

After hatching the alevins were transferred to open topped
nylon screen alevin chambers. The chambers were cuboidal
with sides approximately 10 cm. long and an open top. The
alevin chambers were placed in the same aquariums which
contained the egg cups. The open tops of the alevin
chambers were kept above the surface of the water.

When the fish reached the swim up fry stage they were
transferred to a nylon net enclosure suspended in the test
aquarium. The nylon net enclosure facilitated the daily
exchange of the test solution.

Two egg cups, containing a total of 60 eggs, were placed in
each of the following test concentrations of CMA: 1
mg/liter, 50 mg/liter, 1000 mg/liter, and 5000 mg/liter, and
the control 0 mg/liter. Three egg cups, containing a total
of 90 eggs, were placed in each of the following test
concentrations of CMA: 100 mg/liter and 550 mg/liter.

The test solutions were changed daily. This was
accomplished by moving the test animals to aquariums
containing fresh solution. After use, the aquariums were
rinsed with American River water and filled with fresh test
solution. Proper amounts of dry calcium acetate and
magnesium acetate were added to water to make up the 500
mg/liter, 1000 mg/liter, and 5000 mg/liter test
concentrations. A stock solution of 10 mg/l of CMA was used
to make up the 1 mg/liter, 50 mg/liter, and 100 mg/liter
test solutions.

Each day the number of dead eggs, hatched alevins, dead
alevins, living swim up fry and dead swim up fry were
recorded. The dissolved oxygen level, pH, and water
temperature were also measured daily. The air temperature
in the environmental chamber was continuously recorded and
remained at $9° + 1°C$ throughout the test period.

b.   Results

(1)   Short Term Static Bioassays - Results of the static acute
bioassays are shown in Table 1. CMA and calcium acetate are
less toxic to Rainbow Trout than NaCl. Magnesium acetate is
more toxic to Rainbow Trout than NaCl. For Fathead Minnows,
magnesium acetate is more toxic than NaCl, while calcium
acetate is less toxic than NaCl.

TABLE 8.1
Results of the Short Term Fish Bioassays (LC_{50}'s in mg/liter)

	NaCl	CMA	Ca Acetate	Mg Acetate
Rainbow Trout	12,200	18,700	16,200	4,300
Fathead Minnow	11,400	21,000	14,300	9,000

(2)   Long Term Bioassays - A fungal infection persisted in the
100 mg/liter and 500 mg/liter concentrations despite
repeated treatment efforts. It is well known that fungi can
affect survival and development in fish. Additionally,
disruption of the air supply in the 100 mg/liter tank caused
10 alevins to die. Consequently, the results are not
quantitative.

Despite the problems, some results were obtained. Some
difficulty in hatching due to egg membrane hardening, was
observed in the 5000 mg/liter concentration. The larval
fish appeared to be less able to free themselves completely
from the egg shell for several days. Two of the developing
fish were unable to penetrate the egg membranes. Healthy
alevins were released from these eggs by rupturing the egg
membrane with slight pressure from a glass rod.

c.   Conclusions

Magnesium acetate is slightly more toxic and calcium acetate is
slightly less toxic to Fathead Minnows and Rainbow Trout than
NaCl. CMA is less toxic to Rainbow Trout than NaCl or either
acetate tested separately.

A continuously maintained concentration of 5000 mg/liter of CMA
impacts the membranes of Rainbow Trout eggs. This suggests that
slightly higher CMA levels could cause increased hatching

mortality. However, a sustained concentration of this level during snow and ice removal is unlikely.

Because this was a laboratory study, it does not completely reflect the actual conditions in the field. A scientifically designed field study should be performed.

(ii)   Zooplankton Bioassays
  a.   Introduction
Zooplankton are small aquatic animals that occupy a position in the food web that is intermediate between some microbes of the lower trophic levels and the carnivores of the higher trophic levels. The Waterfleas of the Order Cladocera form a significant portion of the zooplankton in many waters, and are an important source of food for both aquatic insects and fish (9).

Daphnia magna is a fresh Waterflea found primarily in ponds containing large amounts of suspended organic matter. It feeds on algae, protozoa, bacteria, and organic detritus. D. magna is tolerant of low oxygen levels, and is generally more sensitive to pollutants than fish. It is also relatively easy to culture. If the proper conditions are maintained, females will naturally clone giving rise to parthenogenetic young which are genetically identical to their mother, and to each other.

The ease of culture, sensitivity to pollution, and the ability to produce large numbers of identical test animals makes Daphnia magna an ideal organism for bioassays. D. magna has been used in toxicity studies for many years.

  b.   Materials and Methods
Daphnia magna were purchased from a biological supply company. D. magna was mass cultured in a 20 gallon aquarium and in three liter jars. The individuals actually used in the toxicity tests were cultured in 125 ml wide mouth flint glass bottles. All cultures were kept at room temperature. The culture medium was made by mixing 15 mg of dried sheep manure, 75 gm of dried garden soil, and 3 liters of dechlorinated tap water. This mixture was allowed to stand for 2 days, filtered through cheese cloth, and through a .100 mesh filtering cloth. The filtrate was allowed to stand for

one week before use. The culture medium consisted of 250 ml of the original filtrate added to 650 ml of dechlorinated tap water. The original filtrate was kept until the supply was exhausted, then a new supply was made up. Three times a week .01 mg of yeast and about 100,000 <u>Selenastrum capricornutum</u> cells were added to each culture jar.

The young produced in the 125 ml flint glas jars were removed daily. Neonates to be used in bioassays were placed in a common vessel containing fresh medium from the same batch of medium used in the bioassay. If not needed for a bioassay, the young were either cultured or discarded.

(1)   Short-term Bioassays - Short-term static bioassays were used to determine the median lethal concentration (96 hr LC_{50}) for CMA, calcium acetate, magnesium acetate, and NaCl. The LC_{50} is that concentration of toxicant which kills 50% of the test animals in 96 hours. Tables 8.3, 8.4, 8.5, and 8.6 shows the concentrations for each deicer that was tested. In every test, each concentration and each control was run in triplicate.

The individual tests were conducted in 125 ml wise-mouth flint glass jars to which 100 ml of test solution or control solution was added. The control solution consisted of the growth medium. The test solution consisted of growth medium plus a measured amount of deicer.

After the various deicer concentrations and the control were prepared, 10 neonates (< 24 hours old) were transferred to each bottle using a large bore pipette. The number of nonmotile test animals was determined every 24 hours. The test animals were not fed during the test procedure. After 96 hours the LC_{50} was calculated using the binomial moving average and probit methods.

In the initial short-term bioassays large amounts of bacteria grew in the higher concentrations of CMA, calcium acetate, and magnesium acetate. Tentative results indicated that the bacterial blooms could have been killing the <u>D.magna</u>.

To determine if the bacterial blooms were killing the test animals, the test procedure was modified to minimize the potential for bacterial blooms to develop. All glassware was washed with phosphorus free detergent, rinsed with a 10% solution of hydrochloric acid, rinsed with deionized water, autoclaved for 30 minutes and dried overnight at 110°C. The test medium was autoclaved for 45 minutes either prior to the addition of the tested deicer, or in one case after the addition of the deicer. The D. magna neonates were washed in dechlorinated water, and transferred to a sterile medium prior to placement in the test medium. Table 8.3 shows those tests in which the bacterial growth reducing techniques were used.

(2) Long-term Bioassays - Long-term bioassays were used to test the chronic effects of CMA, calcium acetate, magnesium acetate, and NaCl. Reproductive success was used as a measure of chronic effects in the long-term bioassays because it was assumed that the toxicant level which has no reproductive impairment will have no detectable chronic effects on the individual or the population.

The test medium for the long term bioassays was prepared in the same way that the test medium for the short term bioassays was prepared. The concentrations for deicer tested were chosen based on the results of the short-term bioassays. Table 8.2 shows the concentrations of the various deicers that were tested.

Ten replicates were used for each concentration and control in every test. A neonate D. magna (< 24 hours old) was introduced into each test bottle, and allowed to mature during the course of the experiment. The test was concluded when the controls produced at least six broods of young. The total number of young produced at each test concentration and in the controls was determined. Single factor analysis of variance and the Student-Newman-Keuls test were used to determine if there was any difference among the test concentrations and the control (10, 11). The hypothesis tested in all cases was H_0: the mean number of young produced in the control test bottles equals the mean number of young produced in the test concentrations. All significance tests were done at the 0.5 level.

TABLE 8.2

CMA, NaCl, Calcium Acetate, and Magnesium Acetate Concentrations Tested in Long-Term Static Bioassays (Concentrations in mg/liter)

CMA Run 1	CMA Run 2	CaAcetate
500	500	1000
100	250	100
4	125	10
0.8	62.5	1
0 (Control)	0 (Control)	0 (Control)

MgAcetate	NaCl
1000	2500
100	500
10	100
1	20
0 (Control)	0 (Control)

TABLE 8.3

Results of short-term Daphnia magna bioassays using Calcium Magnesium Acetate

Test Run 1 - Non-sterile media

Concentrations mg/liter	Number of test organisms	24 hrs	Number of test organisms nonmotile at:		
			48 hrs	72 hrs	96 hrs
8355	30	30	30	30	30
1671	30	15	18	22	25
167	30	2	5	6	6
17	30	2	5	8	11
0 (Control)	30	1	1	2	13

Test Run 2 - Non-sterile media

Concentrations mg/liter	Number of test organisms	24 hrs	Number of test organisms nonmotile at:		
			48 hrs	72 hrs	96 hrs
1000	15	1	7	14	15
500	15	0	2	3	5
100	15	0	0	0	0
0 (Control)	15	0	0	0	0

LC_{50} estimated by:

Binomial method	-	-	653.0	572.2
95% confidence limits	-	-	500-1000	100-1000
Moving average method	-	-	653.0	-
95% confidence limits	-	-	533.767.0	-
Probit method	-	-	641.4	-
95% confidence limits	-	-	525.2-785.2	-

Test Run 3 - Non-sterile media

Concentrations mg/liter	Number of test organisms	24 hrs	Number of test organisms nonmotile at:		
			48 hrs	72 hrs	96 hrs
2000	30	0	10	19	30
400	30	1	1	3	25
80	30	0	4	12	12
16	30	0	0	0	0
0 (Control)	30	0	0	0	1

LC_{50} estimated by:

Binomial method		-	-	1386.9	113.5
95% confidence limits		-	-	400-2000	16-400
Moving average method		-	-	1386.9	149.4
95% confidence limits		-	-	943.7-2430.3	28.3-222.7
Probit method		-	-	1337.4	152.9
95% confidence limits		-	-	$0 = \infty$	$0 = \infty$

Test Run 4 - Sterile media

Concentrations mg/liter	Number of test organisms	24 hrs	Number of test organisms nonmotile at:		
			48 hrs	72 hrs	96 hrs
10,000	30	30	30	30	30
2,000	30	0	8	9	14
400	30	0	0	4	4
80	30	0	0	1	1
0 (Control)	30	0	0	0	0

LC_{50} estimated by:

Binomial method		-	3001.4	2861.0	2148.1
95% confidence limits		-	2000-10,000	2000-10,000	2000-10,000
Moving average method		-	-	2149.5	1421.4
95% confidence limits		-	-	1551.3-3002.6	993.2-2113.3
Probit method		-	-	1971.3	1526.6
95% confidence limits		-	-	$0 = \infty$	$0 = \infty$

Test Run 5 - Sterile media

Concentrations mg/liter	Number of test organisms	24 hrs	Number of test organisms nonmotile at:		
			48 hrs	72 hrs	96 hrs
6,000	30	24	30	30	30
2,400	30	2	2	2	9
960	30	1	1	1	1
384	30	0	0	0	0
0 (Control)	30	0	0	0	0

LC_{50} estimated by:

Binomial method		-	4222.9	3519.4	2942.6
95% confidence limits		-	2400-6000	2400-6000	2400-6000
Moving average method		-	4223-0	3290.4	2946.6
95% confidence limits		-	3661.9-4982.3	2801.8-3966.4	2259.1-3125.5
Probit method		-	4138.4	3289.9	2706.2
95% confidence limits		-	$0 = \infty$	$0 = \infty$	2264.9-3239.3

Test Run 6 - Modified Growth Reducing Method/Seeding
     A. Modified bacterial growth reducing techniques

Concentrations mg/liter	Number of test organisms	24 hrs	Number of test organisms nonmotile at:		
			48 hrs	72 hrs	96 hrs
6000	30	20	30	30	30
2400	30	13	27	29	29
960	30	0	8	17	19
384	30	0	0	0	0
154	30	0	0	0	0
0 (Control)	0	0	0	0	0

LC_{50} estimated by:

binomial method	3109.3	1317.9	887.9	830.5
95% confidence limits	2400-6000	960-2400	384-2400	384-960
moving average method	3412.4	1322.6	1001.1	958.5
95% confidence limits	2699.9	1062.1-1683.0	805.5-1209.9	777.4-1227.0
probit method	3647.4	1305.1	949.2	899.1
95% confidence limits	2860.4-4941.8	1086.4-1567.7	792.9-1136.4	750.8-1075.8

B. Modified bacterial growth reducing technique and seeding
with unsterilized culture medium.

Concentrations mg/liter	Number of test organisms	24 hrs	Number of test organisms nonmotile at:		
			48 hrs	72 hrs	96 hrs
6000	30	24	30	30	30
2400	30	21	24	30	30
960	30	1	11	14	14
384	30	0	0	0	0
154	30	0	0	0	0
0 (Control)	30	0	0	0	0

Test Run 7 - Mixed Bioassay
A. Bacterial growth reducing techniques.

Concentrations mg/liter	Number of test organisms	24 hrs	Number of test organisms nonmotile at:		
			48 hrs	72 hrs	96 hrs
6000	30	30	30	30	30
2400	30	26	30	30	30
960	30	0	4	25	25
384	30	0	0	1	10
0 (Control)	30	0	0	0	0

LC_{50} estimated by:

binomial method	721.9	1338.1	676.2	514.3
95% confidence limits	960-2400	960-2400	384-960	384-960
moving average method	-	1220	676.2	514.3
95% confidence limits	-	1041.6-1459.2	596.5-780.0	370.4-644.0
probit method	-	1242.4	699.5	508.1
95% confidence limits	-	$0 = \infty$	589.4-813.9	385.4-628.0

B. Unsterilized culture medium.

Concentrations mg/liter	Number of test organisms	24 hrs	Number of test organisms nonmotile at:		
			48 hrs	72 hrs	96 hrs
6000	30	30	30	30	30
2400	30	29	30	30	30
960	30	18	27	30	30
384	30	27	28	28	28
0 (Control)	30	0	0	0	0

LC_{50} estimated by:

binomial method	1891.6	1259.5	1054.5	999.9
95% confidence limits	960-2400	384-2400	384-2400	384-2400
moving average method	1891.6	1259.5	1054.2	-
95% confidence limits	1625.2-2321.7	808.0-1641.2	899.8-1244.6	-
probit method	2426.7	1315.0	1058.3	-
95% confidence limits	817.2-11071.9	1070.1-1616.0	877.1-1264.9	-

TABLE 8.4

Results of Short-term Daphnia Magna Bioassays Using Calcium Acetate

Test Run 1 - Non-sterile Medium

Concentrations mg/liter	Number of test organisms	24 hrs	Number of test organisms nonmotile at:		
			48 hrs	72 hrs	96 hrs
1500	30	21	28	30	30
500	30	0	1	19	21
100	30	0	1	2	2
20	30	0	0	0	0
0 (Control)	40	0	0	2	15

Test Run 2 - Non-sterile medium

Concentrations mg/liter	Number of test organisms	24 hrs	Number of test organisms nonmotile at:		
			48 hrs	72 hrs	96 hrs
1000	30	13	19	28	30
500	30	2	5	16	16
250	30	0	0	0	1
125	30	0	0	0	1
62.5	30	0	0	0	1
0 (Control)	30	0	0	0	1

LC_{50} estimated by:

binomial method	-	828.6	484.8	481.7
95% confidence limits	-	>500	250-1000	250-1000
moving average method	-	828.6	519.5	481.7
95% confidence limits	-	691.4-1117.7	457.3-592.5	411-643
probit method	-	834.5	520.0	441.4
95% confidence limits	-	705.8-1051.0	450.4-600.7	$0 = \infty$

TABLE 8.5

Results of Short-term Daphnia Magna Bioassays Using Magnesium-Acetate

Test Run 1 - Non-sterile Media

Concentrations mg/liter	Number of test organisms	24 hrs	Number of test organisms nonmotile at:		
			48 hrs	72 hrs	96 hrs
1500	30	8	15	30	30
500	30	0	0	8	9
100	30	0	0	0	0
20	30	0	0	0	0
0 (Control)	30	0	0	2	15

Test Run 2 - Non-sterile Media

Concentrations mg/liter	Number of test organisms	24 hrs	Number of test organisms nonmotile at:		
			48 hrs	72 hrs	96 hrs
1000	30	1	1	1	20
500	30	0	0	0	0
250	30	0	0	0	0
125	30	0	0	0	0
62.5	30	0	0	0	0
0 (Control)	30	0	0	0	0
LC_{50} estimated by binomial method		-	-	-	857.1
95% confidence limits		-	-	-	500-1000

Test Run 3 - Non-sterile Media

Concentrations mg/liter	Number of test organisms	24 hrs	Number of test organisms nonmotile at:		
			48 hrs	72 hrs	96 hrs
2000	30	4	12	20	27
400	30	0	6	7	28
80	30	2	9	9	9
16	30	0	0	0	0
0 (Control)	30	0	0	0	1
LC_{50} estimated by: binomial method		-	-	1090.0	126.6
95% confidence limits		-	-	400-2000	80-400
moving average method		-	-	1090.0	139.8
95% confidence limits		-	-	665.1-2173.5	91.6-208.2
probit method		-	-	888.7	
95% confidence limits					

TABLE 8.6

Results of Short-term Daphnia Magna Bioassays Using Sodium Chloride (NaCl)

Test Run 1 - Non-sterile Technique

Concentrations mg/liter	Number of test organisms	24 hrs	Number of test organisms nonmotile at: 48 hrs
10000	10	10	10
1000	10	0	0
100	10	0	0
0 (Control)	10	0	0

Test Run 2 - Non-sterile Technique

Concentrations mg/liter	Number of test organisms	24 hrs	Number of test organisms nonmotile at: 48 hrs	72 hrs	96 hrs
10000	30	30	30	30	30
5000	30	27	30	30	30
2000	30	1	1	4	11
500	30	0	1	4	13
0 (Control)	30	0	1	1	11

Test Run 3 - Non-sterile Technique

Concentrations mg/liter	Number of test organisms	24 hrs	Number of test organisms nonmotile at: 48 hrs	72 hrs	96 hrs
5000	15	12	14	15	15
3000	15	0	3	8	10
0 (Control)	15	0	0	0	1

Test Run 4 - Sterile Technique

Concentrations mg/liter	Number of test organisms	24 hrs	Number of test organisms nonmotile at: 48 hrs	72 hrs	96 hrs
10000	30	30	30	30	30
2000	30	0	0	0	0
400	30	0	0	0	0
80	30	0	0	0	0
0 (Control)	30	0	0	0	0

Test Run 5 - Sterile Technique

Concentrations mg/liter	Number of test organisms	24 hrs	Number of test organisms nonmotile at: 48 hrs	72 hrs	96 hrs
6000	30	30	30	30	30
2400	30	0	0	0	0
960	30	0	0	0	0
384	30	0	0	0	0
0 (Control)	60	0	0	0	0

Test Run 6 - Sterile Technique

Concentrations mg/liter	Number of test organisms	24 hrs	Number of test organisms nonmotile at: 48 hrs	72 hrs	96 hrs
4000	30	4	5	5	6
3500	30	0	2	2	2
3000	30	0	0	0	1
2500	30	0	0	0	0
2000	30	0	1	1	1
0 (Control)	30	0	0	0	0

Test Run 7 - Sterile technique

Concentrations mg/liter	Number of test organisms	24 hrs	Number of test organisms nonmotile at: 48 hrs	72 hrs	96 hrs
5000	30	17	26	29	29
4500	30	3	7	8	15
4000	30	1	3	5	7
3500	30	0	0	0	0
3000	30	0	2	3	3
0 (Control)	30	1	1	1	1

LC_{50} estimated by:		48 hrs	72 hrs	96 hrs
binomial method	-	4697.3	4640.9	4500
95% confidence limits	-	4500-5000	4500-5000	4000-5000
moving average method	-	4697.3	4640.9	4500
95% confidence limits	-	4588.4-4793.2	4542.9-4718.5	4272.7-4619.1
probit method	-	4666.4	4511.4	4317.6
95% confidence limits	-	$0 = \infty$	$0 = \infty$	$0 = \infty$

c. Results

Short-term Bioassays - The results of the short-term bioassays used to determine the acute median lethal concentration (LC_{50}) for CMA, calcium acetate, magnesium acetate, and NaCl are shown in Tables 8.3 - 8.6. The following is a summary of these results:

        <384 mg/liter - CMA when bacteria are present

        1421 mg/liter - CMA when bacteria are reduced

         482 mg/liter - Calcium acetate when bacteria are present

         127 mg/liter - Magnesium acetate when bacteria are present

        4500 mg/liter - NaCl

Long-term Bioassays - The results of the CMA, calcium acetate, magnesium acetate, and NaCl long term bioassays are shown in Tables 8.7 - 8.9. The maximum concentrations at which no significant decrease in reproduction was noted were:

        125 mg/liter - for CMA

        500 mg/liter - for NaCl

        100 mg/liter - for Calcium acetate

        100 mg/liter - for Magnesium acetate

These levels are the No Observed Effects Concentrations for the various deicers.

d. Discussion

CMA is readily used as a nutrient by a variety of bacteria. The tests indicate that the bacterial blooms which occur in CMA solutions at room temperature are deleterious to Daphnia magna. D. magna in test jars with bacterial bloom experienced heavy and usually complete mortality. The mortality may have been due to oxygen depletion caused by the high oxygen demand of the bacterial bloom. The high bacterial growth rate made it necessary to establish two distinct LC_{50} values, one for CMA and one for the bacterial blooms that occur in CMA solutions.

The laboratory data must be carefully applied to known field conditions before hypotheses about how CMA will affect aquatic ecosystems can be made. In the field, CMA can be expected to behave similarly to NaCl. Because CMA melts snow and ice as it is applied, it will create its own solution which will rapidly flow into the aquatic environment. The amounts of CMA and NaCl required for effective deicing are similar, therefore, similar concentrations of CMA and NaCl should be expected in the aquatic ecosystems.

Goldman and Hoffman (12) measured the levels of chloride entering the

182

TABLE 8.7

Daphnia Magna Long-Term CMA Test Results

CMA Test Run #1 - The number of young produced in each replicate during the second CMA long-term bioassay.

Jar #	1	2	3	4	5	6	7	8	9	10	Mean	St.Dev.
500	0	0	0	4	1	0	0	9	0	0	1.4	2.95
100	53	46	67	53	59	53	52	53	65	62	56.3	6.65
20	53	44	44	49	42	47	48	44	40	55	46.6	4.76
4.0	32	54	47	55	52	61	49	20	52	21	44.3	14.61
0.8	59	37	46	58	40	18	38	51	30	34	41.1	12.75
0*	30	62	30	31	42	36	47	28	46	53	40.51	11.45

*Control
Concentrations (mg/liter)

Comparisons of the means of young produced at each concentration.

Concentration (mg/liter)   500   100   20   4   0.8   0
Mean # of D. magna born   1.5* 56.3* 46.6 44.3 41.1 40.5
*Significant difference from Control at the 0.05 level.

CMA Test Run #2 -   The number of young produced in each replicate during the second CMA long-term bioassay.

Jar #	1	2	3	4	5	6	7	8	9	10	Mean	St.Dev.
500	0	0	0	0	0	0	0	0	0	0	0	0
250	0	3	0	42	0	0	0	0	0	0	4.5	13.21
125	106	119	130	77	113	53	86	0	108	92	88.4	38.27
62.5	5	105	65	82	80	87	59	89	96	63	73.1	28.12
9*	121	73	104	96	95	81	74	93	90	86	91.3	14.34

*Control
Concentrations (mg/liter)

Comparisons of the means of young produced at each concentration.

Concentration (mg/liter)   500   250   125   62.5 0
                            0*   4.5*   88.4 73.1 91.2

*Significant difference from Control at the 0.05 level.

TABLE 8.8

Daphnia Long-term NaCl Test Results

Jar #	1	2	3	4	5	6	7	8	9	10	Mean	St.Dev.
2500	57	77	42	62	48	68	30	74	106	42	60.6	36.35
500	120	82	89	91	127	81	129	104	115	82	102	19.00
100	105	90	104	96	75	83	100	71	102	82	90.8	13.70
20	97	106	103	107	88	80	69	111	90	70	92.1	16.60
0	121	73	104	96	95	81	74	93	90	86	91.3	15.71

Comparisions of the means of the number of _D. magna_ born at each
concentration during the NaCl   long-term bioassay.

Concentration (mg/liter)   2500   500   100   20   0
Mean # of _D. magna_ born   60.6* 102   90.8 92.1 91.3

*Significant difference from control at the 0.05 level.

184

TABLE 8.9

Daphnia Magna Long-term Ca Acetate and Mg Acetate Test Results

The number of young produced in each replicate during the Ca acetate and Mg acetate long-term bioassay

Ca Acetate

Jar #	1	2	3	4	5	6	7	8	9	10	Mean	St.Dev.
1000	13	25	0	0	0	3	0	0	0	0	4.1	8.40
100	75	62	66	66	0	66	73	80	95	88	67.1	25.84
10	45	58	67	65	69	25	32	42	39	88	53	19.58
1	59	58	60	49	45	56	60	79	50	48	56.4	9.67

Mg Acetate

Jar #	1	2	3	4	5	6	7	8	9	10	Mean	St.Dev.
1000	0	0	0	11	1	13	11	133	109	8	28.6	49.29
100	89	49	75	92	80	91	79	71	73	81	78.	12.58
10	70	38	60	67	61	49	61	63	64	53	58.6	9.49
1	60	57	69	53	67	59	61	0	56	42	52.4	19.87

Control

Jar #	1	2	3	4	5	6	7	8	9	10	Mean	St.Dev.
0	57	62	62	48	44	47	71	58	72	0	52.1	20.63

Comparisions of the means of young produced at each concentration.

	MgAcet	CaAcet	MgAcet	CaAcet	MgAcet	CA Acet	Mg/Acet	CaAcet
Contr

Concentration
(mg/liter)   1000 1000 100 100   10   10   1   1   0
                28.6* 4.1* 78   67.1 58.6 53   52.4 56.4
52.1

*Significant difference of control at the 0.05 level.

aquatic ecosystem due to salt use in deicing operations in the Lake Tahoe basin in California and Nevada. Their measurements indicate that large amounts of chlorides are transported into the aquatic ecosystem during the winter, but the amounts vary tremendously with time and place. Runoff samples collected from areas near roads receiving NaCl during the winter had a mean concentration of 270 mg of Cl^-/liter with a maximum concentration of 2051 mg Cl^-/liter concentrations. Concentrations at the entrance to Donner Lake were as high as 163 mg Cl^- /liter. The winter average in Donner Lake was 12 mg Cl^-/liter.

If CMA was used for deicing, concentrations of acetate similar to those of chloride may be expected to occur in the immediate vicinity of the road. Thus, in rare instances, concentrations approaching the LC_{50}'s for CMA to Daphnia magna could occur in the immediate vicinity of the road. The amount of acetate that would likely be found in streams directly affected by road deicing is above that required to produce bacterial blooms under laboratory conditions. However, because of low winter temperature, bacterial growth should be retarded long enough for the acetate to be diluted to concentrations at which a bacterial bloom would not occur. In the cold water, the acetate would be slowly decomposed by bacteria. The amount of bacteria in the receiving waters may increase providing additional food for other organisms.

Use of CMA for deicing would raise the calcium and magnesium levels in receiving waters, thereby increasing water hardness. Increasing the hardness in some streams could cause a change in what species of animals live in the stream. Some organisms live only in soft water, while others live in hard water. Some organisms live in either. No generalized hypothesis can be made concerning the impacts of the increase in water hardness on aquatic biota, due to the diversity of water conditions, probable differing application rates of CMA, and different species found in different geographical areas.

The results of this study are not exhaustive, due to the limitations inherent in single species laboratory studies. The results of this study simply indicate that CMA is probably not highly toxic to zooplankton. Carefully controlled field studies should be done to more thoroughly determine the impacts of CMA on zooplankton.

(iii)   Phytoplankton Bioassays

a.   Introduction

Phytoplankton are the microscopic algae of the aquatic environment. These algae form part of the producer trophic level and provide energy to the higher trophic levels. To determine the impacts of CMA, calcium acetate, magnesium acetate, and NaCl, both unialgal and natural water algae bioassays were performed. Algae bioassays test for both stimulatory and toxic effects of a substance on algae, because they measure the algal growth rate and the maximum standing crop. The unialgal bioassay is a standardized test of growth response of a well characterized organism under standardized laboratory conditions. The method used for this test was a modification of the EPA bottle algal assay test (4) using an artificial growth medium rather than natural receiving water. This method was used so regional water characteristics could be disregarded. The natural water bioassay used natural water with indigenous algae and other naturally occurring micro organisms. This allowed assessment of the deicers on one particular assemblage of organisms.

b.   Materials and Methods

The bioassays were conducted using both the Green Algae <u>Anabaena flos-aquae</u>. Cultures of both species of algae were obtained from the U.S. Environmental Protection Agency, the Environmental Research Laboratory in Corvallis, OR. After reception, the cultures were checked for authenticity and purity. Both species of algae were maintained in 500 ml Erlenmeyer flasks which contained 200 ml of algae medium. Every 7-10 days the <u>S. capricornutum</u> was recultured, while <u>A. flos-aquae</u> was recultured every 2-3 weeks. Additionally, both species were placed on 1% agar so new cultures could be started if the cultures maintained in the liquid medium became contaminated.

The synthetic culture medium that was used for algae cultures and some of the bioassays, was that characterized by the EPA (4). Table 8.10 shows the nutrient concentrations in the medium. The pH of the medium was 7.5+0.1.

To insure that the cultures and bioassays were bacteria free, the medium was autoclaved. All glassware were washed in a phosphate free detergent, rinsed in hydrochloric acid, rinsed in deionized water, autoclaved, and ovendried at 100°C prior to use. All transfers of algae

TABLE 8.10

Final Concentration of Macronutrients in Algae Culture
Medium as Salts and Elemental Concentrations

Compound	Concentration (µg/liter)	Element	Concentration (µg/liter)
NaNO$_3$	25.50	N Na	4.20 11.00
NaHCO$_3$	15.00	C	2.14
K$_2$HPO$_4$	1.04	K P	0.469 0.186
MgSO$_4$.7H$_2$O	14.70	S Mg	1.91 2.90
MgCl$_2$.6H$_2$O	12.164		
CaCl$_2$.2H$_2$O	4.41	Ca	1.20
H$_3$BO$_3$	185.51	B	32.46
MnCl$_2$.4H$_2$O	415.61	Mn	115.37
2nCl$_2$	3.27	Zn	1.57
CoCl$_2$.6H$_2$O	1.428	Co	0.35
CuCl$_2$.2H$_2$O	0.012	Cu	0.004
NaMoO$_4$.2H$_2$O	7.26	Mo	2.88
FeCl$_3$.6H$_2$O	160.0	Fe	33.05
Na$_2$EDTA.2H$_2$O	300.0	--	

were performed using aseptic technique.

Table 8.11 shows the concentrations of NaCl, calcium acetate, magnesium acetate, and CMA in each test run. All test concentrations and controls were run in triplicate. The concentrations for testing were equivalent weights (e.g. 1g NaCl is equivalent to 1.67 g CMA).

The bioassays were conducted in an environmental chamber. The temperature was maintained at 24°C + 1°C, and the relative humidity was maintained at about 50%. Test flasks were 500 ml pyrex Erlenmeyer flasks which contained 200 ml of test solution. Loose fitting aluminum foil was used for flask closures. The flasks were continually shaken by 100 RPM. The cultures were grown under continuous cool white fluorescent light, the light was maintained at 400 ft. candles + 10% 0.5 inches above the shaker table. The <u>Anabaena flos-aquae</u> was covered with cheese cloth to reduce light intensity by 50%.

Algal growth was measured by fluorescence using a Turner Model III fluorometer. Both the specific growth rate and the maximum standing crop were measured. The specific growth rate was calculated by using the

$$\frac{n(\frac{x2}{x1})}{Tx-T1}$$

equation: G =

where G = Growth Rate
x1 = biomass at time one
x2 = biomass at time x
The maximum standing crop is the maximum biomass occurring during the test run. Practically, the maximum standing crop is assumed to be achieved when the biomass increases less than 5% per day. The maximum specific growth rate is the largest rate in growth occurring at any time during the incubation period.

The fluorescence readings were converted to the g/liter of chlorophyll and the number of cells/ml. Replicate means and the standard deviation of each series of replicates was calculated. A one-way analysis of variance was performed for each test run to determine if significant differences existed among treatments. For each treatment, the specific

TABLE 8.11

The Concentration of NaCl, CMA and Calcium and Magnesium Acetate Used for Each Bioassay:

Flask #	Test Toxicant	Concentrations
1	Control	0
2	NaCl	1 mg/L
3	NaCl	10 mg/L
4	NaCl	50 mg/L
5	NaCl	100 mg/L
6	NaCl	1000 mg/L
7	CMA	1.67 mg/L
8	CMA	16.71 mg/L
9	CMA	83.55 mg/L
10	CMA	167.1 mg/L
11	CMA	1671 mg/L
12	Control	0
13	Ca Acetate	15.07 mg/L
14	Ca Acetate	150.7 mg/L
15	Ca Acetate	1507 mg/L
16	Mg Acetate	18.35 mg/L
17	Mg Acetate	183.5 mg/L
18	Mg Acetate	1835 mg/L
19	Control	0

190

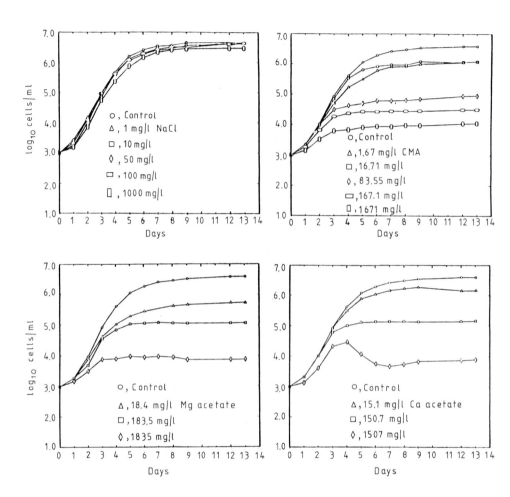

Figure 1. The growth of <u>Selenastrum capricornutum</u> in various
concentrations of NaCl, CMA, calcium acetate and
magnesium acetate (1/19/82 - 2/1/82).

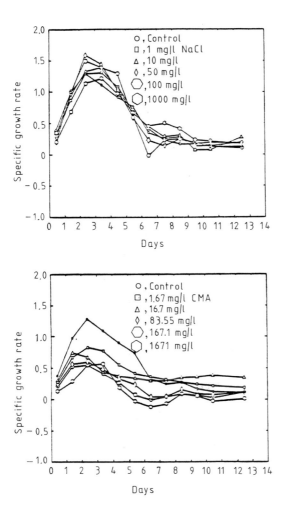

Figure 2. The specific growth rate (two-day moving average) of <u>Selenastrum</u> <u>capricornutum</u> grown in various concentrations of NaCl and CMA (1/19/82 - 2/1/82).

192

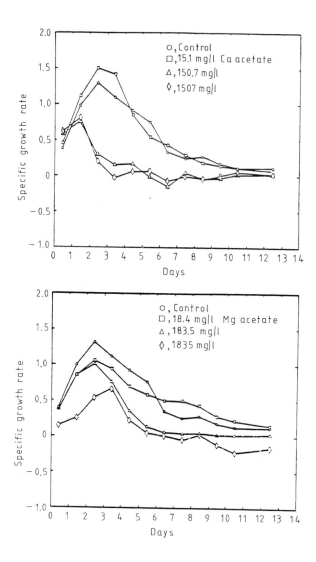

Figure 3. The specific growth rate (two-day moving average)
of <u>Selenastrum capricornutum</u> grown in various
concentrations of calcium acetate and magnesium
acetate (1/19/82 - 1/1/82).

growth rate and standing crop were calculated for each day.

The purpose of the natural-water, algae-bioassays was to determine if
algae in natural receiving water, with an established bacterial
population, would be affected differently by the various chemicals
tested than was algae in the artificial medium. The algae bioassay test
method was modified so that natural water was used rather than
artificial medium. Water from the American and Bear Rivers of
California was used. Nutrients and an inoculum of S. capricornutum were
added to the water. The concentrations tested are shown in Table 8.11.

c. Results
Unialgal Bioassays - Four Selenastrum capricornutum bioassays and three
Anabaena flos-aquae bioassays were run. Of these, two S. capricornutum
and two A. flos-aquae bioassays were successful. The other tests were
rejected because the controls did not grow properly. The results of the
successful bioassays were used to assess the effects of CMA on the
environment. Each successful bioassay will be reported individually.

S. capricornutum bioassay of 1/19/82 - 2/1/82. The growth of S.
capricornutum in various concentrations of NaCl, CMA, calcium acetate,
and magnesium acetate are shown in Figure 1. Figures 2 and 3 show the
specific growth rates observed. A two day moving average was used to
decrease the effect of sampling error.

None of the NaCl concentrations caused a significant difference in
growth during the bioassay. Growth depressions occurred in all CMA
concentrations tested. Both the calcium acetate and the magnesium
acetate treatments showed a growth rate depression similar to that of
CMA. In both treatments, the observable differences occurred later in
time than CMA treatment.

S. capricornutum bioassay of 3/13/82 - 3/25/82. The growth of S.
capricornutum in various concentrations of NaCl, CMA, calcium acetate,
and magnesium acetate are shown in Figure 4 . The specific growth rate
(two-day moving average) is shown in Figures 5 and 6 for the different
treatments.

Except for the first day, the growth of all NaCl concentrations were not
statistically different (at the .05 level of significance) from the

194

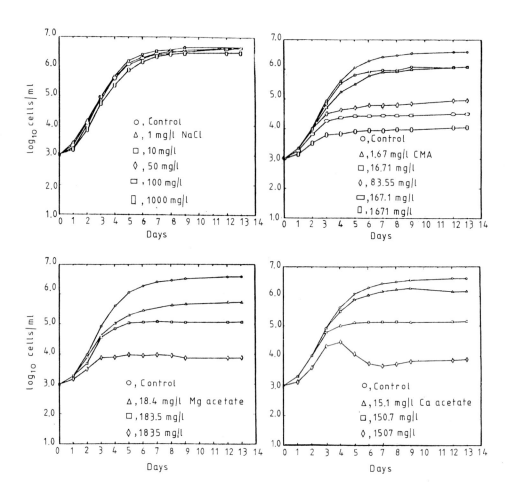

Figure 4. The growth of <u>Selenastrum capricornutum</u> in various concentrations of NaCl, CMA, calcium acetate and magnesium acetate (3/13/82 - 3/25/82).

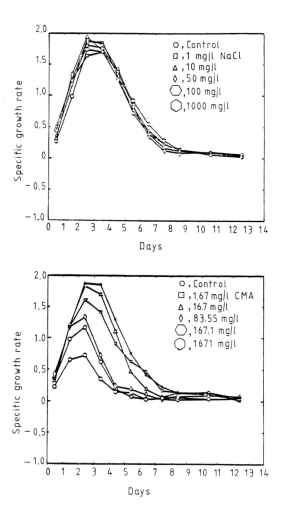

Figure 5. The specific growth (two-day moving average) of <u>Selenastrum</u> <u>capricorntum</u> grown in various concentrations of NaCl and CMA (3/13/82 - 3/25/82).

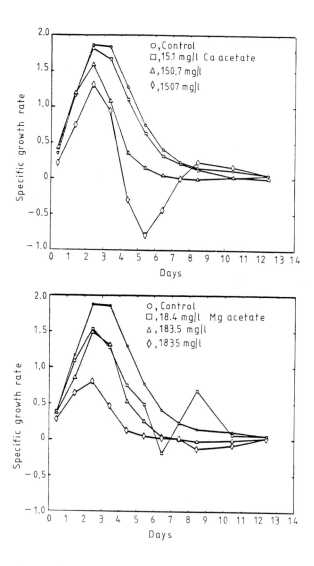

Figure 6. The specific growth rate (two-day moving average) of <u>Selenastrum capricornutum</u> grown in various concentrations of calcium acetate and magnesium acetate (3/13/82 - 3/25/82).

control. All concentrations of CMA exhibited depressed growths.

The specific growth rates for NaCl were indistinguishable from the controls. CMA treatments showed a consistent pattern of depressed growth rates. The most severe depression occurred at the highest concentration. Although less regular than the CMA results, both the calcium acetate and the magnesium acetate treatments showed growth rate depressions. Generally, the higher concentrations were more depressed than the lower concentrations.

A. flos-aquae bioassay (7/13/82 - 7/28/82). The A. Flos-aquae bioassays grew more slowly than the S. capricornutum bioassays, and the maximum standing crop and the daily growth rates attained were lower. The results are shown in Figure 7 . The daily specific growth rates (two-day moving average) for the treatments are shown in Figure 8.

The maximum standing crop in the NaCl treatments was not significantly different from that of the controls. Despite some inconsistencies in the data, the higher concentrations of CMA had a lower standing crop than the controls.

The data for calcium acetate and magnesium acetate are more consistent than those for CMA. All concentrations of both calcium acetate and magnesium acetate exhibited a significantly smaller standing crop than the controls.

The data on specific growth rates is too inconsistent to draw any conclusions.

A. flos-aquae bioassay (1/18/83 - 2/7/83). The standing crop and the daily specific growth rate of various concentrations of NaCl, CMA, calcium acetate and magnesium acetate are shown in Figures 9 and 10 .

The NaCl treatments showed a growth pattern similar to the previous A. flos-aquae bioassay. By the end of the experiment the highest NaCl concentration had a significantly smaller standing crop than the controls and the lower NaCl concentrations.

The CMA treatments showed a similar reduction in standing crop at the higher concentrations to the reductions seen in other algae bioassays.

198

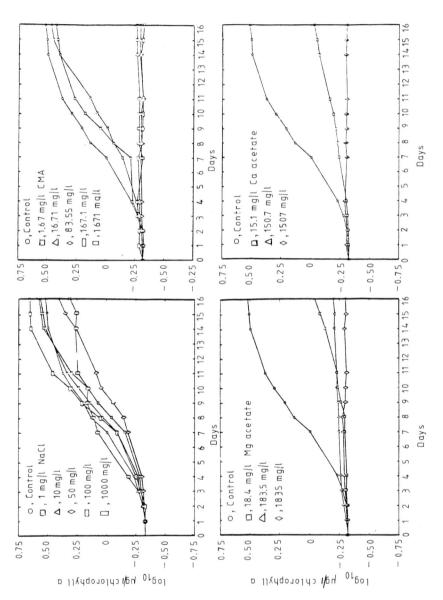

Figure 7. The growth in *Anabaena flos-aquae* in various concentrations of NaCl, CMA, calcium acetate and magnesium acetate (7/12/82 - 7/28/82).

199

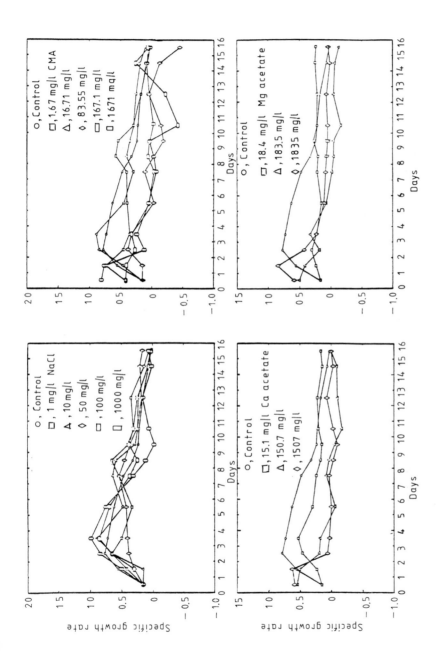

Figure 8. The specific growth rate (two-day moving average) of *Anabaena flos-aquae* ride, CMA, calcium acetate and magnesium acetate (7/12/82 - 7/28/72).

200

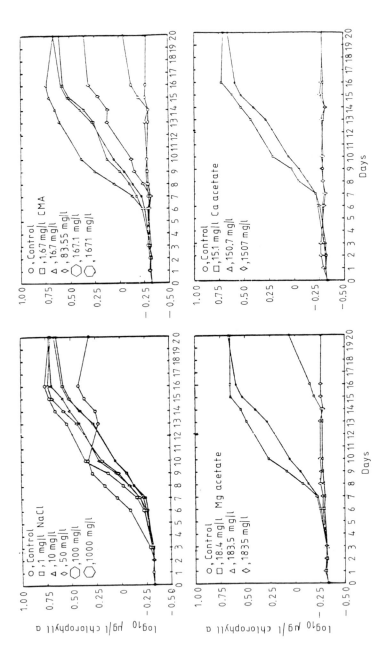

Figure 9. The growth of Anabaena flos-aquae in various
concentrations of NaCl, CMA, calcium acetate
and magnesium acetate (1/18/83 – 2/7/83).

201

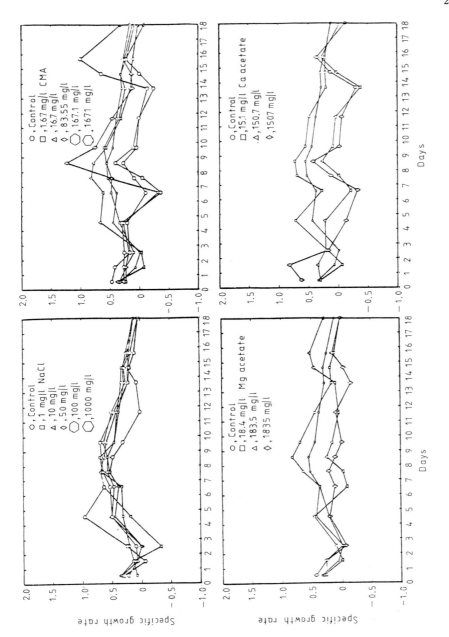

Figure 10.   The specific growth rate (two-day moving average)
of Anabaena flos-aquae grown in various concentrations
of NaCl, CMA, calcium acetate and magnesium acetate
(1/18/83 - 2/7/83).

By the end of the bioassay, the standing crop in the control, the 1.67 mg/liter and the 16.71 mg/liter were all significantly larger than the 167.1 mg/liter and 1671 mg/liter concentrations.

The calcium acetate and magnesium acetate treatments exhibited a reduced standing crop in the higher treatment levels compared to the controls. The maximum standing crop in the control and in the 15.1 mg/liter treatment of calcium acetate developed significantly larger standing crops than either the 150.7 mg/liter and 1507 mg/liter concentrations of calcium acetate. There was no significant difference between the control and the 15.1 mg/liter treatment. For magnesium acetate, all treatment levels exhibited a depressed maximum standing crop.

Natural Water Bioassay (12/14/82 - 12/24/82) - The maximum standing crops of the combined S. capricornutum and indigenous algae in various concentrations of NaCl, CMA, calcium acetate and magnesium acetate are shown in Figure 11. The specific growth rates (two day moving average) are shown in Figure 12 for the different treatments. The results of this bioassay are somewhat different from the unialgal S. capricornutum bioassays.

During the experiment, all of the NaCl treatment showed similar maximum standing crops. The highest concentration of CMA (1671 mg/liter) treatment showed significantly less growth than the controls.

The highest concentration of CMA (1671 mg/liter) treatment showed significantly greater growth than lower level sodium chloride treatments, but was not significantly different from the controls.

For magnesium acetate, there was no significant difference among any of the treatments or the control by the end of the experiment. There was a large variance among the replicates which made the analysis difficult.

For calcium acetate, the 15.1 mg/liter treatment's maximum standing crop was indistinguishable from the control. The other treatments had distinctly lower maximum standing crop than the controls.

Except for the highest NaCl level, there is little difference in maximum specific growth rate among the NaCl treatments and the control. CMA treatments exhibit a maximum specific growth rate depression at the

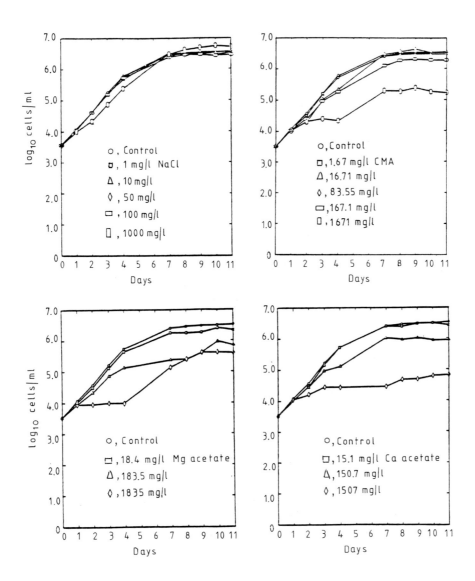

Figure 11. The growth of the combined <u>Selenastrum</u> <u>capricornutum</u> and indigenous algae (American River) in various concentrations of NaCl, CMA, calcium acetate and magnesium acetate (12/14/82 - 12/24/82).

highest concentration. Both calcium acetate and the magnesium acetate show specific growth rate depressions at the higher treatment levels, however, the magnesium acetate seem to recover towards the end of the experiment.

Natural Water Bioassay (12/28/82 - 1/6/83) - The maximum standing crops attained in various treatments of NaCl, CMA, calcium acetate, and magnesium acetate is shown in Figure 13. The specific growth rates (two day moving average) is shown in Figure 14 for different treatments.

The results of this bioassay for NaCl were similar to the first natural water bioassay. From the 4th day to the next to the last day there were no differences among standing crops of the various NaCl treatments or the control. On the last day of the experiment, the lowest treatment level (1 mg/liter) exhibited a significantly lower standing crop than the other treatments.

For CMA the control and the lower level CMA treatments had significantly greater standing crops at the end of the experiment than did the 1671 mg/liter treatment, but there was no statistical difference among them.

For calcium acetate, the control and the lowest level treatment were significantly different from the two highest treatment levels from the third day of the test to the last day of the test. On the last day of the test, the highest treatment had a significantly lower biomass than the other treatments and control which were not significantly different among themselves.

Magnesium acetate showed a significantly depressed standing crop for the highest treatment for the last day of the test. There was no significant difference among the other treatments and controls.

The specific growth rates (two day moving average) show results similar to those of the first natural water bioassay. The NaCl treatments did not show any significant differences, except for an initial specific growth rate depression for the 1000 mg/liter treatment, and a specific growth rate depression in the 1 mg/liter treatment. The CMA, calcium acetate and magnesium acetate specific growth rates all show an initial growth rate depression for the two highest levels followed by an increase in the specific growth rate in the last days of the experiment for the next to highest treatment level.

d. Conclusions

The S. capricornutum unialgal bioassays indicated that NaCl, up to 1000 mg/liter had no apparent statistically significant effect on maximum standing crop or specific growth rate. The A. flos-aquae 1000 mg/liter, NaCl treatment showed a statistically significant growth difference. Initially, the treatment exhibited increased growth, but by the end of the experiment this treatment exhibited decreased growth. In the natural water bioassays the treatments, except for the 1000 mg/liter treatments, NaCl had no statistically significant effect on algal growth. In the 1000 mg/liter treatments there was an initial depression in the standing crop and the growth rate followed by an increase toward the end of the experiment.

For CMA the unialgal and natural water bioassays exhibited a significant depression in maximum standing crop and maximum growth rate at the highest treatment levels. The 83.55 mg/liter, 167.1 mg/liter, and 1671 mg/liter treatments all significantly affected growth in the unialgal bioassays. In the natural water bioassays, only the 1671 mg/liter CMA treatment showed a significant reduction of standing crop and growth rates. Calcium acetate and magnesium acetate had effects similar to the CMA treatments.

The algal bioassays demonstrate that CMA, calcium acetate, and magnesium acetate are more deleterious to algae growth than NaCl. The effect of CMA, calcium acetate, and magnesium acetate was decreased in the natural water bioassays, but algal growth was still delayed even if not statistically different except at the highest treatment levels. A conservative estimate of the maximum concentration at which little effect from CMA, calcium acetate, or magnesium acetate would probably be less than 50 mg/liter.

8.4.3 Terrestrial Ecology

The impacts of CMA on terrestrial ecology were examined by limited laboratory studies on vegetation and soil. The abbreviated time period and the lack of industrially produced CMA precluded an in depth field evaluation on either the vegetation or the soils. The vegetation study for this investigation measured the impacts of CMA in the irrigation water, and in foliar sprays. The soils study measured the nutrients and metals that leached out of selected soils when a 1 N solution of CMA was passed through the soil.

(i) Roadside Vegetation Study

a. Introduction

The short time period allotted for the CMA study precluded an in depth field evaluation of CMA's impacts on roadside vegetation. Therefore, this study was limited to a pot study. Testing was conducted on eighteen woody species selected from plants found adjacent to highways in snowy areas. Species selected for this study are listed in Table 8.12. All testing was by the University of California at Davis (13).

b. Materials and Methods

During this study, it was assumed that some of the CMA applied to control ice and snow would leave the roadway as runoff, or as traffic generated aerosols. To test for the impacts of CMA leaving the roadway as runoff, plants were irrigated with CMA solutions. To test for the impacts of CMA leaving the roadway as aerosols, other test plants had their tops sprayed with CMA solutions. NaCl was also applied to test plants as irrigation solutions or as aerosols. Additionally, one plant of each species was used as a control.

The concentrations of CMA and NaCl used in the irrigation water were selected to bracket the CMA concentrations expected to occur, or the NaCl concentrations known to occur, within 25 feet of the highway. The aerosol concentrations approximated the concentrations expected or found in snowmelt on highway surfaces.

The plants that were used in testing were obtained either as bare root stock or in one and two gallon containers in various soil mixtures. Plants received as bare root stock were inspected, soaked in a water bath, and root pruned to remove damaged or diseased roots. Bare root plants were planted in appropriate sized containers in a potting mix of sand, redwood bark and peat moss. Plants from the Western U.S. were received in 1 and 2 gallon plastic containers containing a potting mix of Douglas Fir bark and sawdust.

All plants were maintained in the nursery can yard of the Environmental Horticultural Department of U.C. Davis. Test solutions were made by dissolving appropriate amounts of CMA in tap water.

Soil irrigation test solutions contained 5, 10, 50 or 100 milliequivalents (meq) of CMA or NaCl; the control consisted of tap water. Sufficient solution was applied at each application to allow

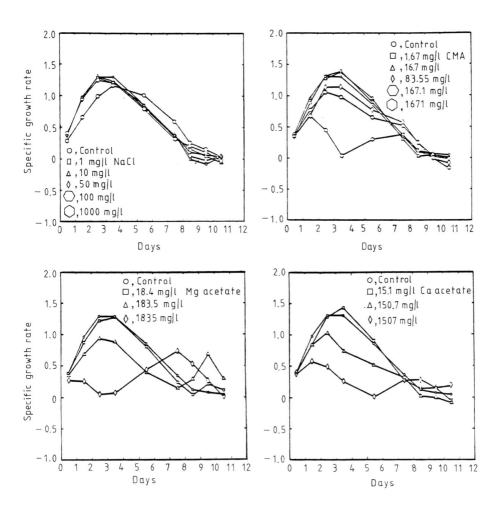

Figure 12 . The specific growth rate (two-day moving average) of
the combined <u>Selenastrum capricornutum</u> and indigenous
algae (American River) grown in various concentrations
of NaCl, CMA, calcium acetate and magnesium acetate
(12/14/82 - 12/24/82).

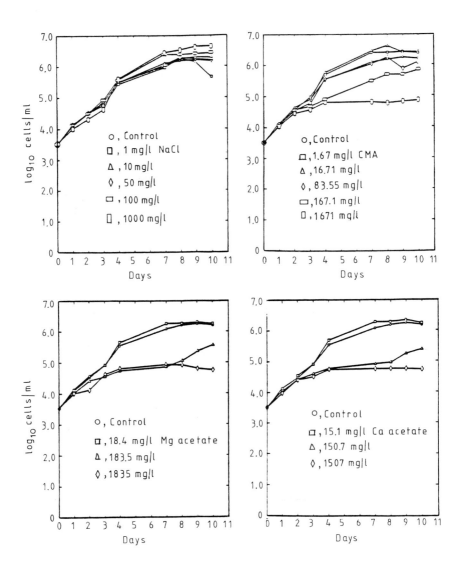

Figure 13. The growth of the combined <u>Selenastrum capricornutum</u> and indigenous algae (Bear River) in various concentrations of NaCl, CMA, calcium acetate and magnesium acetate (12/28/82 - 01/06/83).

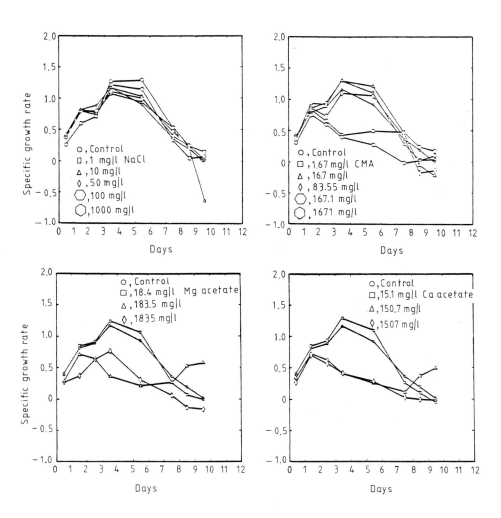

Figure 14. The specific growth rate (two-day moving average) of the combined _Selenastrum capricornutum_ and indigenous algae (Bear River) grown in various concentrations of NaCl, CMA, calcium acetate and magnesium acetate (12/28/82 - 01/06/83).

about 25% of the solution to leach through the soil. Throughout testing we endeavored to provide the plant's soil an ionic concentration equal to the test solution applied. Three plants of each species were treated with each concentration of CMA or NaCl. One plant of each species was used as a control.

Between February and June of 1982 a total of five applications were made. The application dates were determined by monitoring the amount of salts remaining in the soil.

The test solutions for spray applications were made by dissolving the appropriate amounts of CMA or NaCl in tap water. The concentrations used were 0.1, 0.5, 1.0, and 2.0 N. Between February and May 1982 four spray applications were made.

The early 1982 results suggested that CMA was less damaging than NaCl in 14 of the species tested, while four showed about the same degree of damage as NaCl. However, various technical problems prevented a good quantitative examination.

A second season of treatment was conducted to improve the quantitative nature of the data. Rabbit Brush (Chrysothamnus nauseosus) was not tested in the second season.

The type, extent and progression of injury was determined by visual observation. The treatment damage was analyzed by comparing replicates for each treatment level with the control plants used as a standard.

c. Results

Tables 8.12 and 8.13 show numerical comparisons of the damage caused by NaCl and CMA. Table 8.12 rates the treatment caused damage on a scale from one to five. One (1) indicates no treatment damage, while five (5) indicates severely damaged plants. Table 8.13 groups the four separate treatment levels together so that for both soil and spray treatment application modes, the treatment related damage for NaCl and CMA may be directly compared. In Table 8.13 a low number corresponds to minimal damage, while a high number indicates maximum damage. Nine species were more severely damaged by NaCl, one species was more severely damaged by CMA, and in eight species the degree of damage was too low to allow a comparison. In general, the spray treatments produced greater damage than the irrigation treatments.

TABLE 8.12

Plant Treatment Numerical Evaluation

SPECIES		SOIL				SPRAY			
		5 meq	10 meq	50 meq	100 meq	0.1N	0.5N	1.0N	2.0N
Abies concolor	NaCl	1[b]	1	2	2	2[c]	2	4[b]	4[b]
White Fur	CMA	1	1	2[d]	1	1	1	1	3[b]
Acer saccharum	NaCl	1	2	3	4	1	2	3	3
Sugar maple	CMA	1	1	1	2	1	1	1	2
Amelanchier canadensis	NaCl	1	1	2	5	1	2	3	4
June Berry	CMA	1	1	1	2	1	1	1	2
Arctostaphylos patula	NaCl	1	1	2	4	3	5[ab]	4[a]	5a
G. Manzanita	CMA	1	1	1	2	3	5[ab]	1	2
Betula papyritera	NaCl	1	1	4	5	2	2	3	4
Paperbark Birch	CMA	1	1	1	2	1	1	1	2
Calocedrus decurrens	NaCl	1	1	2	3	2	2	3	4
Incense Cedar	CMA	1	1	2	3	1	2	1	2
Chrysethamnus nauseous	NaCl	1	1	2	2	1	1	2	2
Rabbit Brush	CMA	1	1	2	1	1	1	1	2
Cornus florida	NaCl	1	2	4	5	1	2	3	4
Flowering Dogwood	CMA	1	1	1	2	1	1	2	2
Elaeagnus angustifelia	NaCl	1	1	1	2	1	2	3	3
Russian Olive	CMA	1	1	2	4	1	1	1	1
Fraxinus pennsylvania	NaCl	1	1	2	2	1	3	2	3
White Ash	CMA	1	1	1	1	1	2	2	2
Malus hopa	NaCl	1	1	3	3[e]	2	2	3	5
Flowering Crabapple	CMA	1	1	1	2[e]	1	1	1	2
Pinus jeffreyi	NaCl	1	1	2	2	1	2	3	4
Jeffrey Pine	CMA	1	1	1	2	1	1	1	2
P. lambertiana	NaCl	1	1	3	5	1	2	4	5
Sugar Pine	CMA	1	1	1	2	1	1	1	2
Quercus alba	NaCl	1[f]	1[fg]	2[f]	3[g]	2	2	2	3
White Oak	CMA	1	1[g]	1[g]	4[dfg]	1	1	2	2
Q. boresalis	NaCl	1	2[g]	2	5[b]	2	4	3	4
Northern Red Oak	CMA	1	1[g]	4[d]	5[f]	1[f]	1[g]	2[fg]	2
Salix sp.	NaCl	1	1	1	2	2	2	4	5
Willow	CMA	1	1	1	1	1	1	2	2
Thuja occidentalis	NaCl	1	1	2	4	1	2	2	5
Arboruitae	CMA	1	1	1	1	1	1	1	1
Viburnum lantana	NaCl	2	3	4	5[a]	2	3	5	5[2]
Wayfaring Tree	CMA	1	2	1	2	1	1	2	2

(TABLE 8.12 Continued)

		SOIL				SPRAY			
Species		5 meq	10 meq	50 meq	100 meq	0.1N	0.5N	1.0N	2.0N
Thuja occidentalis	NaCl	1	1	2	4	1	2	2	5
Arboruitae	CMA	1	1	1	1	1	1	1	1
Viburnum lantana	NaCl	2	3	4	5[a]	2	3	5	5[a]
Wayfaring Tree	CMA	1	2	1	2	1	1	2	2

LEGEND

Damage %	Numerical Ranking
0	1
0-25	2
25-50	3
50-75	4
75-100	5

FOOTNOTES

a. Treatment series includes dead replicate(s).
b. Treatment series includes replicate(s) which died of causes other than being totally treatment related. Environmental stresses other than test treatment suspected as cause of death.
c. Drought/heat stress damage; replicate(s) observed to have been severely wilted.
d. Plants of poor original quality.
e. Includes plants with regrowth foliage; all previous 1982 foliage was destroyed by the treatments.
f. Treatment series includes replicate(s) partially removed by California state pathologists to test for oak wilt (Caratocystis fagacearum).
g. Treatment series includes replicate(s) totally removed by California state pathologists to test for oak wilt (Caratocystis fagacearum).
h. Not used in second season.

TABLE 8.13

NaCl, CMA Damage Rate Comparisons

SPECIES	SOIL		SPRAY	
	NaCl	CMA	NaCl	CMA
Abies	2	1	8	1
Acer	6	1	5	1
Amelanchier	5	1	6	1
Arctostaphylos	4	1	13	7
Batula	7	1	7	1
Calocedrus	3	3	7	2
Chrysothamnus	2	1	2	1
Cornus	5	1	6	2
Klaeagnus	1	4	5	0
Fraxinus	2	0	5	3
Malus	4	1	8	1
Pinus jeffreyi	2	1	6	1
P. lambertiana	6	1	8	1
Quercus alba	3	3	5	2
Q. borealis	6	7	9	2
Salix	1	0	9	2
Thuja	4	0	6	0
Viburnum	10	2	11	1

Plants which were of poor quality masked the treatment related damage. Additionally, the organic potting mix for seven plant species strongly absorbed the applied ions, interfering with the Plant's ability to absorb the ions. The soil absorption of salt ions buffered the treatment related injury.

As with other laboratory studies, this study has its limitations. Different species react to the stress of NaCl and CMA differently. Also, different soil types will influence the impacts of NaCl and CMA.

(ii) Soil Evaluation

In response to FHWA requests, the State transportation agencies of Minnesota, West Virginia and Maryland set soil samples and information about the samples to the TransLab for evaluation. Additionally, two soil types from the Lake Tahoe Basin were sampled by TransLab personnel. Of the eleven soils sampled, seven were selected for testing.

Soils were received in thin-walled tubes, approximately 12-18 inches in depth and were relatively compacted and undisturbed. The compactness, in addition to the high clay contents of some samples, necessitated treating the samples as disturbed soils rather than undisturbed as originally planned. Most of the samples were compacted to the point that no test solution would be able to percolate through the soil during testing procedures.

Soils tested included one Maryland soil from Anne Arundel County, three soils from West Virginia, two soils from Minnesota and a soil from California's Lake Tahoe Basin. Description of the soils sampled are:

Maryland

Monmouth Collington Associate. This soil is an ultisol that developed from unconsolidated marine sediments of various sizes that were deposited in the Atlantic Coastal Plain. The soil is sandy to loamy. The soil sample represents the horizons just below the topsoil.

Minnesota

Clontarf fine sandy loam. The sample was taken in Chippewa County. This is sandy loam found in stream deltas and outwash plains, in areas of gently sloping topography. This soil is moderately well drained.

Maryland's loam is a nearly level, poorly drained, calcareous soil found in channels, on stream deltas and outwash plains. This sample was from Chippewa County. The permeability of this soil is moderate in the surface layers and rapid in the lower sandy levels. Surface runoff is slow while available water capacity is moderate. Organic content is high and the surface layer is mildly to moderately alkaline.

West Virginia

The West Virginia samples were soils from the DeKalb Series in Fayette County. The DeKalb series consists of moderately deep, well drained soils on ridge tops, hillsides, and mountainsides. These soils are formed in acid material weathered from sandstone. Slopes range from 3 to 70 percent, but are generally more than 40 percent.

DeKalb soils have a low to moderate available moisture capacity and rapid permeability. Natural fertility is generally low. In gently sloping areas, non-stony DeKalb soils are suited to crops, hay and pasture. Permanent pasture does poorly on these somewhat droughty soils. Trees grow fair to well and most of the acreage is wooded. The entire soil profile is strongly acid to very strongly acid.

Sample #1 has a coarser texture containing fewer sandstone fragments and is more droughty than previously described. It is described as DeKalb fine sandy loam from 1 to 10 percent slopes. This soil is suited to all crops commonly grown in the two counties. Short-rooted pasture plants do poorly and the hazard of erosion is severe in unprotected areas. This sample apparently had not been contaminated with deicing salts.

Samples #2 and #3 are classified as DeKalb channery loam from 30 to 40 percent slopes and are representative of this series with the exception of the surface being free of stones. Sample #3 may have been affected by some deicing salts during past winters.

Lake Tahoe

The Lake Tahoe sample was taken along Highway 89 approximately 2.5 miles northwest of the Highway 50/89 separation. Triplicate samples were taken 40 feet from the edge of the pavement.

The samples were from Elmira-Gefo Soil Association found from the Tahoe Valley to the California-Nevada state line. The soils of this associate are formed in granitic materials on glacial outwash fans and moraines. The Elmira series are somewhat excessively drained soils underlain by sandy granitic alluvium. Within the Elmira series, slopes are 0-30 percent which is between the elevations of 6200-6500 feet and normally 20-35 inches precipitation per annum. Characteristically, vegetation is an open stand of sagebrush with coniferous forest and perennial grasses.

a. Methods

Lysimeters (soil containers) were used to contain the soil samples for
the percolation testing. The lysimeters were constructed from 20 gauge
seamless brass tubing, 3 inches inside diameter and 12 inches long.
Supports within the lysimeter for holding the filter paper and soil
samples consisted of circles of coarse brass screen cut to fit and
silver soldered in place 1-1/2 inches from the tube bottom. During the
test, lysimeters were held vertical with a wooden rack. Test filtrate
which had passed through the soil column was collected in a 500 ml
Erlenmeyer flask.

Samples selected for testing were removed from the shipping tubes. Only
the upper 12 inches of each sample was used for testing. Each sample
was made up of three replicates. Some soil from each replicate of the
sample was combined to make a test sample. Each test sample was air-
dried, using fans for air circulation and ground by a power mortar and
pestle to pass a 2mm screen. Most soils had one or more distinct layers
within the top foot of soil. This method breaks down the normal
stratification of the upper 12 inches of soil structure producing a
uniformly disturbed sample that is representative of the upper one foot
of the soil horizon.

From each prepared soil sample, three 500 gram subsamples were taken for
the testing procedure. Each 500 gram replicate subsample was placed
into separate lysimeters fitted with a layer of #40 whatman filter paper
placed on top of the retaining screen. Three lysimeters with identical
replicate subsamples were used for each soil type. The 500 gram samples
were added in one motion to minimize particle size segregation. The
lysimeters containing filter and soil on top of the filter were dropped
20 times through a distance of 2.5 cm onto a flat woody block on a bench
surface. The uniform "tapping" of the lysimeters and subsample was
intended to achieve relatively uniform compaction. Once the samples had
been "tapped," another close fitting whatman filter was placed on top of
the sample. This filter was used to ensure minimal soil disturbance
when the cylinders were loaded with liquids to be tested.

Preliminary testing indicated that approximately 750 ml of test solution
should be charged in the lysimeter to ensure 500 ml of filtrate for
chemical determinations on selected parameters.

Because project funding limited testing to only one concentration of calcium-magnesium acetate and distilled water, 1 N reagent grade concentration was selected. The 1 N CMA concentration, while stronger than could be expected in a field situation (except in very unusual cases) would allow an evaluation of the chemical's potential for extracting nutrients and/or metals from the soil environment.

The soil loaded lysimeters were charged with 750 ml of 1 N CMA or the distilled water control and allowed to percolate until they stopped dripping filtrate or adequate sample for chemical determinations was secured. Filtrate was iced and immediately taken to the California Department of Water Resources Chemistry Laboratory for analysis. Parameters analyzed for were: pH, specific conductance, hydrolyzable orthophosphate, nitrate, potassium, and chemical oxygen demand.

b. Results

Tables 13 and 14 summarizes the results of the chemical analysis. The results note a range for the three replicates and a mean in parathesis. Distilled water and 1 N CMA chemical analysis are noted at the bottom of the table. The 1 N CMA used for the testing had a pH of 8.1. The results only compare the amount of metals or nutrients removed by either 1 N CMA or a distilled water control. The relationship between the amount of metal/nutrient removal and the total available in a particular soil was not determined. A general soil chemistry analysis for each soil was not conducted.

The results indicate that iron was removed by 1 N CMA at a significant level when compared to the control in five of the seven soils tested. Likewise, the effect of CMA on aluminum removal is significant in five of the seven soils tested. Three of the seven soil results showed no aluminum in either the control of CMA treatment. It is not known if this reflects absence of aluminum in these soils or that it did not move in response to treatments.

Sodium movement from the soil was noted in some of the soils; however, it was significant in only two cases, one of which (West Virginia #1) was known to have been contaminated by deicing salt. Significant removal of orthophosphate was noted in only one soil but five of the seven soils showed significant removal of hydrolyzable orthophosphate resulting from the leaching of CMA. Nitrate removal was not significant

218

in most of the soils. However, five of the seven soils did exhibit a significant loss of potassium when compared to the controls.

Based on the results noted in Tables 8.13 and 8.14, it appears 1 N Calcium-magnesium acetate has the potential to remove significant amounts of iron, aluminum, sodium, hydrolyzable orthophosphate and potassium from the soil.

This soil analysis was very limited. The results indicate that a 1 N CMA solution can remove some ions from a lysimeter. However, the laboratory conditions used in testing are not the same as the conditions which will be encountered in actual use. Further studies should be performed both in the laboratory and in the field.

TABLE 8.14

Leachate Volume Specific Conductance and COD Summary from the Soil Study

	Leachate Volume ml	Specific Conductance μmhos/cm	COD mg/liter
MD2 - CMA - DH$_2$O	567 582	30,730 115	63,400 233
MN1 - CMA - DH$_2$O	584 568	31,300 130	61,800 67.6
MN^3 - CMA - DH$_2$O	587 580	31,160 247	66,000 54.0
WV1 - CMA - DH$_2$O	547 548	30,930 193	61,100 122.3
WV2 - CMA - DH$_2$O	564 565	30,600 214	62,100 121.0
WV3 - CMA - DH$_2$O	543 553	31,230 352	65,700 48.0
CA1 - CMA - DH$_2$O	577 587	31,000 82	66,500 78.8
IN CMA Distilled H$_2$O		25,500 12.9	66,800 0

TABLE 8.15

Nutrient and Metal Summary from Soil Study (mg/liter)

	Iron	Aluminum	Calcium	Magnesium	Sodium	Orthophosphate	Hydrolyzable Orthophosphate	Nitrate	Potassium
Md2									
CMA	4.0	40.0	9593.0	5900.0	1600.0	0.09	0.1	0.97	23.0
DH$_2$O	0.3	0.0	2.3	2.0	4.0	0.00	0.1	0.65	4.7
MN1									
CMA	0.2	0	9900.0	5783.0	14.3	0.07	0.07	3.2	29.3
DH$_2$O	0.17	0	9.3	2.7	1.0	0.14	0.11*	2.8	3.4
MN3									
CMA	0.2	0	10,433	5267	14.0	0.07	0.05	3.0	17.3
DH$_2$O	0	0	38.7	7.0	2.0	0.007	0.1	2.4	2.3
WV1									
CMA	3.7	12.7	9650.0	5283.0	45.7	0.07	0.007*	1.0	4.83
DH$_2$O	0.9	2.3	5.3	1.0	29.0	0.02	0.02	0.65	2.1
WV2									
CMA	0.6	26.7	9450.0	5783.0	2.0	0	0.06	11.0	2.4
DH$_2$O	0	0	14.0	2.0	5.3	0.02	0.03	10.7	5.2
WV3									
CMA	0.2	0	9933.0	5617.0	16.7	0.03	0.03	15.3	21.3
DH$_2$O	0	0	44.3	4.7	4.7	0.02	0.04	19.3	6.3
CA1									
CMA	0.4	1.0	9867.0	5400.0	22.7	0.05	0.06	1.03	5.1
DH$_2$O	0.1	2.7	4.3	1.3	7.0	0.04	0.13	0.23	6.6
INCMA	0.2	0	10,200.0	5500.0	13.0	0.08	0.02	0.25	7.5
DH$_2$O	0	0	0	0	0	0	0	0	0

REFERENCES

1   D.M. Murray and V.F.W. Ernest. 1976. An Economic Analysis of the Environmental Impact of Highway Deicing. U.S. Environmental Protection Agency. EPA 600/2-76-105.

2   F.S. Adam. 1973. Highway Salt: Social and Environmental Concerns. Highway Research Record No. 425. Highway Research Board, Washington, D.C.

3   R. Field, et al. 1981. Environmental Impact of Highway Deicing. U.S. Environmental Protection Agency. Water Quality Research. Edison, New Jersey.

4   W.E. Miller, J.C. Greene, and T. Shiroyama. 1978. The _Selenastrum capricornutum_ Printz Algal Assay Bottle Test. U.S. Environmental Protection Agency. 600-78-0118. Corvallis, Oregon.

5   M. Sheeler. 1983. Experimental Use of Calcium Magnesium Acetate. Research Project HR-253 Iowa Department of Transportation. Ames, Iowa.

6   DeFoe, J.H. 1984. Evaluation of Calcium Magnesium Acetate as an Ice Control Agent. Research Report No. R-1248. Michigan Department of Transportation. Lansing, MI.

7   F.R. Kopperdahl. 1976. Guidelines for Performing Static Acute Toxicity Fish Bioassays in Municipal and Industrial Wastewaters. California Department of Fish and Game.

8   D.I. Mount. 1968. Chronic Toxicity of Copper to Fathead Minnows, _Pimephales promelas_. Rafinsque. Water Res. 2(3): 215-223.

9   R.W. Pennak. Fresh-water Invertebrates of the United States. Second edition. John Wiley and Sons, New York.

10   R.R. Sokal and J.F. Rolf. 1981. Biometry. 2nd edition. W.H. Freeman, San Francisco.

11   R.R. Sokal and J.F. Rolf. 1969. Biometry 1st edition. W.H. Freeman, San Francisco.

12   G.R. Goldman and R.W. Hoffman. 1975. A Study of the Influence of Highway Deicing Agents on the Aquatic Environment in the Lake Tahoe Basin and Drainages Along Interstate 80. Ecological Research Associates Report, Prepared for the California Department of Transportation, Sacramento, CA.

13   A.T. Leiser and S. John. 1983. Evaluation of the Effects of Calcium Magnesium Acetate on Selected Plant Species. Report prepared for the California Department of Transportation, Sacramento, CA.

Chapter 9

CATALYSIS OF THE COMBUSTION OF CARBONACEOUS PARTICLES (SYNTHETIC CHARS AND COAL) BY ADDITION OF CALCIUM ACETATE*

Yiannis A. Levendis

Mechanical Engineering, Northeastern University, Boston, MA 02115

9.1   ABSTRACT

The catalytic effects of calcium enrichment on the reactivity of carbonaceous particles was investigated in the range 670-3000 K. The char particles were prepared from polyfurfuryl alcohol (PFA) and were spherical and of uniform size. A few additional experiments were carried out with a HVA bituminous coal. Three different methods were used to introduce the calcium additive: precipitation of calcium carbonate, impregnation with calcium acetate, and calcium ion exchange. Electron microscopy showed that the distribution of calcium was remarkably uniform in particles containing a bimodal distribution of micro- and transitional-pores, whereas for particles with micropores only the Ca concentration was high at the surface and low at the center. X-ray analysis indicated that the conversion of the carbonate to the oxide at low temperatures (below 900 K) takes place only after all carbon has been consumed. Combustion studies showed that the calcium catalyzed the oxidation reaction rate at all temperatures investigated by up to two orders of magnitude. The effectiveness of the catalyst introduced by the different methods was comparable, with the calcium ion exchanged chars being, in general, the most reactive.

9.2   INTRODUCTION

An extensive literature has been published dealing with the catalysis of carbon and coal gasification by various metals and metal compounds, especially of alkali and alkali earth metals[1-14]. Calcium, in particular, has been the subject of numerous investigations as a naturally occurring mineral and as a low cost additive which can catalyze the rate of gasification[1-8] and serve as a sulfur scavenger in gasification and combustion[15]. A number of techniques of calcium introduction have been developed, ranging from mixing ground methods that introduce calcium inside the coal matrix, show considerable

---

*Adapted with permission from the *Journal of Energy and Fuels* from "Catalysis of the Combustion of Synthetic Char Particles by Various Forms of Calcium Additives" by Y.A. Levendis, S.W. Nam, M. Loewenberg, R.C. Flagan and G.R. Gavalas. Copyright 1989, American Chemical Society.

promise for capturing sulfur by retention within the ash[18].
limestone with coal and subsequent injection of the mixed powder into the
combustor[11] to incorporation of calcium within the coal matrix by such methods
as ion exchange[6,7,9,15,16,18], $CaCO_3$ precipitation[20], or impregnation[8]. The

The introduction of minerals in the carbon matrix changes both the
physical and the chemical structure of the particles and greatly affects the
reactivity. Kinetic studies at low to intermediate particle temperature
(500-1600 K) have showed that the catalytic activity of calcium depends on its
concentration[6,7], inclusion size[18,21], uniformity of dispersion[13] and chemical
form[3,8]. Hence, char pre-treatment, pyrolysis conditions and maceral
composition may influence the catalytic activity through their impact on the
Ca-treatment process. It is of interest to note that, while for some lignite
chars the reactivity increased monotonically with calcium loading[6], for others
it saturated at a modest calcium loading (4 wt% Ca)[22].

The purpose of this work was to investigate the catalytic effect of
calcium on carbon gasification in a broad range of temperatures, especially at
temperatures above 1200 K which have received limited attention previously.
The spatial distribution of the calcium additive as a function of both the
porous structure of the char and the mode of addition, as well as the effect
of this distribution on the catalytic action were also examined.

Although the technological interest of calcium addition is in coal
combustion or gasification, most of the experiments in this study employed
polymer derived chars. These synthetic chars were obtained by pyrolysis of
polyfurfuryl alcohol (PFA) in the form of monodisperse spherical particles.
Such char particles are ideal for kinetic investigations, especially at high
temperature, because of their precisely known size (50$\pm$1 μm in diameter) and
shape, their well characterized physical and chemical nature[23], and their
homogeneity and lack of residual mineral matter. Use of these synthetic chars
for fundamental studies of the Ca-enhanced carbon reactivity overcomes some of
the problems previously encountered with HCl-HF demineralized chars: (a)
incomplete demineralization[6], (b) substantial alteration of the pore structure
and apparent density (formation of cenospheres[3] etc.), and (c) deactivation of
the calcium catalyst by residual chemisorbed chlorine[6].

The synthetic chars selected for the present study were a plain PFA char
containing only micropores and a high porosity (75% PFA - 25% carbon black)
char containing both micro- and transitional pores. A limited number of
experiments was carried out with an HVA bituminous coal to obtain some
information about the influence of Ca on the combustion of natural fuels under
similar conditions. Both the synthetic char and the coal particles were
treated with calcium by: (i) precipitation of $CaCO_3$ within the pores of the

chars, (ii) impregnation with calcium acetate solution and (iii) calcium addition by ion exchange. The combustion of the calcium treated chars was studied by a number of techniques in the particle temperature ranges of 670-870 K, 1200-1500 K, and 1800-3000 K.

9.3 EXPERIMENTAL

9.3.1 Production of Synthetic Chars

The glassy carbon materials used in this study were produced from a carbon yielding binder (polyfurfuryl alcohol) and a thinning and mixing agent (acetone). To obtain a high porosity char, carbon black particles, about 20 nm in diameter, were suspended in the polymer-acetone mixture to serve as pore forming agents. The mixtures were introduced at constant rate into an acoustically excited aerosol generator using a syringe pump and were subsequently sprayed into an externally heated thermal reactor. The full description of the atomization and the thermal reactor system is given elsewhere[23]. Following atomization, the uniform droplets were cured by heating to a maximum temperature of 650 K in an inert atmosphere. The resulting particles were collected by sedimentation at the bottom of the reactor. The total residence time in the reactor was approximately 4s. To eliminate sticking of the collected particles, all materials underwent a second pyrolysis treatment for 1hr at 800 K in a horizontal muffle furnace in N_2.

To differentiate the chars produced for the present study the following nomenclature will be used: (a) the plain polymer char will be labeled *low porosity* char since its porosity, e, is ~25%; (b) the char containing 25% carbon black filler will be termed *high porosity* char, e = 48%; and (c) the coal will be referred to as PSOC-680.

Partial oxidation enlarges the fine pores and removes pore constrictions in the synthetic chars, thus, making the particle interior accessible to gases and liquids[23]. The calcium treatment processes were, therefore, facilitated by partial oxidation in air for 5min at 800 K resulting at about 15% conversion. For this oxidation, the particles were spread in a thin layer inside porcelain boats, thereby minimizing bed diffusion resistance. The boats were then introduced into a hot muffle tube furnace for the 5 min. exposure.

9.3.2 Calcium Treatment Techniques

Calcium was added to the synthetic char by $CaCO_3$ precipitation, acetate impregnation, and calcium ion exchange. $CaCO_3$ precipitation involved the following ionic reactions:

(1) $Ca(CH_3COO)_2 + H_2O + CO_2 = CaCO_3 + 2CH_3COOH$ $\Delta H_{298}^0 = 3.6\ kcal/mole$

(2) $2CH_3COOH + Ca(OH)_2 = Ca(CH_3COO)_2 + 2H_2O$ $\Delta H_{298}^0 = -2.69\ kcal/mole$

A small amount of char or coal was evacuated in a 10ml reactor vessel at $70°C$ for 1 day. Carbon dioxide was then introduced into the reactor at room temperature and of 200 mbar pressure, and allowed to equilibrate with the char for 30 min. A slurry of calcium acetate solution in water and a predetermined amount of calcium hydroxide was introduced into the reactor which had been cooled in an ice bath for 10min. in order to prevent desorption of CO_2 upon heating by the exothermic reaction (2). The char and the slurry mixture were stirred with a magnetic stirrer for 3 hrs, during which time the pH of the mixture dropped from 12.0 to 6.8, indicating that all of the $Ca(OH)_2$ had reacted. The char was then filtered, rinsed with distilled water, and dried at $80°C$ for 24 hrs.

Chars were impregnated with calcium acetate by the incipient wetness method. The char was evacuated at $70°C$ for 1 day. Calcium acetate solution (1N) was then slowly added to incipient wetness while the sample was stirred vigorously. Since the synthetic char is not readily wetted, the chars were impregnated under vacuum. The treated sample was dried at $80°C$. The procedure was repeated three times. After the third impregnation the sample was washed and dried.

Calcium ion exchange was performed in a 300 ml beaker placed in a water bath maintained at $50°C$. The char samples were first mixed with 10ml of distilled water for 10 minutes in the reactor to ensure that they were wetted thoroughly. Subsequently, 100 ml of 1 N calcium acetate solution, the pH of which was adjusted to 8.5 with calcium hydroxide, were added to the reactor. The temperature was kept at $50°C$. The reactor was sealed quickly and the slurry was maintained at the initial pH value with continuous addition of 0.01 N $Ca(OH)_2$ solution. A stream of N_2 was used to purge the reactor of air in order to prevent the absorption of atmospheric CO_2 in the solution. At the end of the process, the slurry was filtered and washed with distilled water. The treated char was then dried at $80°C$ for 24 hrs.

9.3.3 Characterization of Chars
(i) Physical Properties

The total (internal and external) initial surface area of the chars was measured by N_2 adsorption at 77 K and CO_2 adsorption at room temperature. The results were analyzed by the BET theory and the Polanyi-Dubinin potential theory, respectively. It was found that the BET area of the low porosity char was $2m^2/g$. After the partial oxidation, it increased to $300m^2/g$. The surface area, as measured by Medek's approximation to the Polanyi-Dubinin isotherm came out to be $59m^2/g$ before, and $560 m^2/g$ after partial oxidation. These values indicate the presence of a vast network of micropores in this char. The porosity of this char after partial oxidation, as measured by CO_2 adsorption was 27%, corresponding to a void volume of $0.22 cm^3/g$. The porosity deduced from the helium and the apparent densities was lower, 25%. The apparent density, as measured by low pressure mercury intrusion, was $1.12cm^3/g$. The true density was found to be circa $1.5cm^3/g$ using helium pycnometry. After calcium treatment, the N_2 BET area came out to be 15, 16, and $20m^2/g$ for the ion exchanged, the precipitated, and the impregnated chars, respectively.

The initial N_2 BET surface area of the high porosity char was $184 m^2/g$, its apparent and true densities were $0.88cm^3/g$ and $1.45cm^3/g$, and the porosity was 40%. After the partial oxidation to about 15% burnout, its total surface area rose to $230 m^2/g$ and the porosity to 48%. Calcium treatment reduced the areas to 80 and $75 m^2/g$, for the ion exchanged and the precipitated chars, respectively.

(ii) Calcium Distribution

The effectiveness of the calcium treatment was assessed by measuring the calcium concentration as a function of distance from the surface of the particle. Samples of the various chars were cast in epoxy, polished in a *Buehler Minimet* automatic polisher, and gold coated for examination with a *CamScan* scanning electron microscope (SEM). Particles that had been sectioned near the middle, i.e. those with the largest diameters, were selected for analysis.

The calcium distribution was determined by energy dispersive spectroscopy (EDS) or, in the case of lower concentrations, with a *JEOL Superprobe 733* electron microprobe by wavelength dispersive spectroscopy (WDS).

Figure 1 shows Ca distributions (presented as mass percentage of CaO equivalents) as a function of distance from the surface of the different chars. Each of the profiles shown is an average from three particles. The variability in calcium levels from particle to particle was small ($\pm5\%$), indicating that the treatment processes are highly repeatable. Furthermore, the two analytical techniques (EDS and WDS) were in very good agreement in the regions where they overlapped. The calcium oxide is uniformly distributed in the high porosity chars that contain both transitional- and micro-pores, since the calcium compounds penetrate readily into the interior of the particles. On the other hand, in the low porosity chars that contain only micropores the penetration is not very effective and the concentration of calcium is high close to the surface and very low at the center. A comparison of the three methods of calcium addition indicates that, for the low porosity chars, ion exchange is the most effective; the calcium concentration is approximately constant in a 5μm thick outside layer. In the next 5μm, the calcium level drops rapidly by two orders of magnitude. Thereafter, the level is again flat all the way to the center of the particle. Calcium carbonate precipitation results in a lower concentration everywhere with a thinner region (1-2μm) of constant concentration near the surface. As before, the concentration drops rapidly reaching a plateau at about 10μm from the surface. Calcium acetate impregnation method resulted in a distribution similar to that of $CaCO_3$ precipitation, except that in the latter method the calcium penetrated more effectively the region near the surface of the particle but more poorly the region close to the center.

The overall calcium loadings in both high and low porosity chars treated by ion exchange were comparable, while in the case of $CaCO_3$ precipitation the high porosity chars had considerably higher loading than the low porosity chars. In all cases, the partial oxidation pre-treatment increased the calcium loading of the particles by enlarging the pores, removing pore constrictions, and enhancing the calcium exchange capability of the chars[24].

SEM-BSE pictures also show the calcium distribution in the particles due to the dependence of the electron backscattering coefficient on the mean atomic number of the material, $\bar{Z}$. Micrographs of sections through calcium treated glassy carbon particles shown in Fig.2 reveal the spatial variation of composition. Examination of Fig. 2 reveals, qualitatively, the same features regarding the radial distribution of calcium as the analyses above. The bright color in the periphery of the particles results from high concentrations of calcium. The particle depicted in Fig. 2a is calcium ion exchanged, while that in Fig. 2b contains precipitated $CaCO_3$. The concentrated (bright) regions near the particle surface graphically illustrate the concentration profile of Fig.1. Figures 3a and b depict high porosity particles that have been treated by ion exchange and $CaCO_3$ precipitation respectively. These particles are uniformly bright, as expected for the uniform distribution of calcium in these chars.

The distribution of calcium in the particles is also illustrated by the ash residue after complete combustion. Ashing experiments were performed in air at 800 K. The ash residue from the combustion of low porosity particles containing precipitated $CaCO_3$ consists of thin bubble-like shells while the residue of the ion exchanged low porosity particles has the form of thick rough shells, shown in Fig.4 a and b. The high porosity synthetic chars produce compact ash residues reminiscent of coal ash. Polished sections of these ash residues, shown in Fig. 5.

(iii) <u>X-ray Diffraction</u>

X-ray diffraction studies of low porosity chars treated with calcium by the ion exchange and precipitation methods were conducted in a *Siemens D500/501* diffractometer at 40kV, 30mA using Ni filtered CuKα radiation. No Ca species diffraction peaks were observed in the x-ray diffraction (XRD) patterns of the calcium loaded chars, presumably because of the small size of the crystallites and the high degree of dispersion[6]. Oxidation of the chars resulted in different XRD patterns depending on the soaking temperature and atmosphere. Samples of chars that were partially burned at 900 or 1400 K in 4% O_2 for 2 s exhibited $CaCO_3$ and CaO peaks. The calcium ion exchanged sample showed particularly strong $CaCO_3$ peaks at both temperatures. Ash produced after complete combustion at 1400 K in 4% O_2 possessed only CaO peaks. Samples pyrolyzed at 1400 K in N_2 exhibited weak CaO peaks only. Finally, complete oxidation at low temperatures (773 K, air) resulted in both CaO and $CaCO_3$ peaks. Partial oxidation at the same temperature resulted predominantly in $CaCO_3$ peaks. Therefore, as previously observed in XRD of lignites[8], crystals of CaO appear to form during pyrolysis of chars at elevated temperatures. However, during oxidation at temperatures below 900 K, calcium carbonate is the predominant product due to the presence of CO_2 as verified by a single thermodynamic calculation. After all carbon is consumed, $CaCO_3$ decomposes to CaO. The conversion of $CaCO_3$ to CaO at temperatures above 900 K is expected to proceed even in the presence of carbon and, thence, CO_2. The fact that this was not seen for the ion exchanged chars at 1400 K presently, is probably due to carbonate regeneration at the cooler region of the furnace or in the sampling probe[25].

The average crystallite size for $CaCO_3$ produced in combustion at 1400 K, estimated from the 3/4 peak width of the (104) diffraction line[29] was 22 nm in fair agreement with results obtained elsewhere[15]. Crystallites of that size could have plugged some of the pores in the chars during combustion.

(iv) Combustion Experiments

The combustion of the chars was studied in a drop tube furnace and a thermogravimetric analyzer.

(a) Pyrometry

The chars were burned at moderate (1200-1500 K) and high particle temperatures (1800-3000 K) in an externally heated, laminar flow (drop-tube) furnace[25]. Particle temperatures were measured by near-infrared two color pyrometry for particles that burned at temperatures that were significantly higher than the wall temperature or inferred from heat balance calculations for particles that oxidized at temperatures close to the wall temperature. The combustion apparatus and the pyrometer are described in Ref. 25. The experiments were conducted at a constant furnace wall temperature of 1470 K, either in air or pure oxygen. Using the pyrometer to view the particles along the axis of the reactor, the entire temperature-time histories of individual burning particle were observed. Typical profiles are shown in Figures 6-11.

In Fig. 6a temperature-time traces of several high-porosity calcium-free particles burning in O_2 are superimposed. The average maximum temperature was about 2600 K, and the mean burnout time was 14 ms. The combustion behavior of similar size particles to which calcium was added by ion exchange is shown in Fig. 6b. The maximum temperatures were about 2900 K and mean burnout time was 11.5 ms. Figure 6c shows traces from combustion of high porosity particles in which $CaCO_3$ had been precipitated. Again, the profiles are rather flat, the average maximum temperatures are 2800 K, and the average burnout time was 12 ms. Thus, calcium addition has a modest effect on reducing the burnout time and increasing the combustion temperature of particles.

Figure 7 shows the temperature history of high porosity particles burning in air. Again, particles treated by both ion exchange and precipitation behave similarly. The particles burn for 25 to 30 ms and the average temperatures are in the vicinity of 2000 K. This combustion behavior is strikingly different from that of calcium-free particles. Under identical conditions, the latter particles burned slowly at temperatures close to that of the reactor wall, with about 2 sec being required for complete combustion.

Combustion of low-porosity particles reveals similar features to the combustion of high porosity particles. The presence of calcium reduces burnout times and increases particle combustion temperatures. Figure 8 depicts combustion of low porosity particles in O_2. The untreated particles burn at an average temperature of 2200 K and a burnout time of approximately 22 ms (Fig. 8a). The calcium treated particles burned faster, approximately 12.5 ms, and hotter, approximately 2500 K (Fig. 8b and c). This behavior seemed to be independent of the method of calcium treatment. Combustion in air, however, revealed again large differences in the manner in which the particles burned. The untreated char particles did not ignite and burned slowly at temperatures between those of the gas and the wall, and the conversion at the end of the 2 sec residence time in the combustion chamber was only 60%[25]. On the contrary, the particles that had been treated by calcium acetate impregnation ignited and burned in an average of 30 ms at a temperature of 1900-2000 K (Fig. 9a). Some ion exchanged or Ca precipitated particles ignited and burned in about 25 ms and circa 1900 K (Fig. 9b), while others burned slowly at a lower temperature. This behavior suggests that the particles so treated were at the verge of ignition. Had the oxygen concentration been a little higher they might too have ignited[25].

A few combustion experiments were conducted in air using the PSOC-680 coal. The combustion behavior of the untreated coal is contrasted to the behavior of $CaCO_3$ precipitated coal in Fig. 10 a and b. The untreated coal particles readily ignited in air, unlike the synthetic Ca-free chars, because of the higher reactivity of coal by virtue of its different pore structure as well as its content of heteroatoms and minerals. The burnout times for the coal char particles were about 35 ms with the average temperatures of 2000 K. The particle temperatures were surprisingly uniform (except for one trace) although the burnout times varied. Early in the combustion of these particles, a distinct temperature peak lasting only 1-2 ms was observed, probably caused by the combustion of volatiles. Introduction of $CaCO_3$ by precipitation accelerated the combustion. The resulting burnout times were of order 15ms (Fig. 10b). The particle temperatures exhibited large scatter and on the average they were somewhat higher than those of the untreated particles.

(b) Moderate Temperatures

Experiments at lower particle temperatures where combustion occurred without a temperature jump were also conducted in the same drop tube furnace, by lowering the oxygen partial pressure, Po_2[25]. At Po_2 below about 7% and a combustor wall temperature range of 1200-1500 K the particles did not ignite, but rather burned slowly at approximately the wall temperature. Combustion rates were deduced from particle size reduction and/or sample mass and density change. The calcium laden chars were again significantly more reactive than the untreated chars. For example, at T_w = 1250 K, T_{gas} = 1225 K, Po_2 = 0.04, and residence time t = 2 s, the conversion of 50μm PFA particles was 22% while the conversion of calcium ion exchanged PFA particles was 85%. Under the same conditions, the conversion of $CaCO_3$ precipitated particles was 68%.

(c) Low Temperatures (TGA)

Experiments were conducted in a *DuPont* model *951* thermogravimetric analyzer (TGA) in air, at temperatures between 650-850 K and a gas flow rate of 100 cm^3/min STP. The samples were heated in nitrogen at 30 K/min until the final temperature was reached and held at that final temperature for an additional 20-30 minutes. Small sample quantities (1 mg) were spread in a thin layer (1 or 2 monolayers) on the balance pan to minimize diffusional resistance in the particle layer. To ensure that the measured rates are free of diffusional limitations, film diffusion, layer diffusion, and particle pore diffusion were examined separately as outlined elsewhere[26]. The layer Thiele modulus[27] at 25% conversion was of order 10^{-2} or less, indicative of uniform oxygen concentration in the particle layer. For the low porosity ion exchanged particles burning at 800 K, which presented the strongest pore diffusion limitations in this section of our investigation because of their high reactivity and small pore size, the particle Thiele modulus, ϕ, was about 0.9, yielding an effectiveness factor, η, a little below unity. Hence, the particles were burning approximately under the kinetic regime I^{28} of combustion. For the less reactive chars, the effectiveness factor was essentially one. This conclusion was directly verified for the high porosity chars where the rate was experimentally found to be independent of particle size. Care should be exercised, however, in assessing the intraparticle penetration of the oxidizer gas in the calcium treated microporous chars since, Ca crystallites might be blocking some of the pores, thereby introducing small uncertainties in the calculation of Knudsen diffusivity.

Figu es 11-13 show the apparent reaction rates for glassy carbons and coal chars at 673, 773 and 873 K as functions of conversion (burnout), where the apparent reaction rate, R_m, and the conversion, X, are given by:

(3) $\quad R_m = -\dfrac{1}{m-m_{ash}}\dfrac{dm}{dt}$

(4) $\quad X = \dfrac{m_{initial}-m}{m_{initial}-m_{ash}}$

In Figures 11a and b showing the apparent rate of low porosity particles burning at 673 and 773 K, the calcium treated PFA chars appear to be more reactive than the untreated PFA chars by up to two orders of magnitude. Calcium-free PFA char rates were nearly independent of conversion, the initial transient being due to the changing gas composition over the bed, from pure N_2 to 21% O_2-79% N_2 upon admission of air. In contrast, the Ca-treated PFA char rates were initially high, but subsequently dropped to a level roughly twice that of the plain PFA char.

In order to understand the differences in the various rate versus conversion plots of Fig. 11a,b, it is necessary to recall the radial distribution of calcium in the chars. The ion exchanged chars with the highest concentration and penetration of calcium exhibited the highest reactivity. In those chars the calcium concentration falls sharply at 6-7μm below the surface; this Ca-rich outside layer, however, contains roughly 60% of the carbon mass. This corresponds closely to the high rate portion of the rate curve (up to 60% conversion). The calcium precipitated and impregnated chars showed similar behavior, except that the high rate region extends to only 30% conversion. This conversion is lower than the 40-50% value that one could calculate on the basis of the calcium concentration distribution. It is possible that the precipitated $CaCO_3$ ceases to catalyze the carbon gasification when it loses contact with the adjacent carbon matrix even if there is still unburned carbon in the vicinity. The difference in the dispersion and chemical form of the calcium additive is also responsible for the large difference in the rate of oxidation, with the chemically bound calcium added by ion exchange being the most reactive. However, in this case it is difficult to separate the effects of the concentration and the chemical form of the catalyst.

Figures 12a, b show the oxidation rate versus conversion for the high porosity chars. The rates are overall higher than those of the low porosity chars (compare Figs. 11b with 12a) and are due primarily to the complete penetration of the calcium additive. For the calcium free chars, the rate at 873K exhibits a maximum at about 15% conversion. As in the low porosity chars, calcium enhances the rate of oxidation of the high porosity chars, with the calcium ion exchanged particles being the most reactive. In all but one case, the rates are monotonically increasing with conversion reflecting the constant calcium distribution in the particles and the progressive opening and enlarging of pores with burnout. The ion exchanged char, exhibits an anomalous behavior at 773 K, undergoing a maximum in the rate. The cause of this behavior is uncertain.

Figure 13 shows oxidation rate versus conversion for the untreated and calcium treated bituminous coal. All samples were devolatilized in N_2 at 773 K for 30 min prior to oxidation. The untreated coal is more reactive than the polymer chars in the high temperature experiments. The oxidation rate exhibits a maximum at about 20% conversion, in a manner similar to the high porosity synthetic char. Both calcium laden chars exhibit overall higher rates without undergoing an early maximum in rate. The rate drop at high conversions (circa 80%) may be due to burnout of all carbon in proximity to the catalyst, and/or the existence of small regions where the catalyst had not penetrated since, the coal had not been subjected to any pyrolysis or pre-oxidation prior to calcium addition.

9.4   ESTIMATION OF RATE PARAMETERS
9.4.1 Medium and High Temperatures

Kinetic rate parameters were estimated using the approach of Loewenberg and Levendis for particle temperatures at or above 1200 K. Their approach applies to the case in which particle combustion occurs at nearly constant temperature and apparent density, conditions which were satisfied for at particle temperatures at or above 1200-1300 K. The analysis assumes first order reaction and solves the pseudosteady film heat and mass transfer equations to obtain an apparent rate constant, k_a, and particle temperature, T_p, in terms of the observed burnout time, conversion and initial particle radius for each temperature-time trace.

Comparison of calculated and measured particle temperatures was used to check whether the primary oxidation product was CO or CO_2. Carbon monoxide is generally accepted as the dominant product at temperatures above 1300 $K^{31,32}$. However, in view of the presence of calcium known to catalyze oxidation to CO_2 at lower temperatures[31,33], it was thought useful to reexamine the question. Calculations with CO being the only product (burning far away from the particle) yielded temperatures within 100 K of the experimental values. On the other hand, the temperatures calculated by assuming that CO_2 is the only product were much higher (as much as 1000 K) than the measured. Thus, it was concluded that CO was the dominant product of combustion.

An Arrhenious plot of the estimated apparent rate coefficient is shown in Fig. 14 for the low and high porosity synthetic chars used. The estimated rates of the calcium laden particles are one order of magnitude higher than those of the untreated chars at all but the highest temperatures. At the highest temperatures, above approximately 2600 K, the apparent rate of the calcium treated chars is lower than the rate at 2000 K. This negative temperature dependence could be due to decline of the number of active sites by thermal rearrangement of the carbon matrix[25,34] and/or to a decrease in the effectiveness of the catalyst. From the reaction kinetics standpoint it would be interesting to evaluate the catalytic effect of calcium on the basis of intrinsic rather than the apparent rates. The data analysis required for this purpose is practical only for the high porosity chars which acquired uniform calcium distribution upon treatment. Even then, the analysis should take into account the changes in surface area and pore size distribution at various extents of conversion, in the presence of the calcium additive. We did not carry out such detailed char characterization, but we did measure the initial surface area and the area after 20% burnout. In view of the limited data available, we carried out a first order analysis using the following assumptions:

(i)   the effect of the additive is to plug a certain fraction of the pores; the remaining accessible pores have the same size as in the untreated char and are uniformly distributed throughout the particle.

(ii) the fraction of accessible pores is determined by comparing the surface area of treated and untreated chars at 20% conversion.

Having made these assumptions we calculated the intrinsic rate constant from the apparent constant using the high Thiele modulus limiting analysis presented in our earlier papers[35,36,30]. The resulting rate constants are shown in the high temperature region of Fig. 16.

9.4.2 Low Temperatures

The apparent reaction rate, R_m, for various treated and untreated chars, plotted against burnout, was shown earlier in Figures 11-13.[1] To calculate the intrinsic rate, R_i, we assumed the effectiveness factor to be equal to unity, as justified earlier, so that:

$$(5) \quad R_i = \frac{R_m}{A_{tot}}$$

where A_{tot} is the total surface area, here taken as the N_2 BET area at 77 K. For comparison purposes, the reactivities of the chars were listed at 20% burnout, where the apparent rates of most chars had reached plateaus. The N_2 BET surface areas for these chars at 20% conversion were 320 m^2/g for the low porosity char and 35 to 55 m^2/g for the calcium treated chars. The area of the high porosity char at 20% burnout was about 380 m^2g, while that of the calcium treatment chars was between 140 and 150 m^2/g. The areas vary little with combustion temperature in the range of 673-873 K. The apparent and intrinsic reaction rate constants k_m and k_i respectively, can be calculated by dividing R_m and R_i by the surface oxygen pressure (for first order kinetics). At these low temperatures, the surface oxygen concentration was assumed equal

---

[1]The apprarent rate per unit mass, R_m is related to the apparent rate per unit external area, R_a by: $R_a = R_m \gamma \sigma_a$[38], where γ is the characteristic dimension of the particle (ratio of particle volume to external area) and σ_a is the apparent density.

to the ambient. The estimated rate parameters for these chars are shown in Figures 15 and 16.

Figure 15 shows that the apparent rate constant, k_m, increases with addition of calcium by up to two orders of magnitude at the lowest temperature and up to five times for the highest temperature. As expected, the catalytic effect of calcium decreases with increasing temperature. The intrinsic reaction rate constant, k_i, follows similar trends, Fig. 16. Here the differences between the various chars become even more pronounced because of the lower surface area of the calcium-containing chars. This higher intrinsic reactivity of the low porosity chars can be explained on the basis of their overall higher calcium loading, 2.5-3.0 wt% Ca in contrast to 1.5-2.0 wt% Ca of the high porosity chars. It can be noted that both the apparent and intrinsic activation energies E_a and E_i, of the chars appear to decrease with Ca-loading.

9.5   CONCLUSIONS

Monodisperse synthetic char and coal particles have been treated with calcium using the following methods: (i) precipitation of $CaCO_3$ within the pores of the chars, (ii) impregnation with calcium acetate solution and (iii) calcium addition by ion exchange. The distribution of calcium inside low porosity synthetic chars was high in the vicinity of the surface but dropped rapidly to negligible values at 5 to 10μm depth below the surface depending on the method of calcium treatment. Ion exchange produced the deeper penetration than the other two methods of calcium addition. In the high porosity chars the distribution of calcium was very uniform throughout the particles for all methods of treatment. X-ray diffraction shows that for all methods of treatment calcium carbonate is the predominant form of calcium during combustion at temperatures below about 900 K at the presence of carbon.

Combustion experiments showed that calcium enhanced the rate of oxidation in the range of 673-3000 K. At the lower temperatures the enhancement factor was about two orders of magnitude while at the highest temperatures it was only a few percent. All three modes of calcium introduction enhanced the combustion rate with calcium ion exchange producing the highest reactivity. The catalytic rate enhancement is manifested despite a certain extent of pore plugging and loss of surface area caused by the calcium additive.

REFERENCES

1   R.G. Jenkins, S.P. Nandi, P.L. Walker, Jr. *Fuel* **1973**, *52*, 288-303.
2   E.J. Hippo, P.L. Walker, Jr. *Fuel* **1975** *54*, 245-248.
3   A. Tomita, O.P. Mahajan, P.L. Walker, Jr. *Fuel* **1977**, *56*, 137-144.
4   A. Linares-Solano, O.P. Mahajan, P.L. Walker, Jr. *Fuel* **1979**, *58*, 327-332.
5   T. Takarada, Y. Tamai, A. Tomita, *Fuel* **1985**, *64*, 1438-1442.
6   T.D. Hengel, P.L. Walker, Jr. *Fuel* **1984** *63*, 1214-1220.
7   A. Linares-Solano, E.J. Hippo, P.L. Walker, Jr. *Fuel* **1986** *65*, 776-779.
8   Y. Ohtsuka, A. Tomita, *Fuel* **1986** *65*, 1653-1657.
9   M.L. Jones, D.P. McCollor, and B.J. Weber, "Combustion Behavior of Ion Exchanged Coal Chars." Presented at the Fall 1987 Meeting of the Western States Section of the Combustion Institute. Hawai, November 1987.
10 H. Marsh, R.R. Adair, *Carbon* **1975** *13*, 327-333.
11 M. Letort, G. Martin, *Bull. Soc. Chim. France* **1947**, *14*, 400-405.
12 P.L. Walker, Jr., M. Shelef, R.A. Anderson *Chemistry and Physics of Carbon* (Edited by P.L. Walker), Vol. 4 p. 287. Edward Arnold, London, 1968.
13 W.M. Tudenheim, G.R. Hill, *Ind. Engng. Chem.* **1955**, *47*, 2129-2134.
14 E.T. Turkdogan, and J.V. Vinters, *Carbon* **1972**, *10*, 97-111.
15 H.N.S. Schafer, *Fuel* **1970**, *49*, 197-213.
16 P.L. Case, M.P. Heap, C.N. McKinnon, D.W. Pershing, R. Rayne, *Am. Chem. Soc. Div. Fuel Chem. Prepr.* **1982**, *27*(1), 158-165.
17 H. Freund, R.K. Lyon, *Comb. & Flame* **1982**, *45*, 191-203. And references therein.
18 K.K. Chang, R.C. Flagan, G.R. Gavalas, P.K. Sharma *Fuel* **1986**, *65* 75-80.
19 L.R. Radovic, P.L. Walker, Jr, R.G. Jenkins *Fuel* **1983**, *62*, 209-212.
20 P.K. Sharma, G.R. Gavalas, R.C. Flagan, *Fuel* **1987**, *66*, 207-209.
21 A. Oya, H. Marsh, *Jour. of Mater. Sci.* **1982**, *17*, 309-322.
22 E.J. Hippo, R.G. Jenkins, P.L. Walker, Jr. *Fuel* **1979**, *58* 338-345.
23 Y.A. Levendis, R.C. Flagan, Submitted to *Carbon*, **1988**.
24 W.S. Kalema, "A Study of Coal Oxidation" *PhD Thesis*, California Institute of Technology, 1984.
25 Y.A. Levendis, R.C. Flagan, G.R. Gavalas, Submitted to *Comb. & Flame*, **1988**.
26 R. Sahu, Y.A. Levendis, R.C. Flagan, G.R. Gavalas, *Fuel* **1988**, *67*, 275-283.
27 P.B. Weitz, C.D. Prater, *Advan. Catalysis* **1954**, *6*, 143.
28 M.F.R. Mulcahy, I.W. Smith, *Rev. Pure and Appl. Chem.* **1969**, *19*, 81-108.
29 M.A. Short, P.L. Walker, Jr. *Carbon* **1963**, *1*, 3-9.
30 M. Loewenberg, Y.A. Levendis, Submitted to *Comb. & Flame*, **1988**.
31 B.C. Young, S. Niksa, *Fuel* **1988**, *67*, 155-164.
32 I.W. Smith, *Nineteenth Symposium (International) on Combustion*, The Combustion Institute, Pittsburgh, PA, **1982** 1045-1065. And references therein.
33 D.W. McKee, "Chemistry and Physics of Carbon", (Ed. P.L. Walker, Jr. and P.A. Thrower), Marcel Dekker, New York, 1981, Vol. 16 pp.1-118.
34 J. Nagle, R.F. Strickland-Constable, Proceed. of the Fifth Carbon Conf. 1:154, Pergamon Press, Oxford, 1961.
35 G.R. Gavalas, *Comb. Sci. Tech.* **1981**, *24*, 197-210.
36 M. Loewenberg, J. Bellan, G.R. Gavalas, *Chem. Eng. Comm.* **1987**, *58*, 89-97.

NOTATION

SYMBOL	DESCRITPION	UNITS
a	particle radius	cm
A_a	apparent pre-exponential factor	g/cm^2 s $(atm)^n$
A_{in}	intrinsic pre-exponential factor	g/cm^2 s $(atm)^m$
A_{tot}	specific total area	m^2/g
C_∞	ambient oxygen concentration	g/cm^3
C_s	oxygen concentration at particle surface	g/cm^3
E_a	apparent activation energy	kcal/g
E_{in}	intrinsic activation energy	kcal/g
k_i	intrinsic rate coefficient	$g/(cm^2sec(atm)^m)$
m	mass of carbon	g
P_{o2}	ambient partial pressure of oxygen	
r_p	pore radius	cm
R_a	apparent reaction rate	g/cm^2-sec
R_m	apparent reaction rate	g/g-sec
R_i	intrinsic reaction rate	g/cm^2-sec
T_g	ambient temperature	K
T_p	particle temperature	K
X	conversion	
ϵ	total porosity	
η	effectiveness factor	
ϕ	Thiele modulus	

TABLE 9.1

Physical Properties of Partially Oxidized Chars before Calcium Treatment

CHAR	Low Porosity plain polymer	High Porosity polymer + 25% carbon black
Apparent Density (g/cm^3)	1.12	0.75
Helium Density (g/cm^3)	1.5	1.45
Porosity	25%	48%
Average Pore Diameter	~ 15 Å	150Å and 15Å bimodal
BET area (m^2/g)	300	184
CO_2 area at 298 K (m^2/g)	560	-

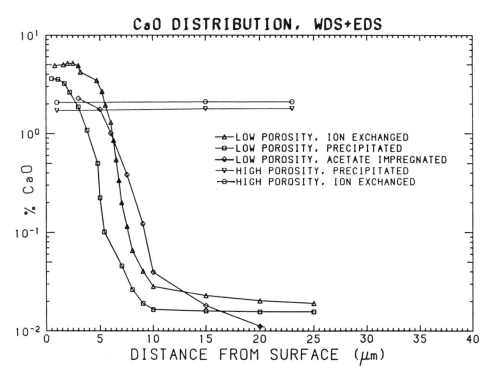

1. Distribution of calcium in CaO equivalents along the radius of calcium treated synthetic chars

238

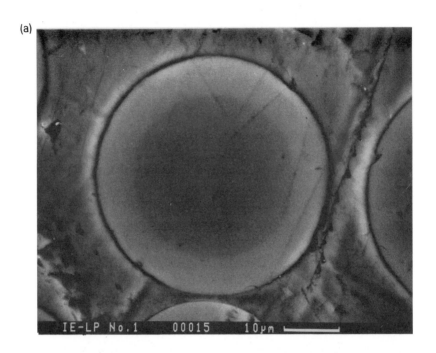

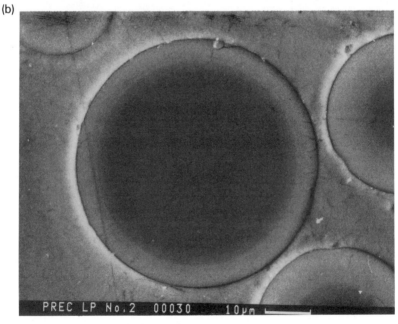

Figure 2. SEM-BSE micrographs of sections through calcium treated synthetic char particles: (a) low porosity ion exchanged; (b) low porosity $CaCO_3$ precipitated.

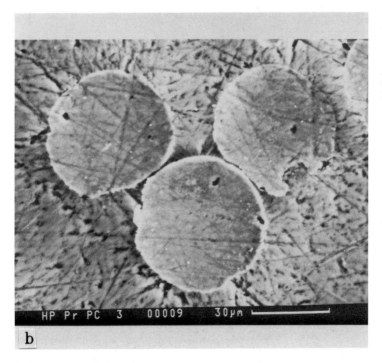

Figure 3. SEM-BSE micrographs of sections through high porosity calcium treated synthetic char particles: (a) ion exchanged; (b) $CaCO_3$ precipitated.

240

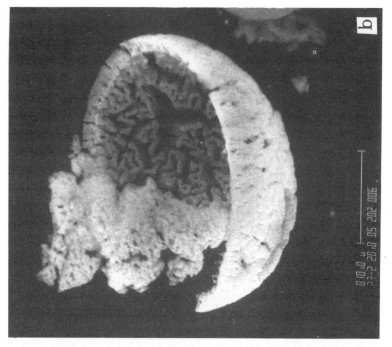

Figure 4: SEM-SE micrographs of residual ash after complete combustion of low porosity synthetic char particles in air at 800 K. The particles were treated with ion exchanged calcium: (a) outside view (b) inside view.

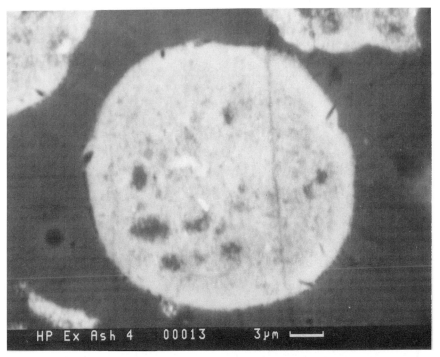

Figure 5. SEM micrographs of polished sections of ash particles resulting from combustion of high porosity chars.

242

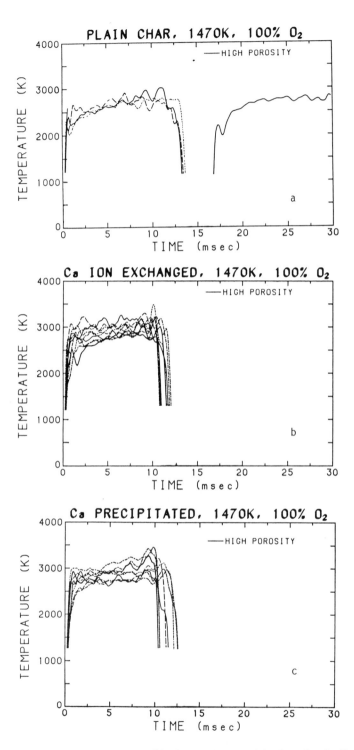

6. Temperature-time traces of high porosity particles burning in O_2: (a) plain char, (b) ion exchanged, (c) $CaCO_3$ precipitated.

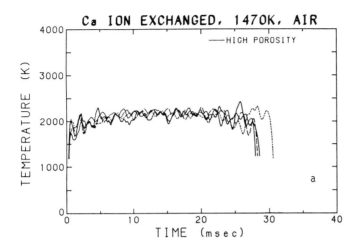

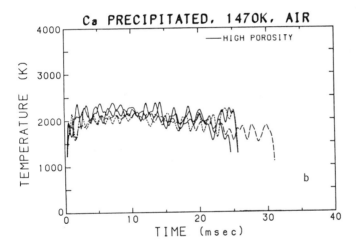

7. Temperature-time traces of high porosity particles burning in air: (a) ion exchanged, (b) CaCO₃ precipitated.

244

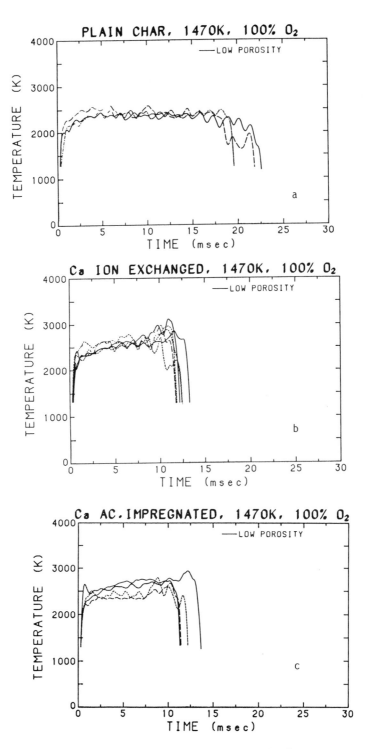

8. Temperature-time traces of low porosity particles burning in O_2: (a) plain char, (b) ion exchanged, (c) calcium acetate impregnated.

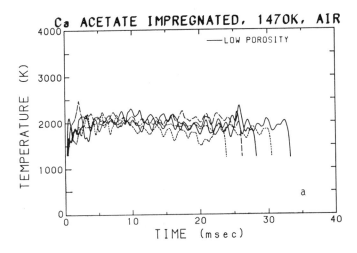

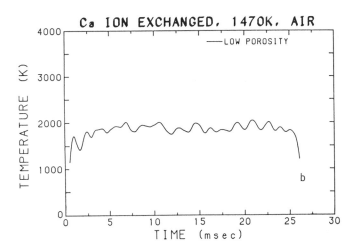

9. Temperature-time traces of low porosity particles burning in air: (a) ion exchanged, (b) calcium acetate impregnated.

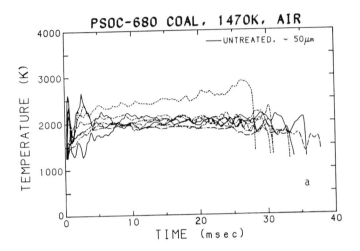

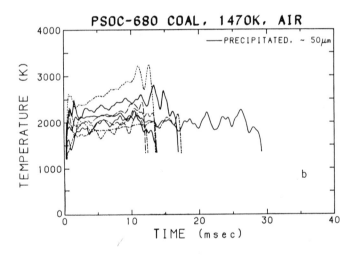

10. Temperature-time traces of PSOC-680 coal particles burning in air: (a) untreated coal, (b) CaCO₃ precipitated.

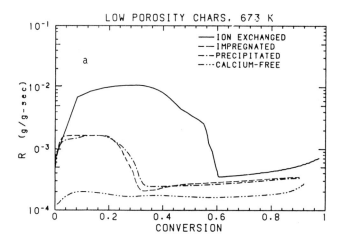

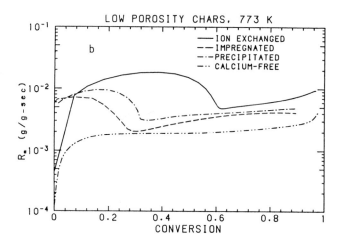

11. Reaction rate R_m vs. conversion for oxidation of low porosity chars in the
TGA at (a) 673 K and (b) 773 K.

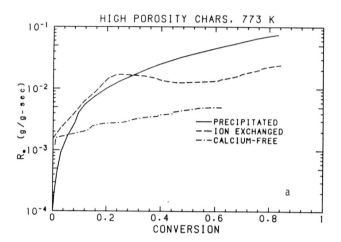

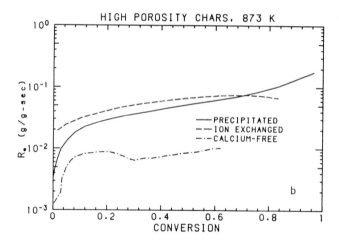

12. Reaction rate R_m vs. conversion for oxidation of high porosity chars in the TGA at (a) 773 K and (b) 873 K.

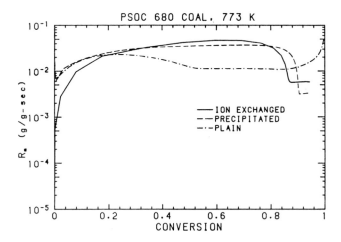

13. Reaction rate R_m vs. conversion for oxidation of PSOC-176 coal-char in the TGA at 773 K.

APPARENT RATE CONSTANT

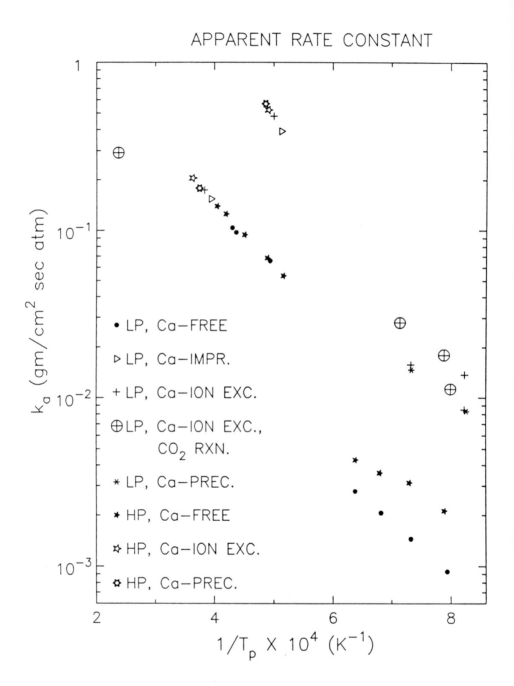

14. Apparent reaction rate coefficient k_a at intermediate to high temperatures, for low porosity (LP) and high porosity (HP) chars.

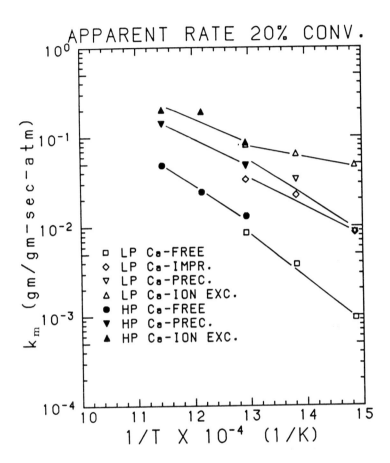

15. Apparent reaction rate coefficient k_m at low temperatures, for low porosity (LP) and high porosity (HP) chars.

INTRINSIC RATE CONSTANT

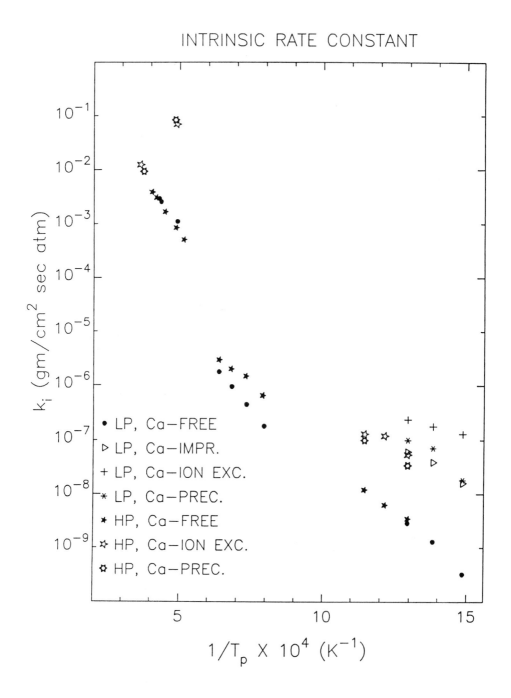

16. Intrinsic reaction rate coefficient k_i, for low porosity (LP) and high porosity (HP) chars.

Chapter 10

CATALYTIC GASIFICATION OF LOW-RANK COALS WITH CALCIUM ACETATE

Yasuo OHTSUKA[1] and Akira TOMITA[2]

[1]Coal Chemistry Laboratory

[1,2]Chemical Research Institute of Non-Aqueous Solutions, Tohoku University, Katahira, Sendai 980 (Japan)

10.1 INTRODUCTION

Catalytic coal gasification has attracted increasing attention, because a lowering of the gasification temperature has several advantages, such as the direct production of methane, less severe material problem and the use of waste heat. Therefore a number of studies on the catalytic gasification have been carried out, as is reflected by the increasing research effort (refs. 1,2). Among many catalysts reported so far (refs. 3,4), it has been generally accepted that alkali metal compounds like K_2CO_3 are the most effective catalysts for the steam gasification of coal.

In previous works (refs. 5-7), we have found that nickel catalyst markedly promotes the steam gasification of brown coal at low temperatures, around 800 K. K_2CO_3 is inactive in this temperature region. The continuous fluidized bed gasification of brown coal with nickel has shown that more than 80 wt% of the coal is converted to very clean gas containing neither tarry materials nor H_2S (ref. 8) and methane-rich gas is directly produced under pressure (ref. 9). We believe that the realization of such a low temperature gasification with less expensive catalysts than nickel should be a final target for the catalytic coal gasification. Therefore we have recently started the study on the development of cheap and disposable catalysts in place of nickel. It has been found that iron catalyst in hydrogen exhibits a high activity comparable to that of nickel catalyst (refs. 10,11), and active, chlorine-free alkali catalysts can easily be prepared from aqueous solutions of NaCl and KCl (refs. 12,13).

Cheap calcium compounds are also one of the promising disposable raw materials as gasification catalysts. With low rank coals, some calcium are inherently present with carboxyl functional groups and these exchangeable calcium cations promote the steam gasification (refs. 14-16). When the catalyst effectiveness of externally-added calcium is examined, these raw coals are usually demineralized with HCl-HF aqueous solution before the incorporation of calcium cations into the acid sites. However, the residual halogen in the demineralized coals may deactivate the calcium catalyst (ref. 17). In most studies, calcium-loaded coals are devolatilized at >1000 K prior to the gasification, but the pretreatment at high temperatures accelerates the agglomeration of catalyst particles (refs. 18,19). Consequently the activity of calcium catalyst may be deteriorated.

We have reported that, without both demineralization and heat treatment, the calcium impregnated on coal is very effective for the steam gasification even at low temperatures of around 950 K (refs. 20-22). In the present work, we will investigate the steam gasification of low rank coals with $Ca(CH_3COO)_2$ in detail in order to clarify some factors controlling the catalyst effectiveness of this compound, for example, the conditions of catalyst preparation, the co-existence of $Mg(CH_3COO)_2$, the coal type, the amount of inherent calcium, and the reaction temperature.

10.2 EXPERIMENTAL

10.2.1 Coal sample

Nine low rank coals from four countries were used, the starting size of coal particles being 1-2 mm in diameter in every case. Among them an Australian Yallourn brown coal was investigated in detail. It was received in a briquette form and then crushed into particles of 1-2 mm in diameter. The ultimate and proximate analyses of all coals used are given in Table 10.1. The amount of exchangeable calcium ions inherently present in coal was determined by leaching with 1 N ammonium acetate solution (ref. 16). The details of the procedure have been described elsewhere (ref. 23).

TABLE 10.1

Ultimate and proximate analyses of coals.

Coal	Code	Country	Ultimate analysis (wt%, daf)					Proximate analysis (wt%, dry)		
			C	H	N	S	O	VM[a]	Ash	FC[a]
Rhein Braun	RB	F.R.G.	65.8	5.5	0.8	0.3	27.6	54.8	2.9	42.3
Yallourn	YL	Australia	67.1	4.8	0.8	0.3	27.0	55.2	0.9	43.9
Morwell	MW	Australia	67.9	5.0	0.5	0.3	26.3	51.6	1.5	46.9
Velva	VL	U.S.A.	69.1	4.9	1.4	0.6	24.0	47.7	8.8	43.5
South Beulah	SB	U.S.A.	71.6	4.8	1.4	2.9	19.3	38.6	13.7	47.7
Colowyo	CW	U.S.A.	74.0	5.0	1.9	0.4	18.7	36.3	6.3	57.4
Wandoan	WD	Australia	75.8	6.8	1.0	0.3	16.1	53.9	16.9	29.2
Taiheiyo	TH	Japan	77.0	6.3	1.5	0.3	14.9	49.4	11.5	39.1
Illinois No.6	IL	U.S.A.	77.0	5.2	1.5	3.6	12.7	38.9	10.9	50.2

[a]VM, volatile matter; FC, fixed carbon.

10.2.2 Catalyst material and addition

$Ca(CH_3COO)_2$ was usually used as a catalyst precursor among different calcium compounds examined. Water-soluble salts like $Ca(CH_3COO)_2$, $Ca(NO_3)_2$ and $CaCl_2$ were impregnated on raw coal by grinding coal particles in their aqueous solutions. For water-insoluble salts like $Ca(OH)_2$, $CaCO_3$, CaS and $CaSO_4$ they were kneaded with coal in water. The resultant mixture was then dried at 380 K in a N_2 stream. In order to make clear the gasification activity of calcium magnesium acetate (CMA), $Ca(CH_3COO)_2$ and $Mg(CH_3COO)_2$ were simultaneously impregnated on coal in the same way as above. Furthermore, for comparison of the effectiveness of calcium catalyst, K, Mg, Ba, Fe and Ni as well as Ca were impregnated on coal from aqueous solutions of nitrates. In every case a nominal catalyst content is expressed as wt% of metal in the dried sample.

10.2.3 Steam gasification

The gasification experiments were carried out with an atmospheric thermobalance (Shinku-Riko, TGD-3000). The coals (about 20 mg) mounted onto a quartz cell were rapidly heated at a rate of about 300 K/min in a H_2O (66 kPa)/N_2 stream, and soaked for 2 h at a constant temperature, usually 923 K. Since raw coal was used in all runs instead of devolatilized char as a starting material, the reaction consisted of the coal devolatilization and subsequent char gasification stages. The reactivity of char in the latter stage will be discussed throughout the present paper. Data is processed as follows. Char conversion is expressed as wt% on a dry, volatile-free, ash-free, catalyst-free basis. The specific rate of char, the gasification rate per the unit weight of residual char, was almost independent of the conversion up to 50–60 %. Thus the average rate in this range was used as an index for the reactivity of char.

10.2.4 Characterization of catalyst

X-ray diffraction analysis (XRD) was carried out by using Cu-Kα radiation (45 kV x 30 mA) to characterize calcium catalyst in the coal devolatilization step prior to char gasification. The samples for XRD were prepared in a thermobalance by heating calcium-loaded coals in pure N_2 and then quenching the resulting chars to room temperature. Unless otherwise stated, the heating conditions were as follows: heating rate, 300 K/min; devolatilization temperature, 923 K; soaking time, 5 min. Temperature changes in chemical form of $Ca(CH_3COO)_2$ physically-mixed with YL coal were followed *in situ* by using a high temperature X-ray diffraction (HTXRD; Shimadzu XD3A/HX3) technique (refs. 21,24). The HTXRD measurements were carried out during heating the samples up to 923 K at a rate of 10 K/min in a N_2 stream. The average crystallite size of calcium species was determined by the Debye-Scherrer method.

256

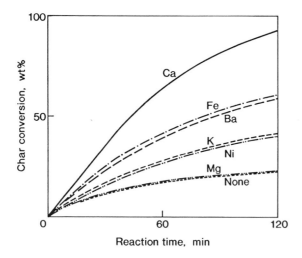

Fig. 10.1 Gasification profile of Yallourn brown coal with various metal nitrates at 923 K.

10.3 RESULTS

10.3.1 Calcium and other metals

Figure 10.1 illustrates the gasification profile of YL coal with several metals at a loading of 1 %. Magnesium was inactive, but all the other metals exhibited the catalytic effect. Among active catalysts the calcium was most effective; char conversion at 923 K reached more than 90 % in 120 min. Interestingly, iron was more active than nickel, and barium was less active than calcium. The rate enhancement by nickel or potassium was small at a low loading of 1 %.

10.3.2 Calcium compounds

Figure 10.2 illustrates the influence of the kind of calcium salts on the gasification reactivity of YL coal. The calcium loading was 5 %. The catalysts from $Ca(CH_3COO)_2$, $Ca(NO_3)_2$, $Ca(OH)_2$ and $CaCO_3$ enhanced the gasification rate to a quite similar extent. Coal was completely gasified within 60–70 min in every case. It is noteworthy from a practical point of view that cheap $Ca(OH)_2$ and $CaCO_3$ are effective. The details of the coal gasification with $Ca(OH)_2$ have been described elsewhere (refs. 20,22). $CaCl_2$ and CaS also promoted the gasification, and the degree of rate enhancement by these compounds was slightly lower than that of above four salts. Since some Cl- or S-containing compounds evolved during the gasification with $CaCl_2$ or CaS would cause some problems like corrosion, the practical use of these compounds seems not so easy. The catalyst from $CaSO_4$ showed only a slight activity.

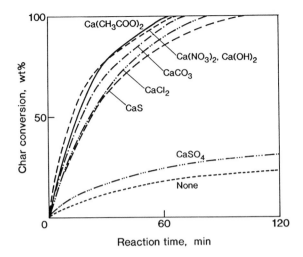

Fig. 10.2 Gasification reactivity of Yallourn coal with different calcium compounds at 923 K.

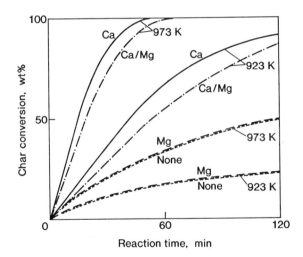

Fig. 10.3 Reactivity of Yallourn coal in the co-existence of $Ca(CH_3COO)_2$ and $Mg(CH_3COO)_2$ at 923 and 973 K.

10.3.3 Coexistence of calcium acetate and magnesium acetate

In order to clarify the catalyst effectiveness of CMA, YL coal was gasified in the co-existence of $Ca(CH_3COO)_2$ and $Mg(CH_3COO)_2$ at a loading of 1 % Ca and 1 % Mg. Figure 10.3 illustrates the gasification profiles at 923 and 973 K. $Mg(CH_3COO)_2$ alone showed no catalytic effect in this temperature range, similarly as in the case of $Mg(NO_3)_2$ (Fig. 10.1). However, the co-impregnated $Ca(CH_3COO)_2$ and $Mg(CH_3COO)_2$ markedly promoted the gasification. At 923 K char

conversion reached 85 % in 120 min, and at 973 K the coal was completely
gasified within 70 min. The effectiveness of this mixture was a little smaller
than that of $Ca(CH_3COO)_2$ alone. Figure 10.3 reveals that CMA can be used
instead of pure $Ca(CH_3COO)_2$ as a catalyst raw material for coal gasification.

10.3.4 Coal particle size

The catalytic effectiveness in coal gasification is often dependent on the
size of coal particles, in other words, the degree of contact between coal and
catalyst. The method of catalyst addition used in the present study reduced the
starting particle size to <0.02 mm. Thus the contact between coal and catalyst
was very intimate. In order to examine the effect of particle size on the
gasification reactivity, coarse particles of YL coal with the size of 0.25-0.50
mm in diameter were immersed in an aqueous solution of $Ca(CH_3COO)_2$ and the
resulting particles were dried at 380 K in a N_2 stream. The reactivity of this
sample was almost equal to that for fine powders of <0.02 mm; the average
gasification rate at 5 % Ca and at 923 K was 3.0 and 3.4 h^{-1} for coarse
particles and fine powders, respectively. The size effect was not so
appreciable under these conditions. This can be attributed to the great
adsorptive ability of low rank coals.

10.3.5 Coal type

The catalytic effectiveness of $Ca(CH_3COO)_2$ in the gasification of nine coals
with different carbon and sulfur contents was examined at a loading of 1 %. The

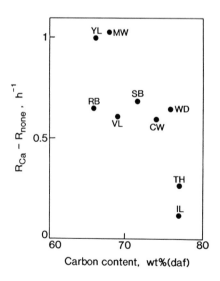

Fig. 10.4 Effectiveness of calcium catalyst for the gasification of several
coals at 923 K.

difference in gasification rates with and without calcium, $(R_{Ca}-R_{none})$, is used as an index of the effectiveness. A plot of $(R_{Ca}-R_{none})$ versus carbon content in coal is illustrated in Figure 10.4, where the symbols designate the names of the examined coals (see Table 10.1). The effectiveness of calcium depended on the coal type; the larger effectiveness was observed for the coals with lower carbon contents. Since the effectiveness for VL and SB coals with high sulfur contents was almost similar as that for RB and CW coals with low sulfur contents, there may be little dependence on the sulfur content.

10.3.6 Calcium loading

Figure 10.5 illustrates the influence of the amount of added calcium on the gasification rates for YL, MW, SB and CW coals. The rates for YL and CW coals increased almost linearly with the calcium loading, whereas the rates for MW and SB coals tended to level off at around 5 %. Walker and co-workers (refs. 15,17) also observed that the effect of calcium loading on the gasification rates of American lignites depended on the coal type. At 5 % Ca all these coals were completely gasified within at least 100 min; the rates for YL, CW, MW and SB coals with calcium were 28, 27, 6 and 3 times those without catalyst, respectively. The degree of rate enhancement by calcium addition was much larger for YL and CW coals which showed lower reactivities without catalyst.

10.3.7 Temperature dependence

The gasification rates of $Ca(CH_3COO)_2$-impregnated YL and SB coals were determined at different temperatures. In both coals the calcium showed no catalytic activity at a low temperature of 773 K, in contrast with an extremely

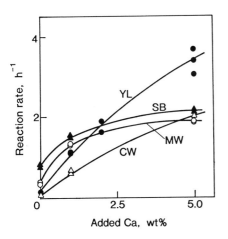

Fig. 10.5 Effect of calcium loading on the gasification rates of several coals at 923 K.

260

high activity of nickel catalyst in this temperature region (refs. 5-7). The
rate enhancement by calcium was appreciable at > 850 K. The gasification rate
increased with temperature and the complete conversion of YL or SB coal with 5
% Ca was achieved within 25 or 40 min at 973 K.

Figure 10.6 illustrates the Arrhenius plots. The apparent activation
energies and frequency factors for YL coal were 170, 170 and 140 kJ/mol, and
4.7×10^8, 7.7×10^9 and 3.2×10^8 h^{-1} at 0, 1 and 5 % Ca, respectively. These
values for SB coal, though the data at 1 % Ca were not shown in the figure,
were 160, 160 and 130 kJ/mol, and 1.4×10^9, 1.5×10^9 and 2.4×10^7 h^{-1} at 0,
1 and 5 % Ca, respectively. For both coals, the activation energy remained
unchanged and only the frequency factor was increased by the addition of 1 %
Ca, and a slight decrease in the activation energy was observed in the presence
of 5 % Ca. Otto et al. (ref. 25) also observed that the apparent activation
energy for the steam gasification of bituminous coal char slightly decreased at
a higher calcium loading. The catalysis of calcium at low loadings may be due
to an increase in the reaction-site density (ref. 25).

Figure 10.6 also shows that, in order to attain a practically sufficient
reaction rate, say 1 h^{-1}, with YL coal, a temperature of 1020 K is required

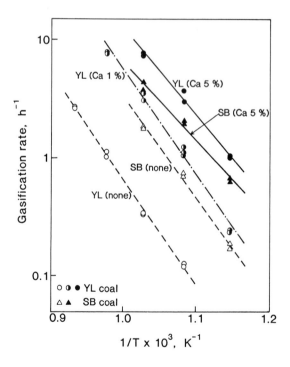

Fig. 10.6 Arrhenius plots for the gasification of Yallourn and South Beulah
coals.

without catalyst, whereas 920 and 870 K are enough with 1 and 5 % Ca, respectively. In other words, the presence of 1 and 5 % Ca could lower the gasification temperature by 100 and 150 K, respectively. With SB coal the addition of 5 % Ca lowered the temperature by 50 K only.

10.3.8 XRD profiles of calcium-bearing chars

Figure 10.7 illustrates XRD patterns for different chars prepared from $Ca(CH_3COO)_2$-loaded coals. In every case the calcium loading was 5 % on the coal before char preparation. For YL coal impregnated with $Ca(CH_3COO)_2$, no XRD lines from calcium species were detectable for the char devolatilized at 923 K (Fig. 10.7A). This observation shows that the calcium catalyst is very finely dispersed on the surface of char. When the devolatilization temperature was raised to 1023 K, the small and broad peaks attributable to CaO appeared (Fig. 10.7B), the average crystallite size being determined to 21 nm. Thus the heat treatment at higher temperatures accelerated the crystallization of calcium species.

When the $Ca(CH_3COO)_2$ physically-mixed with YL coal was used instead of the impregnated-$Ca(CH_3COO)_2$, the diffraction peaks from this compound were detectable even at room temperature. Thus the change in chemical form with temperature could be followed by HTXRD. The dehydration of $Ca(CH_3COO)_2$ completed at <570 K. At around 700 K the XRD lines of anhydrous $Ca(CH_3COO)_2$ disappeared, and instead the peaks of $CaCO_3$ appeared. The diffraction intensity of $CaCO_3$ peaks increased with increasing the temperature. At a constant temperature of 923 K, the decomposition of $CaCO_3$ to CaO proceeded gradually with the soaking time and completed in 40 min. Figure 10.7C shows the XRD profile for the char prepared from the physical mixture of YL coal and $Ca(CH_3COO)_2$. The strong and sharp lines of $CaCO_3$, together with the small peaks of CaO, were observed. The crystallite size of $CaCO_3$ was as large as 40 nm. The comparison with XRD patterns for the chars from $Ca(CH_3COO)_2$-impregnated coals reveals that the physical mixing of $Ca(CH_3COO)_2$ with coal gives the poorly dispersed calcium.

Figures 10.7D and 7E illustrate XRD profiles for calcium-bearing chars from $Ca(CH_3COO)_2$-impregnated SB and TH coals, respectively. No XRD lines of calcium species were observed with SB char, similarly as the case of YL char (Fig. 10.7A), whereas the signals from $CaCO_3$ crystallites were detectable with TH char. These observations show the decreased dispersion of calcium catalyst on the surface of TH char.

The chemical form and dispersion of the catalysts from other calcium salts than $Ca(CH_3COO)_2$ were examined with YL coal in the same manner as in Fig. 10.7A. Strong and sharp peaks due to $CaSO_4$ were observed in the case of $CaSO_4$

262

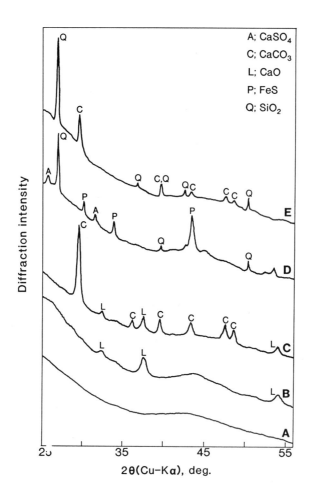

Fig. 10.7 XRD profiles of different calcium-bearing chars.
The conditions of char preparation are as follows; A, D, E, from $Ca(CH_3COO)_2$-impregnated YL, SB, and TH coals at 923 K, respectively; B, from $Ca(CH_3COO)_2$-impregnated YL coal at 1023 K; C, from the physical mixture of YL coal and $Ca(CH_3COO)_2$ at 923 K.

(ref. 21). With the CaS-loaded coal, the small peaks of $CaCO_3$ and CaO as well as CaS were detected, indicating that some CaS are converted to $CaCO_3$ and CaO. On the other hand, for $Ca(NO_3)_2$-, $Ca(OH)_2$- and $CaCl_2$-loaded coals, no XRD lines from calcium species were detectable on these chars prepared at 923 K (ref. 21), similarly as the case of $Ca(CH_3COO)_2$ (Fig. 10.7A). Calcium species derived from these compounds were found to be highly dispersed. Although the absence of XRD signals gave no information on their chemical forms, the XRD results for $Ca(CH_3COO)_2$ physically-mixed with YL coal or impregnated on TH coal suggest that $CaCO_3$ is the predominant species at 923 K (Figs. 10.7C and 7E). When the devolatilization temperature was raised to 1023 K, the broad peaks attributable to CaO appeared with these three chars. The decomposition of $CaCO_3$ to CaO proceeded easily in an inert atmosphere at 1023 K.

10.3.9 Catalyst preparation

Table 10.2 summarizes how both the method of catalyst addition and the heat treatment prior to steam gasification, that is, the devolatilization of calcium-loaded coal, affect the catalyst dispersion and subsequent char reactivity. When YL coal impregnated with $Ca(CH_3COO)_2$ was heated rapidly (300 K/min) up to 923 K and soaked for 5 min, the specific rate of the resulting char, 3.3 h^{-1}, was almost the same as that (3.4 h^{-1}) without heat treatment. The calcium on the surface of this char was very finely dispersed because no XRD lines attributable to calcium species were detected. The dispersion of calcium catalyst lowered with increasing severity of conditions of heat treatment, in other words, with decreasing the heating rate, increasing the temperature, and increasing the soaking time. The severer conditions resulted in lower gasification reactivities of chars; the reaction rate decreased from 3.3 h^{-1} for the char of 300 K/min - 923 K - 5 min to 1.8 h^{-1} for the char of 10 K/min - 1023 K - 30 min. The devolatilization of the physical mixture of YL coal and $Ca(CH_3COO)_2$ at 923 K gave the formation of larger $CaCO_3$ particles (40 nm), which resulted in the considerable decrease in the rate. Thus a good

TABLE 10.2

Influence of conditions of catalyst preparation on the dispersion of calcium catalyst and the gasification rate of char.

Method of catalyst addition[a]	Conditions of heat treatment			Calcium catalyst		Specific rate at 923 K (h^{-1})
	Heat. rate (K/min)	Temp. (K)	Soaking time (min)	Chemical form	Crystal. size (nm)	
Impregnation	No treatment			–	–	3.4
Impregnation	300	923	5	n.d.[b]	–	3.3
Impregnation	10	923	30	n.d.	–	2.6
Impregnation	300	1023	30	CaO[c]	21	2.1
Impregnation	10	1023	30	CaO	24	1.8
Physical mix	300	923	5	$CaCO_3$[d]	40	0.5

[a]YL coal; $Ca(CH_3COO)_2$ (Ca, 5 %).
[b]Not detectable by XRD (Fig. 10.7A).
[c]Shown in Figure 10.7B
[d]With the small peaks of CaO (Fig. 10.7C).

correlation was observed between the catalyst dispersion and the char reactivity.

10.4 DISCUSSION
10.4.1 Reaction scheme

In a previous study using the HTXRD technique (ref. 21), it has been suggested that the calcium-catalyzed gasification at low temperatures of around 950 K may proceed through the following carbonate-oxide cycle.

$$CaCO_3 + C = CaO + 2CO \tag{10.1}$$

$$H_2O + CO = H_2 + CO_2 \tag{10.2}$$

$$CaO + CO_2 = CaCO_3 \tag{10.3}$$

McKee (ref. 26) investigated the catalysis of alkaline earth salts in the steam gasification of graphite at high temperatures of 1100-1300 K, and suggested that the catalytic mechanism may involve a carbonate-oxide oxidation-reduction cycle. The mechanism proposed by McKee has been modified as the above reaction scheme for explaining the catalysis of calcium in the low temperature region.

The occurrence of eqn. 10.1, that is, the promotion of the decomposition of $CaCO_3$ by carbon, was confirmed also in the present study; the TPXRD experiment of the physical mixture of YL coal and $Ca(CH_3COO)_2$ revealed that the decomposition of $CaCO_3$ to CaO at 923 K completed in 40 min on the surface of char, whereas it took much more than 90 min for the completion of $CaCO_3$ decomposition on inert Al_2O_3 (ref. 21). Since the calcium catalyst was more finely dispersed with the impregnated $Ca(CH_3COO)_2$ than with the physically-mixed one (Figs. 10.7A and 7C), the reaction rate of eqn. 10.1 would be larger in the former case. With eqn. 10.2, a high partial pressure of steam (66 kPa) in the present study would favor the shift conversion of CO formed, and further calcium catalyst considerably promotes this reaction during steam gasification (ref. 27). The regeneration of $CaCO_3$ from CaO (eqn. 10.3) proceeded readily in the presence of CO_2 and at temperatures of <1000 K (ref. 21), though the regeneration was not so easy at higher temperatures even in the presence of CO_2. Thus the catalytic gasification of low rank coals with $Ca(CH_3COO)_2$ at around 950 K may proceed via a carbonate-oxide cycle mechanism.

10.4.2 Importance of catalyst dispersion

Walker and co-workers (refs. 18,19) reported that the heat treatment of calcium-exchanged lignite decreased the dispersion of calcium and consequently lowered the catalytic activity toward the subsequent reaction of char with air. As is shown in Table 10.2, the degree of catalyst dispersion was well correlated with the gasification rate of calcium-bearing char. This finding points out that the catalyst dispersion is an important factor which controls the rate enhancement by calcium also in the gasification of coal with steam. When $Ca(CH_3COO)_2$ was impregnated on YL coal and then devolatilized at 923 K, calcium species were not detectable by XRD (Fig. 10.7A) and thus very finely dispersed on the surface of char. Since some exchange of calcium ions with oxygen functional groups in YL coal takes place in the impregnation step, such a high degree of dispersion would be brought about by the presence of ion-exchanged calcium. The increase in severity of conditions of devolatilization of calcium-loaded coal accelerated the agglomeration of calcium particles, and the decreased dispersion of calcium catalyst resulted in the small rate enhancement (Table 10.2). Although the ion exchange or impregnation method provides essentially atomic dispersion in the step of catalyst addition, the heat treatment prior to the gasification decreases the catalyst dispersion and thus lowers the inherent catalytic activity. Therefore the gasification catalyst should be used without such the pretreatment so that the inherent activity can be maximally exhibited in the gasification step.

For TH coal with a smaller content of surface oxygen groups than YL coal, the formation of $CaCO_3$ crystallites was observed on the devolatilized char (Fig. 10.7E). The sintering of $CaCO_3$ particles would arise from a lower exchange extent of calcium ions with these acid cites. The smaller catalyst effectiveness of calcium observed for TH coal, as shown in Figure 10.4, is ascribed partly to the decreased dispersion.

Table 10.2 shows that the catalyst dispersion is affected also by the method of catalyst addition. In contrast with the highly dispersed calcium observed for the impregnated $Ca(CH_3COO)_2$, the physically-mixed $Ca(CH_3COO)_2$ gave large $CaCO_3$ particles in the devolatilization step (Fig. 10.7C). Since no ion exchanges occur in the catalyst addition like physical mixing, the absence of exchanged calcium would be the reason for the poor dispersion. The catalytic effect of poorly dispersed calcium was quite small (Table 10.2).

10.4.3 Effectiveness of calcium catalyst

The effectiveness of various catalysts in the steam gasification of coal has been studied by many investigators, and found to depend on several factors, such as the catalyst loading, the kind of precursor salt, the coal type and the reaction temperature. In this section the catalyst effectiveness of calcium will be discussed from these point of view.

(i) <u>Comparison with other metals</u>. As shown in Figure 10.1, at a low loading of 1 %, the calcium was the most active catalyst among different metals examined in the present study. Although alkali-metal compounds have been generally accepted to be most effective, the potassium at 1 % was less active than calcium. When the catalyst loading was increased up to 10 %, however, the effectiveness of potassium was very large for the gasification of many coals from lignite to anthracite (refs. 28,29). The activity of nickel was not so high in this work, but the nickel at a high loading of 10 % exhibited an extremely high activity toward the gasification of brown coals with low sulfur contents even at low temperatures of < 800 K (refs. 5-7). Other catalysts than nickel, potassium, calcium and iron, were inactive in this temperature region.

With the activity sequence of alkali earth metals, it has been reported that barium is more active than calcium at temperatures of >1000 K (refs. 25,26,30). In the present work, however, the reverse sequence was observed (Fig. 10.1). Calcium catalyst was still more active than barium catalyst even when the activity at 923-973 K was compared at the same atomic per cent, say 0.7 at.% (1.5 wt% Ca and 5.0 wt% Ba) (ref. 21). The difference in activity sequence between the earlier studies and the present work can be explained on the basis of the above-mentioned reaction mechanism. The mechanism suggests that, if alkaline earth metal can readily cycle between carbonate and oxide, the metal is more active. The interconversion between $CaCO_3$ and CaO is easy at around 950 K, but it is difficult at high temperatures of >1000 K because CaO is very stable. On the other hand, the barium shows the reverse behavior. Consequently the difference would be observed. The detailed discussion has been described

TABLE 10.3

Lowering in gasification temperature, ΔT, by catalyst addition in the steam gasification of Yallourn brown coal.

Catalyst			ΔT
Metal	Precursor Salt	Amount (%)	(K)
Fe[a]	$Fe(NO_3)_3$	10	50
Na[b]	NaCl	5	120
Ca	$Ca(CH_3COO)_2$	5	150
K[b]	KCl	10	170
Ni[c]	$Ni(NH_3)_6CO_3$	10	300

[a]Ref. 10.
[b]Ref. 12; metal ions alone are incorporated into coal by the ion exchange
        method using a pH adjusting agent.
[c]Ref. 5.

elsewhere (ref. 21).

Figure 10.6 shows that the gasification temperature for YL coal can be lowered by 100 and 150 K with the aid of 1 and 5 % Ca, respectively. The degree of the lowering in gasification temperature by catalyst addition is a convenient index that represents the catalyst effectiveness quantitatively. This index (ΔT) in the steam gasification of YL coal with different catalysts is summarized in Table 10.3, where the values for nickel, iron, sodium and potassium catalysts have already been shown elsewhere (refs. 5,10,13). On this criterion, the calcium prepared from $Ca(CH_3COO)_2$ was more effective than sodium and iron, but less effective than nickel and potassium. With the most effective nickel catalyst, however, the expensive nickel of a large amount of 10 % is essential for such the large effectiveness, and 15 wt% of coal remains ungasified because of the rapid deactivation of nickel catalyst (refs. 5,7). With potassium catalyst, KCl as a catalyst precursor is inexpensive, but the use of a pH adjusting agent like NH_3 or $Ca(OH)_2$ is unavoidably necessary for preparing the active, chlorine-free catalyst (ref. 13). The application of this catalyst to the coals with high ash contents seems not so easy because of the catalyst deterioration by the reaction with alumino-silicate compounds in minerals (ref. 31). On the other hand, with calcium catalyst, the complete conversion of coal can easily be achieved (Fig. 10.2), and this catalyst is applicable to the coals with high ash and high sulfur contents (ref. 22). Furthermore, there is the advantage that cheap raw materials like $Ca(OH)_2$, $CaCO_3$ and CMA can be used (Figs. 10.2 and 10.3). Therefore it can be concluded that the calcium is the most promising catalyst for coal gasification.

(ii) Influence of catalyst precursor. Figure 10.2 shows that $Ca(NO_3)_2$, $Ca(OH)_2$, $CaCO_3$ and $CaCl_2$ exhibit similar catalyst effectiveness as $Ca(CH_3COO)_2$. Large effectiveness of calcium catalysts from these precursors would be ascribed to the high degree of catalyst dispersion. Interestingly even CaS promoted the gasification to a large extent (Fig. 10.2). The XRD measurement for the CaS-loaded coal revealed the presence of $CaCO_3$ and CaO as well as CaS on the devolatilized char. This observation means that some CaS are converted to $CaCO_3$ and CaO by the reaction with H_2O and CO_2 evolved during devolatilization according to the following equations.

$$CaS + CO_2 + H_2O = CaCO_3 + H_2S \qquad (10.4)$$

$$CaS + H_2O = CaO + H_2S \qquad (10.5)$$

Since the concentrations of H_2O and CO_2 around CaS particles become higher during steam gasification, the reaction rates of eqns. 10.4 and 10.5 would be larger. $CaCO_3$ and CaO formed would promote the gasification through the carbonate-oxide cycle. With the $CaSO_4$-loaded coal, no XRD lines from $CaCO_3$ and CaO were detectable on the char, indicating that no change in the form of $CaSO_4$ takes place. Therefore little effectiveness was observed with $CaSO_4$.

(iii) <u>Influence of coal type</u>. As is seen in Figure 10.4, the effectiveness of calcium catalyst for high sulfur coals (VL, SB) was almost the same as that for low sulfur coals (RB, CW). This suggests that calcium catalyst is hardly deactivated by the sulfur compounds like H_2S evolved during the devolatilization and subsequent gasification. However, Otto et al. (ref. 25) observed that calcium catalyst was severely poisoned by externally-added H_2S in the steam gasification of coal chars. This difference can be explained from the different reaction conditions. The conditions of Otto et al., an extremely low partial pressure of steam (2.7 kPa) and a relatively high temperature of 1023 K, favor thermodynamically the reactions of $CaCO_3$ and CaO with H_2S, that is, the reverse reactions of eqns. 10.4 and 10.5. However, the present conditions of both a high partial pressure of steam (66 kPa) and a low temperature of 923 K are unfavorable to the poisoning of calcium catalyst by H_2S. This would be supported by the fact that the catalyst derived from CaS exhibits a high activity (Fig. 10.2).

It is well known that alkali and alkali earth metal cations exchanged with carboxyl groups naturally exist in low rank coals. Takarada et al. (ref. 23) have shown that the amount of such metal ions, especially Ca and Na, controls the reactivity of low rank coals in steam. The presence of inherent, exchangeable calcium ions may also affect the degree of rate enhancement by

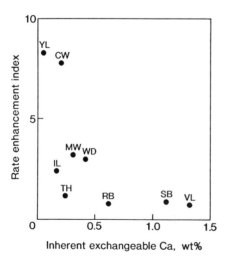

Fig. 10.8 Influence of the amount of inherent exchangeable calcium on the rate enhancement indexes for various low rank coals.

externally-added calcium in the gasification of these coals. A rate enhancement index, $(R_{Ca}-R_{none})/R_{none}$, may be convenient for expressing quantitatively the difference of rate enhancement among several low rank coals. In Figure 10.8, this index at a loading of 1 % Ca is plotted as a function of the amount of inherent, exchangeable calcium. A relatively good correlation between the two, except for TH and IL coals, was observed; the enhancement index decreased with an increase in the amount of inherent calcium. The indexes for TH and IL coals deviated from the expected correlation curve. This deviation would be ascribed to the poor dispersion of calcium catalyst on these chars, as indicated by the fact that the catalyst dispersion was poorer on TH char than on YL and SB chars (Fig. 10.7).

10.5 CONCLUSION

Calcium catalyst prepared from $Ca(CH_3COO)_2$ promoted the steam gasification of low rank coals at low temperatures of around 950 K. The catalytic effectiveness of $Ca(CH_3COO)_2$ depended on the coal type; the larger rate enhancement by calcium was observed for the coals with smaller amounts of inherent exchangeable calcium, such as Yallourn, Colowyo and Morwell coal. The gasification rate increased with the calcium loading, the degree of rate increase being also dependent on the coal type. With Yallourn coal, calcium catalyst was effective even at a low loading of 1 %, and most active among the examined metals like K, Mg, Ba, Fe and Ni. The presence of 5 % Ca lowered the gasification temperature of Yallourn coal by 150 K. Cheap and effective calcium is found to be the most promising catalyst for coal gasification. The co-impregnated $Ca(CH_3COO)_2$ and $Mg(CH_3COO)_2$ also enhanced the reactivity of Yallourn coal to a similar degree as the case of $Ca(CH_3COO)_2$ alone. This finding shows that CMA can also be used as a catalyst raw material. The dispersion of calcium catalyst was affected by the conditions of catalyst preparation. The devolatilization of calcium-loaded coal, that is, the heat treatment prior to the gasification, brought about the decreased catalyst dispersion, which resulted in the lowering of gasification activity. Therefore the gasification catalyst should be used without the pretreatment to gain the maximal activity.

ACKNOWLEDGMENT

The authors gratefully acknowledge the assistance of Miss Fumie Hamaoka, Miss Meiko Nishiyama (presently Mrs. Meiko Kuramoto) and Miss Naoko Yoshida in carrying out experiments.

270

REFERENCES

1 J.L. Figueiredo and J.A. Moulijn (Editors), Carbon and Coal Gasification, NATO ASI Series, Martinus Nijhoff Publishers, Dordrecht, 1986.
2 J.A. Moulijn and F. Kapteijn (Editors), Special Issue of the International Symposium on Fundamentals of Catalytic Coal and Carbon Gasification, Fuel, 65 (1986) 1324-1478.
3 J.L. Johnson, The Use of Catalysts in Coal Gasification, Catal. Rev. Sci. Eng., 14 (1976) 131-152.
4 D.W. McKee, The Catalyzed Gasification Reactions of Carbon, Chem. Phys. Carbon, 16 (1981) 1-118.
5 A. Tomita, Y. Ohtsuka and Y. Tamai, Low Temperature Gasification of Brown Coals Catalysed by Nickel, Fuel, 62 (1983) 150-154.
6 K. Higashiyama, A. Tomita and Y. Tamai, Action of Nickel Catalyst during Steam Gasification of Bituminous and Brown Coals, Fuel, 64 (1985) 1157-1162.
7 Y. Ohtsuka, A. Tomita and Y. Tamai, Catalysis of Nickel in Low Temperature Gasification of Brown Coal, Appl. Catal., 28 (1986) 105-117.
8 A. Tomita, Y. Watanabe, T. Takarada, Y. Ohtsuka and Y. Tamai, Nickel-Catalysed Gasification of Brown Coal in a Fluidized Bed Reactor at Atmospheric Pressure, Fuel, 64 (1985) 795-800.
9 T. Takarada, J. Sasaki, Y. Ohtsuka, Y. Tamai and A. Tomita, Direct Production of Methane-Rich Gas from the Low-Temperature Steam Gasification of Brown Coal, Ind. Eng. Chem. Res., 26 (1987) 627-629.
10 Y. Ohtsuka, Y. Tamai and A. Tomita, Iron-Catalyzed Gasification of Brown Coal at Low Temperatures, Energy & Fuels, 1 (1987) 32-36.
11 K. Asami and Y. Ohtsuka, Gasification of Low Rank Coals with Chlorine-free Iron Catalyst from Ferric Chloride, Proc. 1989 Int. Conference on Coal Science, Tokyo, October 23-27, 1989, pp. 357-360.
12 T. Takarada, T. Nabatame, Y. Ohtsuka and A. Tomita, New Utilization of NaCl as a Catalyst Precursor for Catalytic Gasification of Low-Rank Coal, Energy & Fuels, 1 (1987) 308-309.
13 T. Takarada, T. Nabatame, Y. Ohtsuka and A. Tomita, Steam Gasification of Brown Coal Using Sodium Chloride and Potassium Chloride Catalysts, Ind. Eng. Chem. Res., 28 (1989) 505-510.
14 J.L. Johnson, Relationship between the Gasification Reactivities of Coal Char and the Physical and Chemical Properties of Coal and Char, Am. Chem. Soc. Div. Fuel Chem., 20 (1975) 85-101.
15 E.J. Hippo, R.J. Jenkins and P.L. Walker, Jr., Enhancement of Lignite Char Reactivity to Steam by Cation Addition, Fuel, 58 (1979) 338-344.
16 R.J. Lang and R.C. Neavel, Behaviour of Calcium as a Steam Gasification Catalyst, Fuel, 61 (1982) 620-626.
17 T.D. Hengel and P.L. Walker, Jr., Catalysis of Lignite Char Gasification by Exchangeable Calcium and Magnesium, Fuel, 63 (1984) 1214-1220.
18 L.R. Radovic, P.L. Walker, Jr. and R.G. Jenkins, Importance of Catalyst Dispersion in the Gasification of Lignite Chars, J. Catal., 82 (1983) 382-394.
19 L.R. Radovic, P.L. Walker, Jr. and R.G. Jenkins, Effect of Lignite Pyrolysis Conditions on Calcium Oxide Dispersion and Subsequent Char Reactivity, Fuel, 62 (1983) 209-212.
20 T. Nabatame, Y. Ohtsuka, T. Takarada and A. Tomita, Steam Gasification of Brown Coal Impregnated with Calcium Hydroxide, J. Fuel Soc. Japan, 65 (1986) 53-58.
21 Y. Ohtsuka and A. Tomita, Calcium Catalysed Steam Gasification of Yallourn Brown Coal, Fuel, 65 (1986) 1653-1657.
22 Y. Ohtsuka and K. Asami, Steam Gasification of High Sulfur Coals with Calcium Hydroxide, Proc. 1989 Int. Conference on Coal Science, Tokyo, October 23-27, 1989, pp. 353-356.
23 T. Takarada, Y. Tamai and A. Tomita, Reactivities of 34 Coals under Steam Gasification, Fuel, 64 (1985) 1438-1442.

24 Y. Ohtsuka, Y. Kuroda, Y. Tamai and A. Tomita, Chemical Form of Iron
   Catalysts during the CO_2-Gasification of Carbon, Fuel, 65 (1986) 1476-1478.
25 K. Otto, L. Bartosiewicz and M. Shelef, Effects of Calcium, Strontium, and
   Barium as Catalysts and Sulphur Scavengers in the Steam Gasification of Coal
   Chars, Fuel, 58 (1978) 565-572.
26 D.W. McKee, Catalysis of the Graphite-Water Vapor Reaction by Alkali Earth
   Salts, Carbon, 17 (1979) 419-425.
27 T. Takarada, Y. Ohtsuka and A. Tomita, Pressurized Fluidized-Bed
   Gasification of Catalyst-Loaded Yallourn Brown Coal, J. Fuel Soc. Japan, 67
   (1988) 683-692.
28 T. Takarada, Y. Ohtsuka, A. Tomita and Y. Tamai, Activity Sequence of
   Catalysts on Catalytic Coal Gasification, J. Fuel Soc. Japan, 62 (1983) 414-
   420.
29 T. Takarada, Y. Tamai and A. Tomita, Effectiveness of K_2CO_3 and Ni as
   Catalysts in Steam Gasification, Fuel, 65 (1986) 679-683.
30 T. Yamada and T. Homma, Catalytic Effect of Alkaline Earth Salt on the
   Reaction of Char from Phenol-Aldehyde Resin with Carbon Dioxide, J. Fuel
   Soc. Japan, 58 (1979) 11-17.
31 D.W. McKee, C.L. Spiro, P.G. Kosky and E.J. Lamby, The Catalysis of Coal
   Gasification, Chemtech, (1983) 624-629.

Chapter 11

CALCIUM IMPREGNATION OF COALS AS A MEANS FOR SULPHUR EMISSIONS CONTROL IN COMBUSTION

P.K. Sharma

Jet Propulsion Laboratory, California Institute of Technology Pasadena, CA 91109

11.1 ABSTRACT

Calcium impregnation of coals prior to combustion in an industrial or utility furnace shows promise toward efficient control of sulfur emissions. Recent research carried out in the area is reviewed. Two techniques are known for impregnating coals with calcium: (a) ion exchange, and (b) precipitating calcium carbonate within coal's pores using an ionic reaction. Both techniques use calcium acetate in achieving calcium impregnation. In combustion of calcium treated coals, the release and capture of sulphur oxides in the presence of finely dispersed calcium oxide is an intriguing phenomenon. Combustion of calcium-exchanged coals in a laminar flow reactor under strongly fuel-lean conditions is reviewed. A similar, but limited study made in combustion of a bituminous coal containing precipitated $CaCO_3$ in its pore structure is also reviewed. Limited tests carried out in fluid bed combustion of calcium impregnated coals are also described. Altogether, these tests indicate various advantages of using calcium impregnated coals over conventional methods of sulphur emissions control.

11.2 INTRODUCTION

Sulphur emissions control in pulverized coal combustion is currently achieved by post-combustion flue gas scrubbing which involves capital and operating costs as high as 25% of the total power plant costs. Because of high cost and complex operation of those scrubbing processes, alternative methods of sulphur removal have been under active investigation. The most successful of these, fluidized combustion in a bed of calcined limestone or dolomite, is already commercial in industrial boilers and might soon become commercialized in utility boilers. At the relatively low temperatures of fluidized combustion sulphur oxides combine with the calcium oxide particles to form stable sulphate.

Fluidized combustion also has several other advantages, but its application is limited to new power plants. Other methods of sulphur control suitable for retrofitting existing pulverized coal-fired boilers rely on injecting the calcium sorbent along with the coal and removing the spent sorbent with the ash. At high temperatures and oxidizing conditions prevailing in a pulverized coal furnace, calcium sulphate is not thermodynamically stable, hence sulphur oxide removal must exploit either rate limitations in the release of sulphur oxides or the subsequent capture of sulphur oxides in downstream cooler sections of the furnace, where the equilibrium of sulphur oxide capture becomes favorable.

In line with above reasoning, one mode of calcium sorbent addition involves the injection of ground limestone along with the coal either in a conventional furnace or in a furnace employing staged combustion[1]. This method has the merit of extreme simplicity and low cost but has yielded sulphur removal in the range of 40-60%, insufficient to meet existing environmental standards. Evidently, the kinetics of sulphur oxide capture in the post-combustion region of the furnace are not sufficiently rapid, perhaps due to slow diffusion within the partially sintered calcium oxide particles. The mode of calcium sorbent introduction of interest here is one where calcium is introduced in closer association with the coal particles, i.e. to have calcium incorporated into the micro pore structure of coal. Such dispersion of calcium in the coal can be seen to result in several advantages. First, the kinetic and diffusional resistances to combination of sulfur oxides with calcium are reduced. Secondly, the fuel-rich conditions prevailing in the interior of the coal particle favor capture of sulphur. This feature can be exploited by modifications in the design of existing coal combustors.

11.3 CALCIUM ADDITION TO COAL

There are two methods for incorporating calcium in the coal's pore structure. One version involves impregnating coal with a calcium solution and precipitating calcium carbonate within the pores of particles. The other version involves exchanging acidic groups in coal with calcium cations, using a suitable solution. This exchange may be represented by:

$$RH + R'H + Ca^{++} \rightarrow R^-Ca^{++}R'^- + 2H^+$$

where RH, R'H are carboxylic or phenolic groups in coal. Lignites have sufficient acidic groups for accepting as much as 6-8 wt% calcium. Bituminous coals do not possess sufficient acidic groups but they can acquire the required ion exchange capacity by mild oxidation ($\approx$ 200°C) at the cost of a 10-15% loss in heating value[2]. Adding calcium by ion exchange produces very fine, atomic-scale, dispersion in close association with the sulphur source resulting in more rapid kinetics of sulphur capture.

The alternative method for calcium addition (in the form of $CaCO_3$) to the coal utilizes a novel process not based on the ion exchange capacity of coal. This process utilizes a liquid phase ionic reaction between Ca^{2+}, CO_2 and H_2O to lodge submicron size particles of $CaCO_3$ in coal pore space. Such reactions generally require a basic pH(>7.0) to proceed forward.

Various schemes are possible for contacting coal with CO_2 and an aqueous solution containing Ca^{2+} ions in order to precipitate $CaCO_3$ in the coal matrix. The scheme described by Sharma et al[3] involves mixing coal carrying preadsorbed CO_2 with a slurry consisting of calcium hydroxide in a 1-normal aqueous solution of calcium acetate. The reaction chemistry may be summarized in terms of the following reactions:

$$Ca(CH_3COO)_2 + H_2O + CO_2 \rightarrow CaCO_3 + 2CH_3COOH$$
$$2CH_3COOH + Ca(OH)_2 \rightarrow Ca(CH_3COO)_2 + 2H_2O$$

The overall reaction is:
$$Ca(OH)_2 + CO_2 \rightarrow CaCO_3 + H_2O$$

In principle, calcium acetate acts as an intermediate and is not consumed in the process. In practice some acetate will be lost by adsorption on the coal surface.

The adsorption of CO_2 on several bituminous coals has been studied by several investigators. Walker and Kini[4] have shown that typical bituminous coals adsorb CO_2 at room temperature in amounts in excess of a monolayer. The presence of CO_2 within the pore structure prior to contact with the liquid suspension facilitates $CaCO_3$ precipitation within the pores. As the liquid penetrates the pores, CO_2 will be dissolved in the liquid phase and diffuse outwards. But because of the high rate of the ionic reactions described above, a major part of the $CaCO_3$ is expected to be formed within the pores.

11.4 SULPHUR CAPTURE IN THE COMBUSTION OF CALCIUM-EXCHANGED COALS UNDER FUEL-RICH CONDITIONS

Freund and Lyon[5] studied the combustion of several bituminous and lower rank coals enriched in calcium by ion exchange under fuel-rich conditions. In their studies, calcium addition was made by soaking the coals in a solution of calcium acetate in water.

In the investigations of Freund and Lyon, sulphur capture during combustion was studied as a function of fuel equivalence ratio and carbon conversion. They found sulphur removal as high as 90% at sufficiently high equivalence ratios. At equivalence ratios used in conventional combustion the sulphur removal fell to 60% or lower. In view of this, they proposed this mode of calcium addition in connection with staged combustion. In the first fuel-rich stage sulphur would be captured in the form of sulphide. If this fuel-rich stage is followed by an oxygen-rich stage where combustion is completed at sufficiently low temperatures, the sulphide would be converted to sulphate and remain in that form.

11.5   SULPHUR CAPTURE IN THE COMBUSTION OF CALCIUM-EXCHANGED COALS UNDER EXTREME FUEL-LEAN CONDITIONS

This section describes the studies of Chang[6] et al. on two coals of different rank from the Pennsylvania State University Coal Bank, PSOC 623, a Texas Darco Lignite, and PSOC 680, an Indiana No. 6 high volatile bituminous coal. The elemental analyses for the two coals are presented in Table 11.1. The coal samples were ground to 230x325 US mesh size (44-70µm) and treated with dilute acid to remove minerals from the solid. A procedure developed by Schafer[7,8] for determining the carboxylic acid group content and the total acidity of low-rank coals was adapted to incorporate calcium ions into the coal samples by ion exchange. Each acid-washed coal sample was mixed with 0.2 N calcium acetate solution at 25°C for 4-24 h. The pH of the mixture was kept within 7.0-11.0 with $Ca(OH)_2$ to control the amount of calcium ion exchanged. After calcium exchange the slurry was filtered and washed with demineralized water to remove excess calcium. The treated coal was then vacuum dried at 105°C for 24 h. The ion-exchange capacity of the bituminous coal was insufficient to incorporate enough calcium into the coal for a meaningful Ca/S ratio. Thus, the coal was retreated by oxidation in air at 160°C for 65 h in a fluidized-bed reactor. The oxidized coal was then acid-washed and ion exchanged.

The analysis of the treated coal samples is given in Table 11.2. Comparison of Tables 11.1 and 11.2 shows that the sulphur content of the ion-exchanged coals is substantially lower than that of the starting coals, evidently due to the various pretreatment steps. For the bituminous coal oxidative pretreatment would convert pyrite to sulphate which would be largely removed in the subsequent acid washing. The pulverized combustion tests were conducted in a laminar flow reactor. Details of the reactor design are described in Ref. 9. The test conditions are provided in Ref. 6. The summary of test results is given below.

Fig. 1 presents the extent of sulphur capture as a function of residence time for the bituminous coal PSOC 680. This extent increases rapidly with residence time at all oxygen levels. Similar results were obtained in the combustion of lignite with Ca/S=2.5. For example, at 1600 K wall temperature and 40% O_2, the sulphur capture increased from 16 to 60% as the residence time increased from 0.6 to 5.5 s.

11.5.1 Effect of Oxygen Concentration

Figure 1 also shows the effect of oxygen concentration on sulphur capture. Although the runs with 13 and 21% oxygen show no significant difference, the sulphur capture increases rapidly as the oxygen concentration increases beyond 21%.

11.5.2 Effect of Calcium Content

Tests involving two lignite samples with Ca/S ratios of 2.5 and 0.8, respectively, were conducted at conditions similar to those of the bituminous coal. Figure 2 referring to combustion in 40% oxygen, shows the level of sulphur capture to be approximately proportional to the Ca/S ratio. Similar dependence was observed in experiments conducted with 31% oxygen.

11.5.3 Effect of Reactor Temperature

Experiments with the bituminous coal were also conducted at reactor wall temperature of 1400K. The resulting lower particle peak temperatures (Table 11.3) and longer residence times in the region favouring $CaSO_4$ formation were expected to improve sulphur retention. As shown in Figure 3, sulphur retention increases with increasing residence time and oxygen level. Comparison of Figures 1 and 3 shows that at 21% oxygen and low residence times sulphur capture is substantially higher in the case of 1400 K wall temperature. At 40% oxygen, on the other hand sulphur retention is somewhat higher at 1600 K wall temperature.

The results can be discussed in terms of the following general reaction scheme describing the production and capture of SO_2:

Reducing conditions

$$R-S \rightarrow H_2S + R' \qquad (1)$$
$$R-S + CO \rightarrow COS + R'' \qquad (2)$$
$$H_2S + CaO \rightarrow CaS + H_2O \qquad (3)$$
$$COS + CaO \rightarrow CaS + CO_2 \qquad (4)$$

Oxidizing conditions

$$R-S + 3/2\ O_2 \rightarrow SO_2 + CO + R' \qquad (5)$$
$$CaS + 2O_2 \rightarrow CaSO_4 \qquad (6)$$
$$CaS + 3/2\ O_2 \rightarrow CaO + SO_2 \qquad (7)$$
$$CaO + SO_2 + 1/2\ O_2 \rightarrow CaSO_4 \qquad (8)$$
$$CaSO_4 \rightarrow CaO + SO_2 + 1/2\ O_2 \qquad (8')$$

where R, R', R" indicate organic groups in coal. The set of Reactions (1)-(8) is not necessarily complete and additional reactions have been proposed in recent reports[10,11].

Retention of SO_2 in the ash can result either from reaction sequence (1)-(4), (6) or from recombination of released SO_2, Reaction (8). The dependence on residence time shown in Figure 1 indicates that, under the conditions of the figure, recombination is the dominant mechanism, for retention becomes negligible at residence times <1s. In particular, any $CaSO_4$ formed by Reaction (6) would have to decompose by Reaction (8').

When the wall temperature is reduced to 1400 K, significant sulphur retention is observed even at residence times < 1s (Figure 3). The difference is explained by the different temperature-time histories in the two tests.

The higher sulphur retention for the case of 1400 K wall temperature could also be due to the direct formation of $CaSO_4$ by Reaction 6. Any sulphate formed will be less likely to decompose at lower temperatures.

Considering next the effect of oxygen concentration, Figure 1 shows that sulphur retention is essentially the same at 13 and 21% and increases rapidly beyond 40%. Although the increase at the higher concentrations is consistent with the recombination mechanism, Reaction (8), the similar retention at 13 and 21% concentrations indicates that factors other than recombination may be at work. One possibility examined was the vaporization of elemental calcium by means of the reaction:

$$CaO \ (s) + CO \ \rightarrow \ Ca(g) + CO_2$$

But equilibrium calculations indicate[6] that the amount of calcium vaporized would be no more than 0.3% by weight of coal, so that if fully utilized, it would provide capture of 0.24 wt% sulphur in coal.

A second possible effect of higher particle temperatures arising from higher oxygen concentrations could be introduced via melting and agglomeration of mineral matter in the coal. Calcium has been reported to greatly accelerate the melting of coal ash under reducing conditions between 1200 and 1400 K and under oxidizing conditions between 1500 and 1700 K[12]. The peak particle temperatures reached during the experiments far exceeded the melting temperature of calcium-enriched ash, thus it is reasonable to assume that extensive ash melting and agglomeration took place. Ash melting can have a negative effect on sulphur retention either by depleting calcium due to the formation of calcium silicates and aluminates[13] or by decreasing the total surface area of ash particles, thereby imposing diffusion resistance on Reaction (8). Ash melting, on the other hand, may enhance sulphur retention by encapsulating sulphides or sulphates, thus slowing Reactions (7) and (8') releasing SO_2.

11.6 SULPHUR CAPTURE IN THE COMBUSTION OF COALS WITH PRECIPITATED $CaCO_3$ IN THE
      PORE STRUCTURE

This section describes the studies of Sharma[3] et al. on the bituminous
coal PSOC 680. The calcium impregnation procedure on the coal is described in
Ref.3. The size range of the treated coal was 20-100 μm and Ca/S ratio was 1.0.

A sample of the treated coal was cured in epoxy resin and polished for
study by SEM-EDS to measure the distribution of calcium across the particle cross
section. Another sample was analyzed by XRD to measure $CaCO_3$ particle size from
the broadening of the lines corresponding to the 104 and 116 faces of $CaCO_3$. The
mean particle size was calculated from the Scherrer formula after subtraction of
the instrumental contribution. Figure 4 shows the variation of Ca/S molar ratio
along the particle cross section for the treated PSOC 680 coal as determined by
SEM-EDS analysis. The Ca/S molar ratio at a given location was obtained from the
Ca/S signal intensity ratio by using a calibration factor determined separately
from a coal sample of known calcium and sulphur content. It is seen that with
the exception of the particle centreline and the outer particle surface layer,
the Ca/S ratio within the particle is not far from unity. The high calcium
concentration on the outer particle surface results from a small portion of the
added calcium (less than 10% of total) precipitating on the outer surface of the
particle.   The treated particles were also characterized by X-ray
diffraction-line broadening yielding mean crystallite size 300A°. Freund and
Lyon[5] obtained also by line broadening CaO crystallite size of 250A° for coal
that had been calcium ion-exchanged and devolatilized at 1400°C for 0.5 s.

A limited number of combustion tests were carried out in the laminar flow
reactor[9] at 1600 K wall temperature using a large excess of air. Table 11.4
summarizes the results of two combustion tests. For the case of high oxygen
concentration and high residence time (Test No. 1), the sulphur retention was
86%. This result was confirmed by SEM-EDS analysis on the ash. For combustion
with air and lower residence time (Test No. 2), sulphur retention dropped to 26%.
Micrographs of ash samples obtained by SEM showed that ash particles from Test
No. 1 were approximately twice as large as those from Test No. 2. The enhanced
agglomeration due to higher oxygen concentration may play some role in the higher
retention of sulphur. However, the principal mechanism of sulphur retention
appears to involve SO_2 recapture in the cooler section of the furnace, favoured
by high oxygen content and long residence time.

In the investigations of Chang et al.[6] described in the earlier section,
sulphur retention in the combustion of the same coal pretreated by calcium ion
exchange to Ca/S = 3.4 was measured. Under the combustion conditions of Test No.
1, sulphur retention was also above 80%. This limited comparison suggests that
calcium-precipitated coal may perform as well as calcium ion-exchanged coal in
terms of sulphur retention.

11.7 FLUID BED COMBUSTION

Fluid bed combustion of calcium-exchanged lignite (PSOC 623) and the Indiana bituminous (PSOC 680) coal was investigated[14]. The results for the lignite, which was treated by the ion exchange method are presented in Table 11.5. It is seen that high sulphur retentions (83-95%) are realized. The experimental conditions in the combustion of the pretreated bituminous coal were maintained relatively constant and are summarized in Table 11.6. Sulphur retention obtained under these conditions was about 82%. The relatively high sulphur retentions obtained in fluid bed combustion are consistent with the favorable thermodynamics of $CaSO_4$ formation at the temperature range employed.

11.8 SUMMARY

The studies reviewed here deal with sulphur capture in combustion of calcium impregnated coals. The test conditions included in these studies cover the range from pulverized firing to fluid bed combustion. Under most conditions, sulphur capture with calcium impregnated coals is superior to corresponding combustion processes where physically mixed limestone is used. The economical aspects of calcium impregnation need to be addressed. However, such economics will be favorably impacted in processes where coal is fed as a slurry where it is not necessary to dry the treated coal.

ACKNOWLEDGMENTS

The author wishes to acknowledge several helpful discussions with Professors G.R. Gavalas and R.C. Flagan of California Institute of Technology.

REFERENCES

1   P.L. Case, M.P. Heap, C.N. McKinnon, D.W. Pershing, and R. Payne, Am. Chem. Soc. Div. Fuel Chem., Prepr. 1982, 27(1), 158.
2   W.S. Kalema, "A Study of Coal Oxidation", PhD Thesis, California Institute of Technology, 1984.
3   P.K. Sharma, G.R. Gavalas, and R.C. Flagan, Fuel 1987, 66, 207.
4   P.L. Walker, Jr., and K.A. Kini, Fuel 1965, 44, 453.
5   H. Freund, and R.K. Lyon, Combust. Flame 1982, 45, 191.
6   K.K. Chang, R.C. Flagan, G.R. Gavalas, and P.K. Sharma, Fuel 1986, 65, 75.
7   H.N.S. Schafer, Fuel 1970, 49, 197.
8   H.N.S. Schafer, Fuel 1970, 49, 271
9   C.L. Senior, "Submicron Aerosol Formation During Combustion of Pulberized Coal", PhD Thesis, California Institute of Technology, 1983.
10   A. Levy, E.L. Merryman, and B.W. Rising, Report DOE/PC/30301-11, Battelle, Columbus, 1983.
11   A. Attar, Report EPA-600/7-84-070, North Carolina State University, 1984.
12   G.P. Huffman, F.E. Huggins, and G.R. Dunmyre, Fuel 1981, 60, 585.
13   J.W. Cobb, J. Soc. Chem. Ind. 1910, 29 (5), 250.
14   K.K. Chang, R.C. Flagan, G.R. Gavalas, and P.K. Sharma, presented at the 1984 Annual AIChE Meeting, held in San Francisco, Nov. 25-30, 1984.

TABLE 11.1

Elemental analyses of PSOC 623 (Darco Lignite) and PSOC 680 (Indiana No. 6) on a dry basis provided by the PSU Coal Bank

Wt%	PSOC 623	PSOC 680
Ash	16.6	11.9
Carbon	60.7	68.5
Hydrogen	4.3	4.6
Nitrogen	1.1	1.5
Sulfur	1.1	3.6
Oxygen (by diff.)	14.3	9.5

TABLE 11.2

Analysis of calcium-exchanged coal samples

	Wt%, dry basis			
	Ash	S	Ca	Ca/S
PSOC 680	21.1	1.6	7.1	3.4
PSOC 623 A	22.4	0.8	2.5	2.5
PSOC 623 B	20.2	0.8	0.8	0.8

TABLE 11.3

Calculated particle temperature at two wall temperatures for the combustion of the bituminous coal

Peak particle temperatures (K)		
Oxygen concentration (%)	1400 K Wall temperature	1600 K Wall temperature
13	1493	1740
21	1582	1878
30	1737	2074
40	1929	2278

TABLE 11.4

Sulphur retention in entrained combustion of PSOC 680 (Ca/S = 1.0) at 1600 K reactor temperature

Test no.	Combustion gas flow rate (1/min)	$\%O_2$ in combustion gas	% S in Product gas	% S in ash (by difference)
1	0.50	40	14	86
2	4.0	21	74	26

TABLE 11.5

Fluid-Bed Combustion of Calcium-Exchanged Lignite

Coal: PSOC 623, 0.8% S, (Ca/S) = 1.7
Fluidized Solids: Coal (420-590 μm) 4.3% - Quartz (700 μm) 95.7%

Bed Diameter 1", Bed Height 1.5"

Superficial Velocity (cm/s)	Initial Temperature (°C)	Wt. Addded Coal (g)	Temperature Rise (°C)	Sulfur Capture (%)
45	780	1	20	83.3
	790	1	20	92.9
	805	1	20	93.6
	800	2.6	32	94.1
41	780	1	20	92.3
	800	1	20	93.4
	810	1	20	85.6
	805	3.6	45	95.2

TABLE 11.6

Fluid Bed Combustion of Bituminous Coal (PSOC 680) Treated by Calcium Precipitation

Superficial velocity	41 cm/sec
Bed temperature	775°C
Coal particle size	300-420μm
Ca/S molar ration	~ 1.0
Wt. of coal in bed	~ 1.0 g
% sulfur retention	82.1

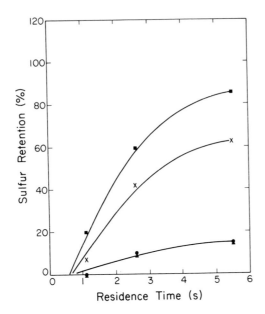

Figure 1. Effect of particle residence time on sulphur retention for the combustion of PSOC 680 (Ca/S = 3.4) at various oxygen concentrations and 1600 K wall temperature. ■, 40; x, 31; ▲, 21; ●, 13% O_2.

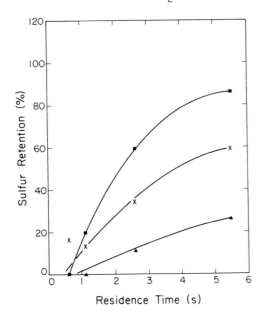

Figure 2. Effect of calcium content on sulphur retention for combustion in 40% O_2 and 1600 K wall temperature. ■, PSOC 680 (Ca/S = 3.4); x, PSOC 623 (Ca/S = 2.5); ▲, PSOC 623 (Ca/S = 0.8).

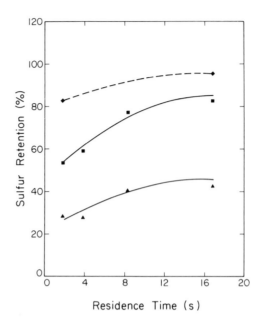

Figure 3. Effect of residence time on sulphur retention during combustion of PSOC 680 (Ca/S = 3.4) at 1400 K wall temperature. ◆, 60; ■, 40; ▲, 21% O_2.

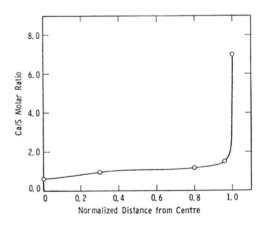

Figure 4. Calcium to sulphur ratio as a function of radial position in a pre-treated coal particle.

Chapter 12

COMBUSTION OF COM WITH ALKALINE ACETATE ADDITIVES

Bruce W. Rising and Herbert R. Hazard

Battelle, Columbus Laboratories, 505 King Avenue, Columbus, Ohio 43201

12.1 INTRODUCTION

This paper describes research in which two additives, barium acetate and calcium acetate, were incorporated into a coal-oil mixture (COM) and a heavy fuel oil for sulfur capture. (1) These additives had been developed previously by Hague International, who injected them as aqueous solutions emulsified with Coal-Oil Mixtures for firing a slot furnace for forging operations. With the moderate temperature conditions of the slot furnace, and the fine dispersion resulting from evaporation of an aqueous solution, Hague reported excellent sulfur capture.(2) The objective of the present research was to explore the use of these additives under conditions simulating those in an industrial boiler. These include a relatively high furnace temperature and a moderate amount of excess air. The COM used for this research was provided by Hague International.

12.2 APPROACH

The use of barium acetate and calcium acetate as additives to coal-oil mixtures (COM) for capture of sulfur during combustion was explored under boiler conditions using a small refractory lined laboratory furnace, fired at a rate of about 22 lb/hr of fuel. The barium acetate was added to the COM (75% oil/25% coal) as a dry powder having nearly the same particle size distribution as the coal in the COM. Two mol ratios of barium to sulfur were used: 1:1 and 3:1. Barium acetate was also mixed as a dry powder with a high-sulfur No. 6 oil at mol ratios of barium to sulfur of 1:1 and 3:1. Calcium acetate was dissolved in water and then blended with COM prior to combustion testing. This method was used because the dry acetate caused the COM to agglomerate and caused seizing of the fuel pump.

Gaseous effluents were measured by standard flue gas analyzers; particulates were collected using EPA Method-5 procedures. A series of liquid impingers was used on the Method-5 as a means of trapping SO_2 for later analysis. An EPA Method-6 sampling train was also used as a means of collecting and measuring SO_2.

Particulate samples collected by the Method-5 sampling unit were weighed and submitted for chemical analysis. Samples were analyzed for carbon, sulfur, total barium or calcium concentration and the fraction of barium soluble in water. The quantity of soluble barium compounds in the flyash was of key concern because of the high toxicity of these compounds. From these analyses, and from analyses of the gaseous effluents, the fate of the fuel sulfur could be traced.

The furnace used was a refractory-lined cylinder 24 inches I.D. and 48 inches long. It was fired at rates from 366,000 to 625,000 Btu/hr, with most observations made at 400,000 Btu/hr; average excess air was 20 percent. As with all small furnaces, much of the coal ash was deposited on the furnace walls as a molten slag, with somewhat less than half appearing as flyash at the sampling positions. The furnace gases passed through water-cooled and air-cooled tube sections for cooling to a temperature near 600 F prior to sampling of particulates and gaseous combustion products.

12.3 EXPERIMENTAL RESULTS

12.3.1 Flue Gas SO_2 Reduction

The SO_2 reduction for each test is determined by comparing the measured SO_2 concentration with the concentration calculated for conversion of all fuel sulfur to SO_2. The SO_2 reduction and sulfur capture data are summarized in Table 12.1. The variation of fuel sulfur accounted for, from 50 to 126 percent, reflects the difficulties of collecting and analyzing the small quantities of stack dust as well as inaccuracies of measuring fuel and air feed rates. Baseline tests reported in Table 12.1 were performed without the use of additives, all other tests were performed using barium and calcium additives in the test fuels indicated.

A 13 percent SO_2 reduction was achieved using a Ba/S mol ratio of 1 in No. 6 fuel oil. Although the fraction of fuel sulfur accounted for in this test is only slightly greater than the basic test, the sulfur content of the flyash increased by a factor of 2. The increased sulfur content of the flyash was also accompanied by an increase in the total particulate emissions.

Additional additive testing continued with barium and No. 6 oil using a Ba/S mol ratio of 3. In this test the SO_2 reduction was measured at different firing rates; the results are illustrated in Figure 1.

The measured SO_2 reduction in the flue gas varied from 77 percent at the lowest firing rate to 43 percent at the highest firing rate and was approximately linear with firing rate. We believe the decline in sulfur capture with increased firing rate is linked to the increase in combustion zone temperatures and a decrease in residence time in the furnace and heat exchanger.

TABLE 12.1

Summary of Test Conditions and Sulfur Capture Data

Test No.	1	2	3	4	5	6A	6B	6C	6D	7
FUEL	No.6	No.6	COM	COM	COM	No.6	No.6	No.6	No.6	COM
ADDITIVE	None	Ba	None	Ba	Ba	Ba	Ba	Ba	Ba	Ca
ADDITIVE/ SULFUR MOL RATIO	0	1	0	1	3	3	3	3	3	6
FIRING RATE (10^3 Btu/hr.)	411	405	365	366	396	397	506	596	531	360
PERCENT SO_2 REDUCTION	–	13	–	0	67	77	54	43	47	0
PERCENT FLUE GAS SULFUR AS SO_2	99.4	91.7	99.4	96.4	69.9	–	–	–	–	89.5
PERCENT FLUE GAS SULFUR IN PARTICULATE	0.6	8.27	0.6	3.6	30.1	–	–	–	–	10.5
PERCENT FUEL SULFUR ACCOUNTED FOR	90	96	105	106	50	–	–	–	–	126

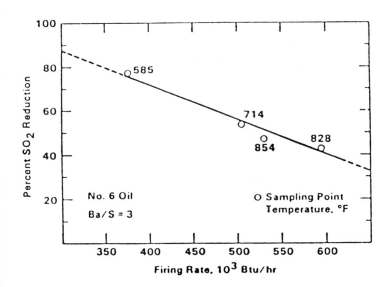

Figure 1. Variation of SO_2 reduction with firing rate.

TABLE 12.2

Particualte Analytical Data

Test No.	1	2	3	4	5	7
Particulate Emissions						
Measured, lb/hr	0.03	0.23	0.11	0.19	0.32	0.93
Theoretical, lb/hr[1]	0	1.17	0.24	0.93	2.45	1.49
Percentage of total ash accounted for	-	19.5	15.8	20.1	13.0	62.4
Particulate Constituents (wt%)						
C	8.9	0.5	6.8	7.5	1.4	7.4
S	4.7	9.2	1.3	3.0	7.2	1.9
$BaSO_4$	-	22.3	-	23.6	42.5	-
Soluble Ba	-	12.9	-	26.3	15.6	-
Ca	-	-	-	-	-	44.0
Sulfur as $BaSoB_4$	-	33	-	100	81	-
Mol ratio Ba/S						
Fuel	0	1	0	1	3	6[2]
Flyash	0	0.66	0	3.1	1.3	18.5[2]

[1]Based on weight of coal ash plus weight of additive metal oxide
[2]Ca/S mol ratio

Barium Acetate + COM. Negligible SO_2 reduction was observed firing COM with a Ba/S mol ratio of 1. However, as in the previous test, there was a significant increase in the sulfur content of the flyash. Also, the flyash emissions increased markedly over the levels measured in the base test. The total sulfur measured in the flue gases exceeded the fuel sulfur by 6 percent in this test.

In the next COM test the Ba/S mol ratio was increased to 3. The SO_2 reduction was 67 percent while the sulfur content of the flyash nearly doubled from that measured in the previous test. As expected, the total particulate emissions also increased in this test. As noted in Table 12.1, the sulfur balance for this test was the lowest for the series. Also, as shown in Table 12.2, the measured particulate emissions account for only a small fraction of the total estimated emissions. It is believed that part of the particulate, including sulfur absorbed from the gas, was deposited on the surfaces of the heat exchanger used to reduce gas exit temperatures.

Calcium Acetate + COM. COM was fired then with calcium acetate added as a solution in water, emulsified with the fuel. With a Ca/S ratio of six the sulfur capture, as determined by flue-gas SO_2 content, was not detectable. However the sulfur content of the flyash was slightly more than that reported in the baseline test. The test conditions here were somewhat different than for the six previous tests, in that the measured CO concentration was greater than 5000 ppm, compared with 100 ppm in other tests. This may have resulted from the high water loading of the fuel mixture, which could reduce flame temperature and may have reduced burner stability. A mass balance of the sulfur indicated that 26 percent more sulfur was detected in the stack than was fired into the furnace.

12.3.2 Flyash Analyses

Flyash samples collected from the flue gases in Tests 1 through 4 and Test 7 were chemically and gravimetrically analyzed. The gravimetric analysis of flyash samples was used to estimate the flyash emission rate. The theoretical emission rate reported in Table 12.2 represents all of the coal ash and metal oxide fired into the furnace.

Barium Acetate + No. 6 Oil. The addition of the sorbent in this test produced a marked increase in the sulfur content of the ash, as was expected; the total particulate loading also increased as expected. The measured emission rate is nearly a factor of 10 greater than that reported in Test 1; however, it only represents 19.5 percent of theoretical value based upon the ash content of the fuel.

Barium acetate effectively combined with the fuel sulfur species to form $BaSO_4$. In Test 2, with a 1:1 Ba/S mol ratio, the fraction of the flue gas sulfur found in the flyash increased an order of magnitude over the previous test.

The flyash samples in Test 2 contained 22.4 percent barium sulfate and 12.9 percent barium in water-soluble form. However, only 33 percent of the total sulfur captured by the flyash was in the form of barium sulfate. Sulfur forms other than barium sulfate were not identified.

In Tests 6a-6d, firing No.6 fuel oil with barium acetate, no flyash samples were collected. Sulfur capture data were based on flue gas analyses alone.

Barium Acetate and COM. Firing COM with a Ba/S mol ratio of 1, the sulfur in the flyash was less than 1 percent of the total sulfur measured in the flue gas. In Test 5, with Ba/S mol ratio of 3, the sulfur in the flyash increased to 30 percent of the total sulfur in the flue gas, but the total sulfur accounted for was only 50 percent, and the sulfur capture as indicated by gas analysis was 67 percent. These differences are accounted for by the fact that dust loading of the flue gas was very low, only 13 percent of theoretical. At least half of the total ash was retained in the furnace as slag, and would not be expected to retain sulfur. Additional flyash could deposit on the inside of the heat exchanger, and could be expected to retain sulfur.

The soluble barium content of flyash samples from Test 4 represented 26.3 percent of the total sample weight. For Test 5 this value was only 15.6 percent, even though more barium acetate was used in Test 5.

Also, as shown in Table Table 12.2, nearly all of the flyash sulfur collected in Tests 4 and 5 is of barium sulfate. In Test 2 only 33 percent of the flyash sulfur is in the form of barium sulfate, although a significant amount of unsulfated barium is available.

Calcium Acetate + COM. In Test 7 a blend of calcium acetate, water and COM was fired into the test furnace. Calcium acetate was used to compare with the sulfur collection efficiency with barium acetate. The COM/water mix was necessary because dry acetate could not be mixed and pumped. Although a Ca/S ratio of 6 was used, analysis of gas samples revealed insignificant SO_2 reduction. As discussed earlier and shown in Table 12.2 the quantity of sulfur contained in the flyash samples was slightly greater than that observed in the baseline test.

Also, shown in Table 12.2 the flyash samples collected contained 44 weight percent calcium and 1.9 weight percent sulfur. The mol ratio of sorbent/sulfur was 18/5 for the flyash samples versus 6 for the fuel. The large quantity of unreacted calcium plus the low level of SO_2 reduction suggest that calcium acetate is not an effective sorbent for SO_2 under these test conditions.

12.4   CONCLUSIONS

Barium acetate proved to be an effective SO_2 sorbent for COM and No. 6 oil At a Ba.S mol ratio of 3 a 67 percent SO_2 reduction was observed for COM; a 77 percent reduction was observed for No. 6 oil at the same firing rate. Flyash emission increased as the Ba/S mol ratio increased. Also, as the Ba/S mol ratio increased from 1 to 3 the quantity of soluble barium compounds in the COM flyash decreased. Increased fuel firing rates produced decreased sulfur capture at a fixed barium/sulfur ratio with No. 6 fuel oil. The decreased sulfur capture is coupled to increased combustion zone temperatures and decreased residence time. When barium acetate was used all flyash samples contained toxic soluble barium compounds. For the COM tests, nearly all of sulfur in the particulates was barium sulfate.

Calcium acetate did not prove to be an effective sulfur sorbent for COM under the test conditions used.

ACKNOWLEDGEMENTS

This work was supported by the Department of Energy, Division of Industrial Energy Conservation, Office of Industrial Programs under Contract No. Eng-92+DEC-AC07-76ID01570 with Ralph Sheneman as DOE program manager. The COM fuel used was provided by Hague International.

REFERENCES

1   B.W. Rising, D.C. Mays, and H.R. Hazard, "Effect of Additives on Emission of SO_x from Combustion of Coal-Oil Mixtures", Report by Battelle Columbus Laboratories to U.S. Department of Energy, Office of Industrial Programs, Contract No. Eng-92+DEC-AC07-76ID01570, April, 1981.
2   J.W. Bjerklie, R.A. Pentry, and S.B. Young, "Demonstration of Sulfur Removal from Combustion Gases of Coal-Oil Mixture", Report by Hague International to DOE Oak Ridge Lab, Subcontract No. 7731, December, 1979.

Chapter 13

MODIFIED LIMESTONE METHOD FOR REMOVING SULFUR FROM FLUE GASES

A. Durych, A. Laszuk, A. Wiechawski
Instytut Inżynierii Chemicznej, Politechnika Krakowska Kraków, Poland

13.1   INTRODUCTION

Flue gas desulfurization in the world, particularly from energetic sources, has been dominated by the methods using calcium compounds for SO_2 bonding. This follows from the fact, that calcium compounds are the cheapest source of alkalinity and the limestone as a mineral is widespread in many countries. Experiments aiming to improve this process are being carried out for many years. Two major trends can be distinguished: double alkali method in which So_2 is absorbed into a clear solution, usually Na_2CO_3 and next the solution is reacted with CaO or $CaCO_3$ in a separate tank to precipitate $CaSO_3$ and regenerate SO_3^{2-} ions. The $CaSO_3$ solids are separated for disposal and clear solution is recycled to the scrubber loop.

Another direction of the improvements leads to the process of slurry scrubbing with soluble additives. Slurry scrubbing with soluble additives gives more rapid rates and larger capacities for SO_2 mass transfer than does a simple slurry scrubbing.

13.2   WORK AT TECHNICAL UNIVERSITY OF CRACOW

In the Institute of Chemical Engineering and Physical Chemistry, Technical University of Cracow initial experiments of SO_2 absorption have been carried out in the laboratory scale using a column with a fluidized bed of plastic spheres (TCA type) with the diameter 35, 70 and 200 mm. Flue gas was simulated by the atmospheric air with 0.5% vol. SO_2 from a gas cylinder. Basic experiments included scrubbing with the limestone slurries of different concentrations.

It has been mentioned above that some soluble additives can improve the process. They can be for example organic acids of the strength between the carbonic and sulfurous acid. Some additional experiments have been carried out with calcium acetate solutions of different concentrations. SO_2 removal efficiency E vs solution concentration X_w is shown in the Fig. 1, exemplary for the column of 35 mm diameter. It can be noticed that SO_2 removal is much higher for the calcium acetate scrubbing. This also means that the zero value of the equilibrium SO_2 concentration over the $CaCO_3$ solutions, often assumed in the literature, is only a very rough approximation.

Generally it can be stated that calcium acetate as a scrubbing liquor for

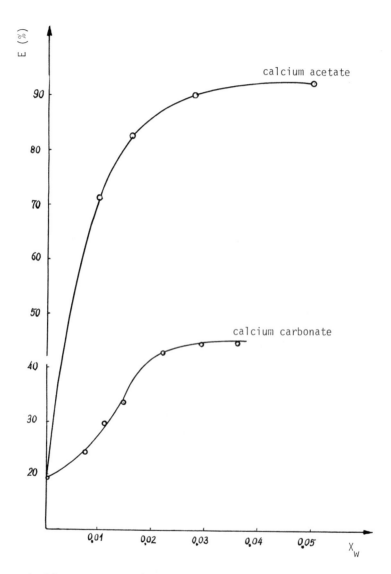

Figure 1. SO_2 removal efficiency E vs. solution concentration X_w.

SO_2 absorption can be very interesting because of the significant intensification of the absorption process as well as the possibility of entire limestone utilization for SO_2 reaction. Because of the good limestone solubility in the acetic acid, the reaction products($CaSO_3$) can be easily separated and the remaining acetate can be recycled to the process. In the classical limestone slurry absorption the products($CaSO_3$ and $CaSO_4$) cannot be separated from the unreacted limestone and this part of reagent is practically lost.

Acedic acid mediates in the process but is continuously regenerated. Only the part of acid which is evaporated, entrained with gas and discharged with the waste product must be made up.

In the literature (G.T. Rochelle, C.J. King, Ind. Eng. Chem. Fund., 16, 1, 1977) the influence of several additives has been widely discussed. Some theoretical dependencies were derived and the most
important soluble additives has been tested as an additive for both lime- and limestone-slurry process. As potential carboxylic acid additives the authors mention acetic, benzoic, glycolic, adipic, isophthalic and sulfopropionic acids.

In our experiments acetic acid was used as the cheapest and easy of access. Reactions of acetic acid with the limestone and sulfur dioxide are presented below:

$$4 \ CH_3COOH + 2 \ CaCO_3 \rightarrow 2(CH_3COO)_2Ca + 2 \ H_2O + 2 \ CO_2$$
$$(CH_3COO)_2Ca + 2 \ H_2O + SO_2 \rightarrow Ca(HSO_3)_2 + 2 \ CH_3COOH$$
$$Ca(HSO_3)_2 + (CH_3COO)_2Ca \rightarrow 2 \ CaSO_3 \downarrow + 2 \ CH_3COOH$$

It can be noticed that acetic acid is fully regenerated in the process.

Some literature data (J.F. Villiers-Fisher, A. Warshaw, US Pat. 3,632,306, B01d 53/34, 1972) and our own experiments showed that the addition of 1% acetic acid (referred to the mass of circulating liquid) is the most appropriate. Greater amount does not improve the absorption and could increase the acid losses.

13.3 EXPERIMENTAL RESULTS

Experiments were carried out in the semi-technical installation in which flue gas from a utility boiler (approximately 2500 m^3/h) was scrubbed in a 2-stage turbulent contact absorber of 500 mm diameter. Rubber hollow spheres of 35 mm diameter were used as a moving packing. Scrubbing liquid was prepared in a 2.5 m^3 stirring tank. In the test No. 1, a 3% limestone suspension without additives was used and as a result 85% SO_2 removal efficiency was achieved with the excess ratio η_w = 1.43 kg added limestone/kg of stoichiometric amount necessary to the absorbed SO_2.

Next test No. 2 was carried out with the 3% limestone suspension and 1% acetic acid additive. Solid concentration in the suspension was maintained on the level 13%. Assuming that the acetic acid additive dissolves the part of limestone with the stoichiometric ratio, the calcium carbonate concentration in the liquid phase was X_w = 0.0217 kg $CaCO_3$/kg water. Mean value of SO_2 concentration in flue gases was Y_I = 1.0 x 10^{-3} kg SO_2/kg gas. At the excess factor η_w = 1.28 limestone consumption was equal to 6 kg/h. During the six day continuous test SO_2 removal efficiency was about 90%.

The third run was eight days long and its aim was to check the influence of the decrease of calcium carbonate concentration in the circulated slurry. In this test the amount of acetic acid was 1% as before (ca 27.5 kg) but the limestone amount was reduced to the stoichiometric quantity. It means that the X_w value was equal zero, η_w = 1 and the solid concentration was increased to the value of 15%. After two days however a little excess (η_w = 1.05) of limestone was applied because the reaction of $CaCO_3$ with the acetic acid could not proceed to the end and some portion of limestone was found in the products.

At the conditions like in the run No. 2 limestone consumption was 4.92 kg/h and the SO_2 removal efficiency could be also maintained on the level of 90%.

13.4   CONCLUSIONS
1)   Small amounts of acetic acid additive improved SO_2 removal efficiency from 85% to 90%.
2)   Limestone consumption due to the additive was decreased from 2.8 to 1.97 g/m^3 flue gas.
3)   Water and acetic acid losses could not be estimated in the experiments. To obtain these data a continuous test lasting some months should be carried out.
4)   The product from tests No. 1 and 2 was easy for centrifuging which is important for the reduction of acid losses.

Chapter 14

A PRELIMINARY EVALUATION OF CMA FOR SULFUR REMOVAL IN COAL-FIRED BOILERS

S. Manivannan and D.L. Wise
Department of Chemical Engineering, Northeastern University, Boston, MA 02115

14.1 ABSTRACT

A plant for the manufacture of CMA-treated coal (calcium, magnesium laden coal) by an ion exchange process with Calcium Magnesium Acetate (CMA) was studied by flowsheeting using the ASPEN PLUS simulator. The process aims at producing CMA-treated coal which reduces sulfur dioxide emissions during combustion and thereby combats acid rain. Also, the process is more effective than the limestone injection method and higher sulfur removal efficiencies are easily obtained[1,2,3]. The simulation study was used to obtain design parameters for the various unit operations involved in the process. Moreover, detailed cost estimation was done using the ASPEN PLUS stand alone costing model. This allowed the calculation of the initial selling price of the product (treated coal) as $41.20/ton (based on $20/ton coal i.e., $21.20/ton for coal treatment), the capital investment as $18,510,000 and the gross operating cost as $61,758,000. These results showed that the process was also economically feasible.

14.2 INTRODUCTION

As oil and gas resources are being depleted rapidly, increased attention is being given to the utilization of coal as the primary fuel. However coal is also a sulfur-containing fuel and combustion of coal produces emissions of sulfur oxides. Based on the information from the National Emissions Data System (NEDS) files as of 1982 total annual emissions of SOx is 29.1 x 10^6 tons and nearly 18.2 x 10^6 tons are from coal fired boilers which constitutes roughly 63 % of the total.

Sulfur oxides when discharged along with the flue gases into the atmosphere can cause serious pollution problems[4]. Sulfur oxides even at low levels of concentration of about 1 ppm can cause eye and lung irritation. At slightly higher levels it can cause bronchospasms and upon prolonged exposure can cause serious damage to the lungs. At concentration levels of 0.25 ppm SO_2 can cause damage to vegetation by a process called chlorosis, by which the green pigment of the leaves is damaged.

Also sulfur dioxide slowly reacts with the atmospheric oxygen to form SO_3 which further combines with moisture to precipitate as "acid rain". Acid rain can cause serious corrosion by deposition on metal structures and also acidifies lakes thereby endangering the lives of aquatic species. In order to prevent the above mentioned serious pollution problems, the Environmental Protection Agency (EPA) has imposed stringent regulations regarding sulfur dioxide emissions. The emission level for the industries is limited to 0.0305 ppm (annual average) and to 0.1394 ppm as the maximum allowable for any 24 hour period in a year. Two different methods of sulfur emission control have been used widely:

a)   Pre-combustion method: This technique attempts to produce a clean coal by the removal of sulfur by physical or chemical methods prior to its combustion.

b)   Post-combustion method: Most of these processes use a lime/limestone slurry to absorb the SO_2 from the flue gases in a scrubbing tower.

The above mentioned methods are expensive and the costs can be as high as 25% of the total power plant costs. A more recent approach is to remove the sulfur during the combustion stage itself. One of these techniques injects a powder of limestone[5] with the coal into the furnace. A more effective method would be to pretreat the coal with calcium additives prior to combustion. The use of calcium treated coal has been investigated by Freund and Lyon[3] and is found to be more effective than the limestone injection technique.

14.3 PROPOSED SOLUTION

It is proposed that Calcium Magnesium Acetate be used to add calcium and magnesium to the coal by ion-exchange[6] with the carboxylic acid groups in the coal. CMA is easily available, cost-effective, is environmentally acceptable and has a high water solubility making it the proposed candidate. A solution of CMA (1 N) in water will be used in the process. The ion-exchanged coal (Calcium, Magnesium laden coal) is then combusted in the boiler and the sulfur is captured from the flue gas and is retained with the ash in the forms of CaS and $CaSO_4$. The following two reactions are known to take place under reducing and oxidizing conditions respectively[1,3,7].

$$Ca + S \quad \rightarrow \quad CaS \tag{1}$$

$$CaO + SO_2 + \tfrac{1}{2}O_2 \rightarrow \quad CaSO_4 \tag{2}$$

which facilitate the conversion of sulfur into CaS and $CaSO_4$ during combustion. During the two-stage combustion of coal, conditions in the first strongly reducing zone may be thermodynamically favorable to sulfur sorption as CaS, being a more stable compound under reducing conditions at higher temperatures. In the low-temperature region, under oxygen rich conditions formation of $CaSO_4$ may be more thermodynamically favored. The overall reaction scheme describing the production and capture of SO_2 can be given by

FIGURE 1: COAL PRETREATMENT SECTION

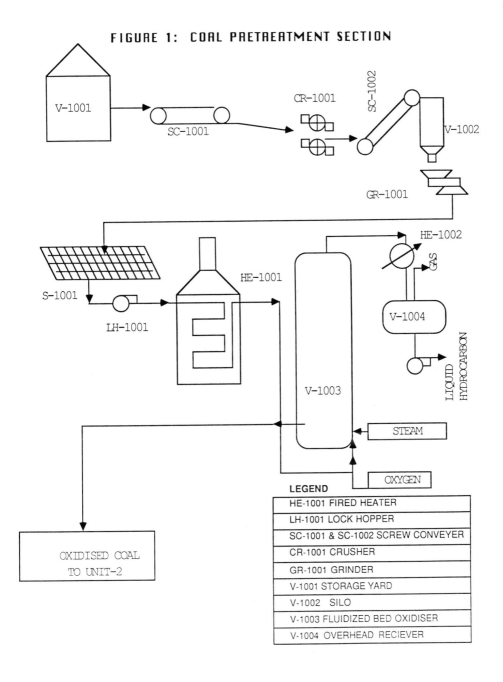

LEGEND

HE-1001	FIRED HEATER
LH-1001	LOCK HOPPER
SC-1001 & SC-1002	SCREW CONVEYER
CR-1001	CRUSHER
GR-1001	GRINDER
V-1001	STORAGE YARD
V-1002	SILO
V-1003	FLUIDIZED BED OXIDISER
V-1004	OVERHEAD RECIEVER

300

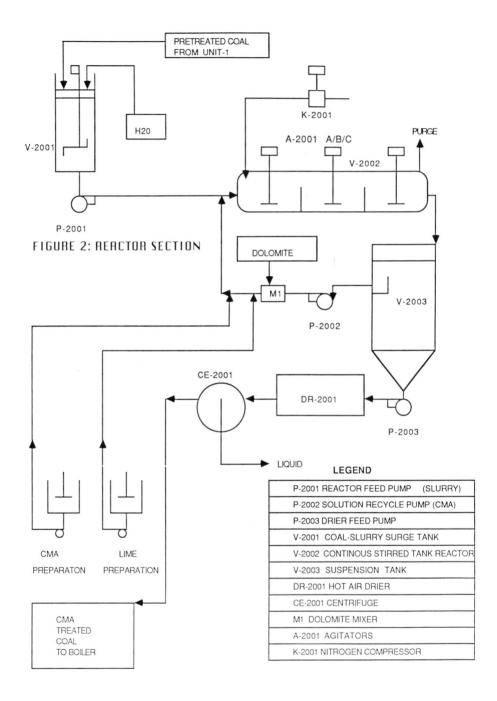

FIGURE 2: REACTOR SECTION

PRETREATED COAL
FROM UNIT-1

H2O

V-2001

P-2001

K-2001

A-2001 A/B/C

PURGE

V-2002

DOLOMITE

M1

P-2002

V-2003

CE-2001

DR-2001

P-2003

LIQUID

CMA
PREPARATON

LIME
PREPARATION

CMA
TREATED
COAL
TO BOILER

LEGEND

P-2001 REACTOR FEED PUMP   (SLURRY)
P-2002 SOLUTION RECYCLE PUMP (CMA)
P-2003 DRIER FEED PUMP
V-2001 COAL-SLURRY SURGE TANK
V-2002 CONTINOUS STIRRED TANK REACTOR
V-2003 SUSPENSION TANK
DR-2001 HOT AIR DRIER
CE-2001 CENTRIFUGE
M1 DOLOMITE MIXER
A-2001 AGITATORS
K-2001 NITROGEN COMPRESSOR

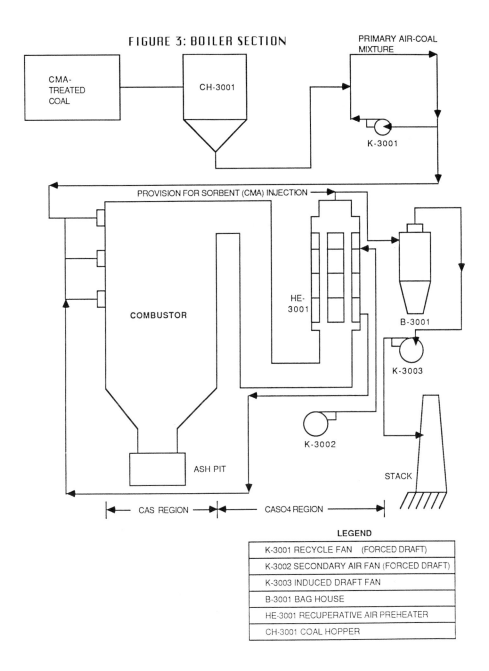

FIGURE 3: BOILER SECTION

LEGEND

K-3001	RECYCLE FAN (FORCED DRAFT)
K-3002	SECONDARY AIR FAN (FORCED DRAFT)
K-3003	INDUCED DRAFT FAN
B-3001	BAG HOUSE
HE-3001	RECUPERATIVE AIR PREHEATER
CH-3001	COAL HOPPER

the following equations[1,3].

Reducing conditions:

$$R\text{-}S \quad \rightarrow \quad H_2S + R' \qquad (1)$$

$$R\text{-}S + CO \quad \rightarrow \quad COS + R'' \qquad (2)$$

$$H_2S + CaO \quad \rightarrow \quad CaS + H_2O \qquad (3)$$

$$COS + CaO \quad \rightarrow \quad CaS + CO_2 \qquad (4)$$

Oxidising conditions:

$$R\text{-}S + 3/2\ O_2 \quad \rightarrow \quad SO_2 + CO + R' \qquad (5)$$

$$CaS + 2\ O_2 \quad \rightarrow \quad CaSO_4 \qquad (6)$$

$$CaS + 3/2\ O_2 \quad \rightarrow \quad CaO + SO_2 \qquad (7)$$

$$CaO + SO_2 + 1/2\ O_2 \quad \rightarrow \quad CaSO_4 \qquad (8)$$

$$CaSO_4 \quad \rightarrow \quad CaO + SO_2 + 1/2\ O_2 \qquad (9)$$

The major design criteria used for the preparation of the flowsheets for the process are summarized as follows:

a) Plant input ranges approximately from 450,000 lb/hr to 550,000 lb/hr of raw bituminous coal.

b) Annual capacity throughput is 1.47 million U.S. tons based upon a 24 hour operating day and 300 operating days per year.

c) The plant is located at the mine mouth and that all resources such as coal, water, power, etc. are assumed readily available.

d) Design of emission control facilities is based upon federal new source performance regulations EPA standards for air and water quality. Two important factors are important in the final evaluation of the process:

   1) As an SO_2 control technology, CMA treated coal should be able to remove above 95% of the sulfur during combustion

   2) The process should be economically feasible to be commercially successful.

14.4 PROCESS DESCRIPTION

The actual process is divided into three major sections. In pretreatment section, the coal is ground to a fine powder using crushing equipment. The oxidation section is required since a mild oxidation of the coal increases the carboxylic acid groups in the coal which in turn increases the ion-exchange capacity of the coal with respect to Ca, Mg[13], but with a slight loss in the calorific value of the coal. Finally in the reactor section the oxidized coal is reacted with Calcium Magnesium Acetate to produce CMA-treated coal. (See Figures 1, 2 & 3.)

A) Pretreatment Section

The coal delivered from the storage is first sent to a crusher. The
outlet diameter size is 800 microns. The coal is then further ground in
a secondary grinder to a size of 150 microns which is sent through a
screen to the downstream section.

B) Oxidation Section

The fine coal is then conveyed by lock hoppers to a preheater where it
is heated up to 450°F. It is then oxidized in a fluidized bed oxidizer
with a stream of oxygen. The overhead vapors from the oxidizer contain
primarily oxides of carbon, sulfur and water vapor. The oxidized coal
is cooled and is sent to the reactor section.

C) Reactor Section

The coal at 80°F is mixed with DM water and agitated to form a slurry.
The slurry is pumped to the reactor which is a series of continuous
stirred tank reactors. The reactant CMA and lime solution to maintain
the pH at 8.5 are added. The reactor is maintained at 130°F by a steam
jacket. Nitrogen is added to maintain an inert atmosphere. Each CSTR
is provided with an agitator for mixing. The liquid-solid products are
sent to the suspension tank where the solids are settled. The settled
solids are pumped by a slurry pump. A fraction is recycled to the
reactor section and the remainder is filtered to remove the liquid. The
liquid with some unreacted CMA and acetic acid is sent to a regeneration
unit where dolomite is added to regenerate CMA. This regenerated stream
is recycled back to the reactor. Fresh CMA is added to make up the
system losses. The filtered solids is sent to a drier where hot air is
blown countercurrently. The dried coal is stored in a secondary storage
yard before it is sent to the boiler.

D) Boiler Section

The treated coal is fed into a hopper from where it is mixed with the
primary air. It is then mixed with the preheated secondary air before
it reaches the burner. The heavy ash in the furnace drops to the ash
pit from where it is removed continuously by a conveyer. But, the fly
ash which is carried by the flue gas is removed mostly just before it
enters the air preheater. The remainder is removed by the bag filter
before the flue gas is discharged to the stack. The $CaSO_4$ formed is
expected mostly with the fly ash and the CaS with the heavy ash.

14.5 PROCESS MODEL USING ASPEN PLUS

ASPEN PLUS is a simulator program that is widely used for process -
flowsheeting[8,9]. The various stream flow rates, type of unit operations,

304

process conditions (temperature, pressure, etc.,) are specified by the user and the program calculates the material and energy balances. It further calculates the design specifications of all the unit operations in the process. A brief description of the program is as follows: (See Appendix-1)

a)   The units of measurement chosen for the input and output reports are both in the English Engineering set of units. The COMPONENTS statement specifies that the components present in the system are hydrogen, oxygen, nitrogen, sulfur, carbon dioxide, acetic acid, water, coal, ash etc.

b)   The NC-COMPS paragraph signifies that COAL and ASH are nonconventional components and that the attributes used to describe these components are proximate anlaysis, ultimate analysis and sulfur analysis. The omission of the second name from the COMPONENTS paragraph for COAL and ASH indicates that the component does not correspond to a component in any of the ASPEN PLUS data banks, so the physical property parameters will not be retrieved and must be entered separately.

c)   The physical property models for nonconventional components use the data provided for the components attributes described by the PROPERTIES SYSOPO statement.

d)   The next step represents the actual flowsheet with the unit operation blocks and to use the appropriate model for each block. The process flow diagram is translated into the various blocks involved. The streams and blocks in the block diagram do not necessarily correspond on a one-to-one basis with the pipes and pieces of equipment in the actual plant. The flowsheet is divided into four sections, namely the pretreatment section, the oxidizer section, the reactor section and the boiler section.

e)   Pretreatment Section: Flowsheet Pretreat represents this section. The coal is reduced to the required size by a two-stage crushing and grinding unit provided with screens of the required sizes. The sizes of the screen opening is specified as 800 and 150 microns.

f)   Oxidation Section: Flowsheet Oxidation represents this section. The block HEATER1 simulates a heater which brings the temperature of the coal to the required oxidation temperature of 450°F. Block SPLIT1 is a fictitious block which splits the main stream of the coal from the stream to be oxidized. Block YIELD converts the nonconventional stream GSCOAL into a stream GSCOALIN which contains conventional components like H_2, O_2, etc. Block OXID is the oxidizer which converts the stream GSCOALIN into hydrocarbon gases. The oxidized coal stream CRCOALIN leaves the unit to the subsequent reactor unit.

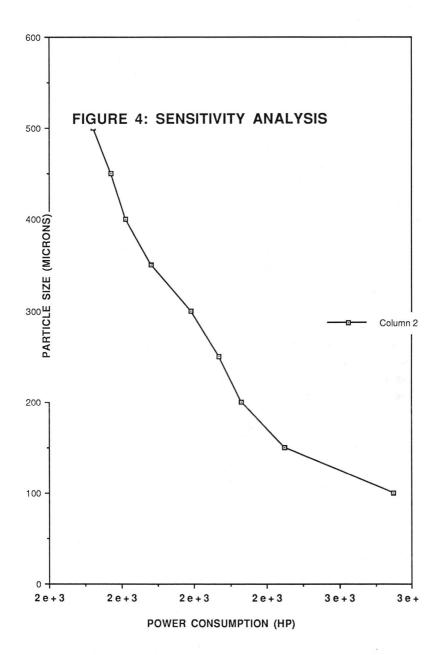

FIGURE 4: SENSITIVITY ANALYSIS

FIGURE 5: SENSITIVITY ANALYSIS

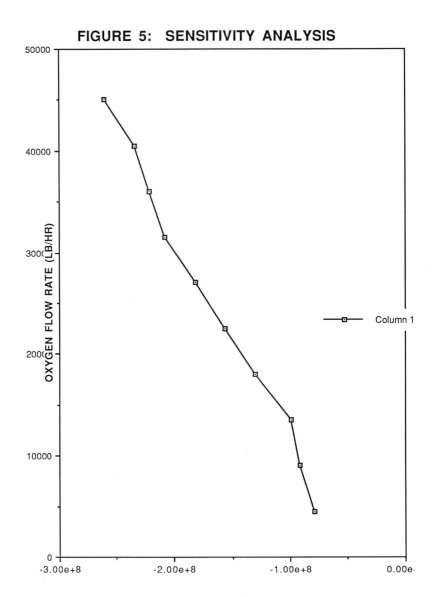

OXIDISER HEAT DUTY (BTU/LB)

FIGURE 6: SENSITIVITY ANALYSIS

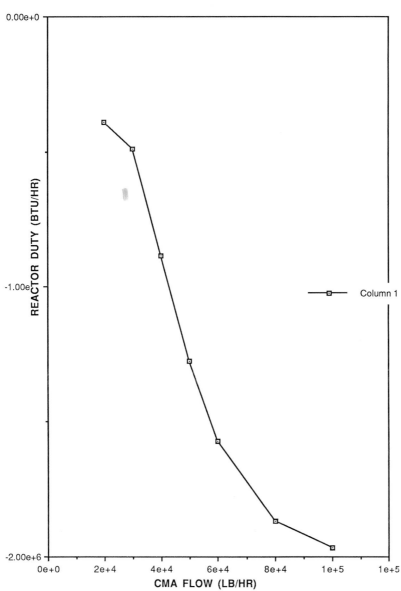

308

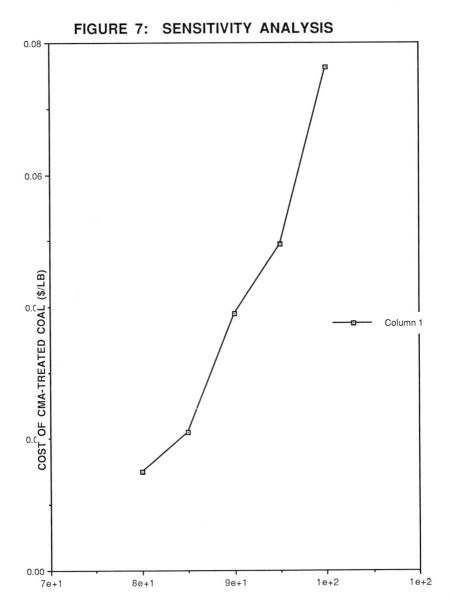

FIGURE 7: SENSITIVITY ANALYSIS

PERCENT REMOVAL OF
SULFUR EMISSIONS

g) Reactor Section: Flowsheet Oxid represents this section. The YIELD1
    block converts the stream OXCOAL into a stream OXCOAL1 with conventional
    components. OXCOAL1 reacts with CMA in the block RCTR to produce
    CALCOAL which is the calcium, magnesium ion-exchanged coal. Block
    CFUGE1 is a solid-liquid separator. The filtrate is sent for CMA
    reclamation using dolomite and is then recycled. The treated coal is
    further dried in a dryer block.

h) Boiler Section: Flowsheet Boiler represents this section. The dry coal
    is mixed with air and is then combusted. The low temperature region of
    the boiler at 2400°F is selected for simulation. The concentration of
    the final combustion products is determined by the stream FLUEGAS which
    leaves the boiler.

14.6 COST ANALYSIS

The ASPEN PLUS stand alone model (See Table 1, 2, & 3) to determine the
capital and the operating costs[10,11,12]. Some of the equipment prices were
obtained from vendor quotes. In other cases, the details of the equipment
were entered and the ASPEN program was requested to calculate the cost. An
interest rate of return of 15% was used as the criteria for profitability.
Thus, the capital cost, gross operating cost, and the initial selling price
were calculated.

The economic evaluation was based on a new 500 MW pulverized coal fired
boiler. The cost comparison with a conventional limestone scrubbing process
is as follows[15]:

	Capital cost	Oper. cost (annual)
a) Limestone Scrubbing[15]	$ 121,953,000	$ 45,190,000
b) CMA-treated coal	$ 18,510,000	$ 4,117,200

This shows that the second process demands lesser capital and operating costs.
Also, the scrubbing system is usually a throwaway system which requires
special raw material handling, feed preparation and waste disposal. This
study has thus confirmed the viability of the CMA-treated coal process in
terms of cost.

14.7 SENSITIVITY ANALYSIS

Often we have a choice of selecting a scheme from several other
alternatives. By sensitivity analysis we can decide the optimal scheme.
Sensitivity is generally referred to as the relative magnitude of the change
in a key variable with respect to an independent variable. If a small change

310

in the independent variable causes a large change in the key variable then it is said to be very sensitive with respect to the independent variable. Thus, sensitivity analyses were carried out on some of the key variables of the plant in order to decide the optimal process scheme. (See Figures 4, 5, 6, 7)

CASE 1:   This analysis is done to determine the optimum particle size for the crushing/grinding section. We can see that the total power consumption increases sharply as the average particle diameter decreases. This may also require increased number of stages which would result in high operating and capital costs for the crushing and grinding units. Notably, the curve rises sharply when the particle diameter drops below 150 microns, which requires sophisticated crushing equipments and fine screens.

CASE 2:   Oxygen rate to the oxidizer increases the oxidizer heat duty and would sharply increase the reaction temperature. This reaction is highly exothermic and this could cause a reaction-runaway situation which could cause serious damage to the internals of the equipments involved. This is overcome by adding a quenching agent to the oxidizer. The rate of quenching vs. temperature is also studied and this enables the design engineer to set the optimum flow rates and also the maximum flowrates needed to shut down the system carefully.

CASE 3:   The reactor heat duty is seen to increase as the flow rate of CMA to the reactor increases due to heat of mixing which would also increase the temperature of the reactor.

CASE 4:   Finally, the cost per ton of the treated coal increases drastically as the percentage of sulfur removal increases.

14.8 CONCLUSION

The calcium-treated coal process has considerable potential for sulfur removal with high efficiency levels[1,2,3,5,14]. The ASPEN PLUS program output gives the results of the process simulation of the plant. The parameters thus obtained from ASPEN PLUS are adequate to project a preliminary design of the plant in a commercial scale. Moreover, the cost analysis shows that the process is competitive with the conventional flue gas desulfurization technology, thus encouraging process development.

TABLE 14.1

Executive Summary

INVESTMENT

Physical Plant	$6,814,000
Nondepreciable Items	177,000
Interest during Construction (18%)	287,000
Start-up	2,272,000
Tax Credit	-886,000
Working Captial	7,802,000
Contingency Allowance	2,044,000
TOTAL	**$18,510,000**

SCHEDULE

Project Start	JUN, 1989
Mechanical Completion	APR, 1990
Commercial Production	JUL, 1990

REVENUE

Capacity Production Rate (lb/hr)	450,000
Normal Production Rate (lb/hr)	450,000
Initial Selling Price ($/lb)	0.0206* Calculated

COSTS

Initial Cost per Unit	
Operating Rate 100% ($/lb)	0.0174
Operating Rate 75% ($/lb)	0.0177
Operating Rate 50% ($/lb)	0.0184

DEBT

Amount to be Financed (40.00%)	$3,543,000
Interest Rate, Long-term	14.00%

PROFITABILITY

Interest Rate of Return on Equity	15.00%
Payout Time -- Years	8.25
Return on Investment (ROI)	12.12%
Net Present Value	$-1,459,000
Venture Worth (Discount rate = 17.0%)	5,460,000
Initial Investment	6,461,000
Venture Worth/Initial Investment	0.85
Break-even Fraction	0.19
Break-even Volume (lb/hr)	78,710

TABLE 14.2

Fixed Capital Estimate Summary

Project Completion: 1990, Quarter II

	MATERIAL COST	LABOR COST	LABOR HRS
Process Units	$2,695,000	$866,000	$33,000
Utility Units	0	0	0
Recvng, Shipping & Stor	0	0	0
Service Building	202,000	155,000	6,000
Service Syst & Distrbutn	166,000	53,000	2,000
Additional Direct	0	0	0
SUBTOTAL	$3,063,000	$1,073,000	$41,000
Site Development	31,000	16,000	1,000
Freight	61,000		
Sales Tax	92,000		
TOTAL DIRECT COST	$3,246,000	$1,089,000	$42,000
Contractor Fld Indirects	294,000	795,000	25,000
SUBTOTAL	$3,541,000	$1,885,000	$67,000
Total Direct and Field Indirect			$5,425,000
Contractor Engineering & Home Office		651,000	
Owner's Cost		304,000	
Fees, Permits & Insurance		434,000	
Additional Depreciable		0	
SUBTOTAL			1,389,000
TOTAL DIRECT & INDIRECT			$6,814,000
Process Basis Contingency		$341,000	
Project Definition Contingency		1,703,000	
TOTAL CONTINGENCY			2,044,000
TOTAL DEPRECIABLE CAPITAL			$8,858,000
Land		$177,000	
Royalty & Expenses		0	
Additional Nondepreciable		0	
SUBTOTAL (NONDEPRECIABLE COSTS)			$177,000
TOTAL FIXED CAPITAL			$9,035,000
Working Capital			7,802,000
Start-up Cost			2,272,000
TOTAL INVESTMENT			$19,109,000

TABLE 14.3

Annual Operating Cost Summary

Production Capacity:	100.0%
Base Period:	1990, Quarter II
Principal Product:	CALMAG-C
Plant Availability:	328/365.25 days
Production Per Year:	3550230. (1000 lb)

TOTAL RAW MATERIAL		$57,688,000
TOTAL UTILITIES		577,000
LABOR - Operators: 5 per shift, 3 shifts		
Operating (39,000 hrs)	$671,000	
Maintenance	319,000	
Supervision	198,000	
Fringe Benefits	475,000	
SUBTOTAL, Labor		**$1,662,000**
SUPPLIES		
Operating	$67,000	
Maintenance	213,000	
SUBTOTAL, Supplies		**$280,000**
GENERAL WORKS		
General & Admin	$712,000	
Property Tax	177,000	
Property Insurance	71,000	
SUBTOTAL, General Works		**$960,000**
DEPRECIATION (15 Years, Straight Line)		$591,000
Gross Operating Cost		$61,758,000
Less: Byproduct Credit		0
NET OPERATING COST		**$61,758,000**

SUMMARY:		
Fixed Cost	$2,902,000	
Variable Cost	$58,265,000	
Gross Cost Excl. Depreciation		$61,167,000
Gross Operating Cost		$61,758,000

314

*** INPUT ECHO(ES) ***

>>CURRENT RUN

ORIGINAL RUN   OCTOBER 9, 1989
3:40:54 P.M.   MONDAY
INPUT FILE: DISKA1:[ASPEN.03]M10.INP;4
OUTPUT PDF: M10   VERSION:1
LOCATED IN: DISKA1:[ASPEN.03]
PDF SIZE: FILE (PSIZE)=99999 RECORDS.   IN-CORE =   200 RECORDS.

```
1     TITLE 'CMA-TREATED COAL MANUFACTURE'
2     ;
3     DESCRIPTION "THIS IS A MODEL OF A PLANT WHICH PRODUCES CALCIUM, MAGNES
4                 ION-EXCHANGED COAL.  THE PLANT COMPRISES OF FOUR UNITS NA
5                 A) PRETREATMENT UNIT, B) OXIDATION UNIT, C) REACTOR UNIT
6                 D) BOILER UNIT"
7     ;
8     IN-UNITS ENG
9     ;
10    OUT-UNITS ENG
11    ;
12    COMPONENTS O2 O2 / H2O H2O / CH4 CH4 / CL2 CL2 / C C / CO2 CO2 /
13              S S / H2 H2 / CO CO / O2S O2S / AA C2H4O2-1
14              CA CA / MG MG / N2 N2 / CASO4 CASO4 / COAL / ASH
15    ;
16    DATABANKS SOLIDS
17    ;
18    NC-COMPS COAL PROXANAL ULTANAL SULFANAL
19    NC-COMPS ASH PROXANAL ULTANAL SULFANAL
20    ;
21    PROPERTIES SYSOPO
22    ;
23    NC-PROPS COAL ENTHALPY HCOALGEN / DENSITY DCOALIGT
24    NC-PROPS ASH  ENTHALPY HCOALGEN / DENSITY DCOALIGT
25    ;
26    FLOWSHEET PRETREAT
27         BLOCK  CRUSH1    IN = FDCOAL1               OUT = SCR1
28         BLOCK  MIX1      IN = FDCOAL OVER 1         OUT = FDCOAL1
29         BLOCK  SCREEN 1  IN = SCR1                  OUT = OVER1 UNDER1
30         BLOCK  CRUSH2    IN = FDCOAL2               OUT = SCR2
31         BLOCK  MIX2      IN = OVER2 UNDER1          OUT = FDCOAL2
32         BLOCK  SCREEN2   IN = SCR2                  OUT = OVER2 UNDER2
33    ;
34    FLOWSHEET OXIDATION
35         BLOCK  HTR1      IN = UNDER2                OUT = CRCOALH
36         BLOCK  SPLIT1    IN = CRCOALH               OUT = GSCOAL OXCOA
37         BLOCK  YIELD     IN = GSCOAL                OUT = GSCOALIN
38         BLOCK  OXID      IN = GSCOALIN OXYGN        OUT = HCGAS
39    ;
40    FLOWSHEET REACTOR
41         BLOCK  YIELD1    IN = OXCOAL                OUT = OXCOAL1
42         BLOCK  PUMP1     IN = OXCOAL1               OUT = OXCOAL2
43         BLOCK  RCTR      IN = OXCOAL2 CMA           OUT = CALCOAL
44         BLOCK  CFUGE1    IN = CALCOAL               OUT = WAT CALCOAL2
45         BLOCK  DRYER     IN = CALCOAL2              OUT = FEEDCOAL
46    ;
```

```
47     FLOWSHEET BOILER
48          BLOCK BOIL        IN = FEEDCOAL AIR              OUT = FLUEGAS
49     ;
50     DEF-STREAMS MIXNCPSD PRETREAT
51     DEF-STREAMS MIXNCPSD OXIDATION
52     DEF-STREAMS MIXNCPSD REACTOR
53     DEF-STREAMS MIXNCPSD BOILER
54     ;
55     DEF-SUBS-ATTR PSD PSD
56          IN-UNITS LENGTH = MU
57          INTERVALS 14
58          SIZE-LIMITS 0/44/63/88/125/177/250/354/500/707/1000/
59                     2000/4000/8000/16000
60     ;
61     BLOCK CRUSH1 CRUSHER
62          IN-UNITS LENGTH = MU
63          PARAM DIAM = 1000 TYPE = GYRATORY
64          BWI NCPSD 11.37
65     ;
66     BLOCK SCREEN1 SCREEN
67          IN-UNITS  LENGTH = MU
68          PARAM OPENING = 800 MODE=DRY
69     ;
70     BLOCK CRUSH2 CRUSHER
71          IN-UNITS LENGTH = MU
72          PARAM DIAM = 200 TYPE = GYRATORY
73          BWI NCPSD 11.37
74     ;
75     BLOCK SCRREN2 SCREEN
76          IN-UNITS  LENGTH = MU
77          PARAM OPENING = 150 MODE=DRY
78     ;
79     BLOCK MIX1 MIXER
80     BLOCK MIX2 MIXER
81     ;
82     BLOCK PUMP1 PUMP
83          PARAM PRES=20 EFF=0.7 DEFF=0.95
84     ;
85     BLOCK HTR1 HEATER
86          PARAM PRES=0 TEMP=450 NPHASE=1 PHASE=S
87     ;
88     BLOCK SPLIT1 SSPLIT
89          FRAC NCPSD GSCOAL 0.1/OXCOAL 0.9
90     ;
91     BLOCK YIELD RYIELD
92          PARAM TEMP=450 PRES=15 NPHASE=1 PHASE=V
93          MASS-YIELD MIXED  H2  0.0501 / O2 0.0939 /
94                     MIXED  C   0.70 / S 0.0442 /
95     ;
96     BLOCK OXID RSTOIC
97          PARAM TEMP=450 NPHASE=1 PRES=15
98          STOIC 1 MIXED C -1 / O2 -1 / CO2 1
99          STOIC 2 MIXED H2 -2 / O2 -1 / H2O 2
100         STOIC 3 MIXED S -1 / O2 -1 / O2S 1
101         CONV 1  MIXED C 1
102         CONV 2  MIXED H2 1
103         CONV 3  MIXED S 1
104    ;
105    BLOCK YIELD1 RYIELD
106         PARAM TEMP=450 PRES=15 NPHASE=1 PHASE=L
```

```
107          MASS-YIELD MIXED H2 0.0501 / 02 0.0939 /
108                    MIXED C 0.70 / S 0.045
109     ;
110     BLOCK RCTR MIXER
111     ;
112     BLOCK CFUGE1 SEP
113          FRAC SUBS=MIXED    STREAM=WAT       COMPS=H2O AA     FRACS=1.  1.
114          FRAC SUBS=NCPSD    STREAM=CALCOAL2  COMPS=COAL      FRACS=1.
115          FRAC SUBS=MIXED    STREAM=CALCOAL2  COMPS=CA MG     FRACS=1.  1.
116     ;
117     BLOCK DRYER HEATER
118     PARAM PRES=15 TEMP=300 NPHASE=1 PHASE=L
119     ;
120     BLOCK BOIL RSTOIC
121          PARAM TEMP=2400 PRES=15
122          STOIC 1 MIXED C -1 / 02 -1 / C02 1
123          STOIC 2 MIXED CA -1 / 02 -1 / 02S -1 / CASO4 1
124          STOIC 3 MIXED S -1 / 02 -1 / 02S 1
125          STOIC 4 MIXED H2 -2 / 02 -1 / H2O 2
126          CONV  1 MIXED C 1
127          CONV  2 MIXED CA 0.9
128          CONV  3 MIXED S 1
129          CONV  4 MIXED H2 1
130     ;
131     STREAM FDCOAL
132          SUBSTREAM NCPSD TEMP=70 PRES=15
133          MASS-FLOW COAL 45000
134          COMP-ATTR COAL PROXANAL (14.21 46.00 42.97 11.03)
135          COMP-ATTR COAL ULTANAL  (11.03 68.68 5.01 1.26 0.21 4.42 9.39)
136          COMP-ATTR COAL SULFANAL (2.09 0.03 2.3)
137          SUBS-ATTR PSD (0 0 0 0 0 0 0 0 .1 .2 .35 .25)
138     ;
139     STREAM OXYGN TEMP=200 PRES=15
140          MASS-FLOW 02 45000
141     ;
142     STREAM CMA TEMP=75 PHASE=L PRES=15
143          MASS-FLOW CA 20000 / MG 20000 / AA 20000 / H2O 40000
144     ;
145     STREAM AIR TEMP=150 PRES=15
146          MASS-FLOW 02 400000 / N2 1600000
```

ACKNOWLEDGEMENT

This is to acknowledge the financial support of Stone and Webster Engineering Corporation to the College of Engineering, Northeastern University. In particular, we wish to thank Mr. B. Brodfeld, Vice President, Manager, Advanced Technologies Department, and Ernest R. Zabolotny, Ph.D. of Stone and Webster, as well as Paul H.King, Ph.D., Dean, College of Engineering, Northeastern University.

REFERENCES

1   K.K. Chang, R.C. Flagan, G.R. Gavalas, and P.K. Sharma, Fuel 1986. 65, 75.
2   P.K. Sharma, G.R. Gavalas, R.C. Flagan, "Calcium Pretreatment of Additives for Sulfur Emissions Control in Combustion," Fuel, 66(2), 207-9, 1987.
3   H. Freund, and R.K. Lyon, Combust. Flame 1982, 45, 191.
4   C.B. Rogers, Morton, Report on Sulfur Oxide Control Technology, U.S.Department of Energy, 1985.
5   P.L. Case, M.P. Heap, C.N. McKinnon, D.W. Pershing, and R. Payne, Am. Chem. Soc. Div. Fuel Chem., Prepr. 1982, 27 (1), 158.
6   W. Bartok, H. Freund, and R. Liotta, "Cation Ion Exchange of Coal" US patent No:4468231, 1984.
7   G.A. Simmons, "Parameters Limiting Sulfation by CaO," AIChE Journal, Vol.34, No. 1 1988.
8   Aspen Plus Introductory Manual (Third Edition), Aspen Technology, Inc., Cambridge, MA, U.S.A.
9   Aspen Plus Solids Manual (Preliminary Edition) Aspen Technology, Inc., Cambridge, MA, U.S.A.
10   Aspen Plus Costing Manual 1985, Aspen Technology, Inc., Cambridge, MA, U.S.A.
11   Project Evaluation in the Chemical Industries (3rd ed) by Valle-Riestra, McGraw Hill, 1986.
12   Conceptual Design of Chemical Processes by James M. Doughlas, McGraw Hill, 1988.
13   H.N.S. Schafer, Fuel 197, 49, 197.
14   G.R. Gavalas, Comb. Sci. Tech. 24:197 (1981).
15   J.L. Hudson (Ed) and G.T. Rochelle (Ed), Flue Gas Desulfurization, ACS Symposium Series 1982.

Chapter 15

CALCIUM MAGNESIUM ACETATE FROM THE BIOCONVERSION OF RESIDUE BIOMASS

Debra J. Trantolo, Joseph D. Gresser, Don C. Augenstein, Donald L. Wise
Cambridge Scientific, Inc. Belmont, MA 02178, USA

15.1 ABSTRACT

This chapter details a project to experimentally investigate a competitive system using residue biomass as feedstock for conversion to calcium magnesium acetate, "CMA", an alternative road salt. This new organic road salt will prevent corrosion of bridge decks, underground cables, and rusting of cars and trucks. "CMA" derived from biomass will be less costly, compared to that derived from petroleum and natural gas. The biomass may be (a) woody biomass residues not suitable for lumber or paper pulp, (b) industrial residues such as whey, (c) municipal solid waste (MSW), and (d) sewage sludge residuals. Experimental work to date has focused on bioconversion of sewage sludge residuals to CMA. The process is based on "suppressed methane" fermentation to produce acetic acid from the biomass, followed by liquid ion exchange to recover acetic acid from the fermenter broth prior to the final production step which occurs by passing the acetic acid over limestone. In the following, project results, focusing on sewage sludge residuals as the biomass of choice, show: (a) confirmation of percent bioconversion and kinetics to acetic acid in small batch fermenters; (b) documentation of equilibrium constants for acetic acid recovery via liquid ion exchange; (c) and determination of the rates of conversion to CMA. Extended work is being directed toward: (a) continuous operation of small lab fermenters to obtain engineering parameters; (b) retrieval of scale-up data on liquid ion exchange; and (c) more definitive cost analysis. These objectives are supportive of the eventual operation of a small pilot plant, intended for the eventual commercialization of this technology. The project work presented in this chapter has shown that selected residue biomass can be readily bioconverted to CMA. Thus, because residue biomass exists in such large quantities as may be used to produce the vast amounts of road salt used each year, the potential for a fermentation-derived CMA is highlighted.

15.2 INTRODUCTION

15.2.1 The Problem

The massive use of chloride salts for roadway deicing is the cause of serious corrosion problems and major environmental problems in the "Frost Belt" areas. These problems include: deterioration of Portland concrete bridge decks through chloride ion corrosion of reinforcing steel; corrosion of underground electrical cables; corrosion of structural members in bridges and other highway appurtenances; corrosion of vehicle chassis; pollution of aquatic habitats and drinking water sources by sodium and chloride ions in runoff; and harm to roadside vegetation. In addition to the corrosion and environmental harms associated with NaCl, the use of sodium chloride for deicing the roads has many other drawbacks because of the many subtle costs of deicing with corrosive salts. These include: gasoline wasted because of slow traffic during repairs to bridges and roads, costs of transporting materials for repair, and labor and material costs for replacement. For example, it is estimated that in New York State alone, the deleterious effects of deicing cost more than $500 million per year [1,2].

Clearly, practical alternatives to the use of NaCl or $CaCl_2$ must be found and use must begin immediately. In one example, the New York State Energy Research and Development Authority (NYSERDA) has opted to explore an alternative road salt [3]. However, commercial production of needed quantities of this material in a practical manner is not yet possible.

15.2.2 The Potential Solution

Calcium magnesium acetate (CMA) has been identified as an effective replacement for sodium chloride as a road salt. While as persistent a highway deicing chemical as NaCl, CMA is an attractive substitute because it carries with it none of the attendant corrosion or pollution problems. At present, it suffers only from limitations brought on by its high cost (minimally 10-20 times the cost of NaCl).

If CMA is to be exploited in meeting the need for safer deicing chemicals, an economic source of material is necessary. Because of the ready availability of limes (to supply the calcium and magnesium parts of the salt), the cost concerns lie primarily in acquiring the acetate portions of the CMA. Acetate feedstocks conventionally derived from petroleum and natural gas are much too costly.

The work detailed in this chapter focused on the production of CMA from residue biomass, specifically sewage sludge. The acetate portion of the salt is derived from the fermentation of biomass organics, producing acetic acid. The acid is then concentrated and isolated using liquid ion exchange with the production of CMA occurring in a final step with the addition of limestone. Thus, a potentially sound CMA process is outlined, the objective being to produce a material which offers an economically competitive alternative to NaCl.

15.3 BACKGROUND

15.3.1 Use of Salts and Alternatives for Deicing Roads

Approximately 10 million tons of deicing salts, primarily calcium chloride ($CaCl_2$) and sodium chloride (NaCl), are spread on roadways in the United States annually. Except for the use of NaCl as a raw material for caustic and chloride production, roadway deicing is the largest use for salt and represents about 20% of annual consumption. Roughly one third of total $CaCl_2$ consumption is for roadway deicing. However, $CaCl_2$ represents only a few percent of the total use of deicing salts. At a delivered cost of $25 to $50/metric ton (1.1 cent/lb to 2.2 cents/lb), NaCl is the present economic choice compared to $CaCl_2$ which costs well over $100/metric ton (4.5 cents/lb). Calcium chloride is used only when low temperature conditions require it. Thus it is clear that economic considerations are a primary concern in selecting a road salt for deicing.

Although the damage resulting from the use of highway deicing salts is very high, it must be balanced against the safety and economic benefits derived from keeping pavements clear during winter. The U.S. Department of Transportation's FHWA-initiated research work during the mid-1970's aimed to develop substitutes for chloride salts in deicing. In FHWA-funded research conducted by Bjorkstein Research Laboratories, calcium magnesium acetate (CMA) and methanol were identified as potentially acceptable, noncorrosive alternative deicing chemicals [4]. The mixture of calcium and magnesium acetates was found a promising substitute for NaCl and $CaCl_2$ because of affordable cost, reasonable effectiveness for deicing, and minimal corrosiveness [4]. Inexpensive, locally available dolomitic lime (a mixture of mostly calcium and magnesium compounds) determined the cations associated with acetate for this prior analysis.

Preliminary field testing of a calcium-magnesium acetate (CMA) began in 1983 in Michigan and Washington and plans are underway to determine the environmental impacts of large-scale use of CMA in New York State [3]. It is supposed that there will be some increase in biological oxygen demand (BOD) when CMA is flushed into natural waters, but that oxygen may not be seriously depleted because this run-off occurs when temperatures are low and biological systems function slowly. Damage to soils is not likely from CMA because organic matter is desirable, and calcium and magnesium ions are already present in large amounts. Oxidation of calcium acetate or calcium magnesium acetate by soil microorganisms would generate calcium or calcium magnesium carbonates which are harmless.

15.3.2 The Selection of CMA as the Alternative Road Salt

As early as 1979, a report to the FHA [4] identified CMA as an effective low cost deicing replacement for sodium chloride (NaCl). CMA was shown to act as rapidly as NaCl and had the same persistency as rock salt, but without the environmental drawbacks of pollution and corrosion. In an earlier review [5] on the subject submitted to FHA, no mention of CMA was made. Since that time voluminous literature has emerged focusing on production, road evaluation, corrosion effects, and environmental compatibility. However, a report describing field tests in the state of Washington in 1983 [6] did identify some problems: CMA was somewhat slower than either salt or urea to react on compact snow and ice; and problems of dust, light weight, and brittleness had to be addressed. In the following years CMA evaluations have been conducted at numerous locations in the snow belt [7-12].

Production of CMA has also been investigated. Scheeler (1983) [13] reacted glacial acetic acid with hydrated dolomitic lime. Reaction and drying was carried out in rotary drum mixers. Sand was added to the reaction mixer to improve road performance and to minimize handling problems. Marynowski (1984) [14,15] of SRI International also used acetic acid in reaction with light burned dolime (dolomitic lime). Reaction conditions suggested use of flocculants and diatomaceous earth to facilitate settling. CMA thus produced was recovered using a Niro fluidized bed-spray drier. In an earlier report [16], Marynowski et al. (1983) screened various CMA production methods and suggested that acetic acid might be derived by anaerobic thermophilic fermentation, using Clostridium thermoaceticum to ferment biomass-derived sugars. Fermentation would be followed by reaction of the entire fermenter stream with dolomitic lime (preferred to dolomite because of a more rapid reaction). Corn grain was identified to be the most favorable biomass feedstock, with high glucose corn syrup slightly less favorable. Spray drying and drum drying were suggested as means for product recovery, with unfermented corn solids and bacterial cells separated as a dried animal feed by-product.

Gancy [17] investigated laboratory versions of industrial pelletizers/agglomerators for synthesis and isolation of hydrated CMA derived from the reaction of commercial acetic acid and unslaked dolime. The product was reported to be satisfactory with respect to friability, particle size distribution, and dust, with a solution pH of 7.4, although case hardening of the product was noted. A drum pelletizer was recommended for product recovery.

A study of 19 methods for CMA drying and isolation by Solash (Nov. 1986) [18] focused on fluidized bed, rotary or tumbling bed dryers, drum dryers, and pneumatic or flash drying. The final recommendation was fluidized bed spray granulation. The product was observed to be slower than rock salt in melting ice and to be softer and less dense. Spray drying was deemed unsuitable because the product is usually a porous low density friable material.

CMA's of varying Ca/Mg ratios may be prepared by reaction of acetic acid and carbonate minerals of magnesium and calcium in appropriate ratios. The mixed salts, of which CMA from dolomite is one example, will differ with respect to melting properties. R.U. Schenk [19] (1986) has investigated the effects of such mixed salts on Portland cement: their eutectic temperatures, heats of solution, stability, solubility, rates of ice melting. The conclusion, based on consideration of these numerous factors, favored production of a 3/7 Ca:Mg ratio, although a 5/5 ratio was not excluded. Indeed, a 1/9 ratio would have been considered acceptable if solubility of the hydrated form permitted.

The studies cited thus far have utilized CMA prepared by reaction of commercially acquired acetic acid with suitable mineral derived substrates. Although this is feasible, it is well to keep in mind alternative sources of the acid such as fermentation processes. As mentioned earlier, SRI International carried out an evaluation of acetic acid production from crop grown biomass [14-16] such as corn or corn syrup, but did not consider industrial or municipal wastes, and did not explore woody biomass.

More recently Bungay and Hudson [1] suggested the use of woody biomass as a feedstock for conversion to acetic acid, but did not consider fermentable industrial or municipal wastes. Clausen and Gaddy [20] have noted that agricultural residues including woody biomass such as corn stover amount to about 400 million tons per year and that the carbohydrate fraction (up to about 80%) can be converted to fermentable hexose and pentoses by acid hydrolysis under relatively mild operating conditions. They have demonstrated up to 90% conversion of glucose and 86% conversion of xylose to acetic and propionic acids by fermentation with the organism Propionibacterium acidi-propionici.

Wise and Augenstein [21] evaluated the use of woody biomass, including pulp and paper wastes, as well as municipal solid wastes, for acetic acid production by thermophilic "suppressed methane" fermentation.

In their work, the ready availability of these potential feedstocks is
stressed:

Source	Quantiy US tons/yr	% Cellulose	% Hemicellulose
Municipal Solid Waste	140×10^6	61	22
Agricultural Residue	300×10^6	25-50	10-30
Forestry Residue	168×10^6	40-55	24-40

In this important paper, Wise and Augenstein [21] describe a process
which links the technologies of waste stream fermentation to dilute acetic
acid (3.5%) and extraction of acetic acid from the fermenter broth in a
concentration process using liquid ion exchange. Product calcium acetate is
then formed by back extraction of the organic solution of ion exchange acetic
acid complex with calcium hydroxide in a step which regenerates the liquid ion
exchanger. The production of acetic acid from fermentable organic wastes thus
represents an attractive method for conversion of community waste to a
community resource, with the associated benefits of reducing waste disposal
problems, reducing ground water pollution from landfills, and reducing the
expense associated with production of crop-based feedstocks solely for
fermentation. These workers have pointed out that not only are many of these
cellulosic wastes immediately available for processing but also are available
in quantities to meet or exceed the requirements for yearly production of CMA.
A further advantage indicated for municipal solid waste is that the lignin
content is lower than in most plant materials because some of the lignin has
been removed by pulping.

Wise has proposed a CMA production process coupling anaerobic suppressed
methane fermentation for acetic acid production with liquid ion exchange
technology for concentrating the dilute acetic acid, followed by direct
reaction with dolomite, dolime, or other calcium/magnesium substrate resulting
in CMA and regeneration of the liquid ion exchange resin. The CMA is produced
as a concentrated aqueous solution from which it may be isolated. The liquid
ion exchange (LIEx) medium may be a high molecular weight secondary or
tertiary amine or a material, such as trioctylphosphine oxide (TOPO), which is
dissolved in an organic phase, such as kerosene. Extraction of the acetic acid
into the organic phase is a consequence of an acid base reaction resulting in
an acetate salt of the LIEx. The process is rapid and efficient. As
discussed later in this chapter, equilibrium constants of 6.7 for a TOPO-HAc
reaction and 10.9 using the secondary amine Adogen 283-D (Ashland Chemical)
have been demonstrated.

Wise has also demonstrated the versatility of fermentation technology; in the work cited above, the suppressed methane fermentation yielded organic acids with acetic acid accounting for over 85% of the product, the remainder being primarily propionic and butyric. In another study [22], Wise demonstrated that suppressed methane fermentation could be operated to increase production of higher acids by use of 2-bromoethane sulfonic acid as an inhibitor of methanogens. Relative yields of C_2 to C_6 acids obtained in a packed bed column fermenter were as follows:

Acetic	1.0
Propionic	0.035
n-Butyric	1.2
n-Valeric	0.064
n-Caproic	2.7

15.3.3 Status of Production for CMA Deicing Salt

At present, no commercial sources of CMA exist. Calcium magnesium acetate is available as a specialty product. However, it is not available in adequate quantities or at a low enough cost to be considered as a potential raw material for highway deicing use.

The availability of limes (dolomitic or high calcium) is not viewed as a problem. Almost all parts of the United States, including the snow belt states, have adequate existing or potential sources. The acetate sources, however, represent a much different situation and form the basis for potential cost reductions.

The estimated price of calcium acetate from conventional sources is likely to be close to $440/metric ton (20 cents/lb). With a price of $22 to $44/metric ton (1 to 2 cents/lb) for NaCl for road use, it is unlikely that calcium acetate or CMA will ever completely replace NaCl use in the United States, if conventional processes for making acetic acid are used. Because of these high cost estimates, the first uses for CMA are most likely to be on bridge decks (and on roadways at some distance on either side of the bridges, so as to minimize NaCl drag-on by vehicles to the bridges from the roadways).

In a recent analysis of acetate feedstock alternatives for the FHWA, Marynowski et al. [14] arbitrarily selected a figure of 10% market penetration for CMA into the highway deicing chemical market at a substitution ratio of 1.5 weight unit CMA/weight unit NaCl. This represents a potential demand for acetic acid of 1 million metric tons/year, or more than 50% of the acetic acid production capacity in the United States in 1980 and 75% of the 1980 consumption.

With FHWA-specified CMA production facilities sized from 90.7 metric tons/day (100 U.S. tons/day) to 907 metric tons/day (1,000 U.S. tons/day), the acetic acid feedstock requirement for the largest CMA plant would be about 240,000 metric tons/yr (265,000 U.S. tons/yr), which is equivalent to the total output from a large modern acetic acid plant. Therefore, each of the larger CMA plants would probably require a dedicated acetic acid plant.

Another FHWA-specified goal was to develop a process that would allow the manufacture of CMA from feedstocks not derived from petroleum or natural gas. Given this goal, a logical first candidate for a feedstock would be coal, with both methanol and CO being produced from the coal, and with use of the Monsanto methanol carboxylation route for conversion to acetic acid. At the time of initial SRI International analysis in 1982 [14], all U.S. methanol plants used methane as the feedstock, except one new plant based on heavy petroleum liquids. Numerous methanol-from-coal projects have been proposed for fuel production over the last decade, but none have been built and the likelihood of any significant number of such plants being reconstructed during this decade appears remote.

With this background and guidelines, evaluations of acetic acid production from biomass were undertaken (see, e.g., Ref. 1 and 14).

15.3.4 Limited Potential of CMA Production from Smaller-Scale Industrial Wastes

Initially, there appears to be substantial amounts of available industrial organic residues available for direct fermentation to acetic acids [23]. For example, the CMA that could be produced from whey, the principal waste from dairy processing, could satisfy a meaningful fraction of requirements. Moreover, unlike corn or woody biomass, which require some pretreatment prior to acetic acid fermentation, these industrial wastes are already "pretreated" in the sense that direct acetic acid fermentation may be carried out. On this basis, NYSERDA, for example began an investigation of the "easy-to-convert" wastes such as whey. However, overall, sufficient amounts of CMA will not be produced.

On the other hand, woody biomass represents a uniquely large and renewable resource for conversion to acetate. Moreover, large scale production in the frost belt areas may provide needed economic stimulus.

15.3.5 Substantial Potential of CMA Production from Larger Scale Residual Biomass

There are three major sources of residue biomass in the U.S.: agricultural waste, municipal solid waste (including sewage sludge), and forestry residues [23]. Most plant materials contain three major components: cellulose, hemicellulose, and lignin. Cellulose is generally the major component, making up to 25 to 61% of all plant materials. Its chemical structure is similar in all plants and consists of repeating 1-4 alpha-D-glucopyranose units. The structures of the other components, hemicellulose and lignin, differ depending on the plant material. The cellulosic material in municipal solid waste is lower in lignin that most plant materials, since some of the lignin has been removed by pulping. Wood and stems of monocotyledons, which represent many agricultural residues, contain 25 to 40% cellulose, 25-50% hemicelluloses, and 10 to 30% lignin. With the appropriate pretreatment all components are fermentable to acetate, including water soluble aromatics formed by pretreating lignin [24]. In the late 1970's it was determined that residue biomass could be used to partially replace petroleum as source of fuel and chemicals. It was estimated that residue biomass could replace 5 to 10% of the petroleum used in our country. This amount would be enough to produce much of the organic chemicals used by industry. From a practical and economical viewpoint, this appears an optimistic prediction. However, residue biomass could provide help in our current problems with sodium chloride road salt if efficient and economical methods would be found for converting the major components into CMA.

Also, it is to be noted that calculations reveal that all $NaCl/CaCl_2$ used as a road salt could be replaced by acetate salts if the available organic fraction of municipal solid waste (MSW) was used as the feedstock for acetic acid production. However, questions on "carry-through" materials such as trace heavy metals have delayed the investigation of converting MSW to acetate.

Reports by Lars G. Ljungdahl on improved bacterial strains for acetate production [25-27] demonstrate the feasibility of production of CMA from starch and glucose by fermentation using the acetogenic bacteria <u>Clostridium thermoaceticum</u> and <u>C. thermoautotrophicum</u>. Basic research is needed to obtain better bacterial strains and to develop more understanding of the acetogenic fermentation process. While basic research on the fermentation is useful, the technology of the CMA production process also has to be researched and worked out.

It is to be noted that there are several single species of bacteria, both mesophilic and thermophilic that stoichiometrically transform glucose to acetic acid according to the equation:

$$C_6H_{12}O_6 \xrightarrow{2H_2O} 2CH_3COOH + 4H_2 + 2CO_2$$

Unfortunately, none of these organisms possess cellulase, i.e., the enzyme needed to hydrolyze crystalline cellulose to its constituent monomeric unit, glucose [28]. However, there are well established pure cultures of individual species that can be combined to effect the bioconversion of cellulose to glucose and stoichiometric conversion of the glucose sub-units to acetic acid. It is essential to use at least one species that contains an active cellulase and that the catabolic activities of the combined species interact in a manner that leads to a combined fermentation with acetic acid as the sole product [28].

In summary, the advantages of using residue biomass for fermentation and conversion to acetate are these:

(a)   amounts of residue biomass meet or exceed the amounts needed to make acetates for all conceivable salt uses;

(b)   the economics of larger scale use of residue biomass as feedstock must necessarily be more advantageous than the buying of corn from the Midwest;

(c)   larger scale processing costs will necessarily be less than fermentation of isolated industrial wastes streams, which cannot supply sufficient amounts of acetate;

(d)   community goodwill, as well as the economics, should be enhanced in that residue biomass will be perceived as being put to good use (the use of feed corn for road deicing, for example, may actually incur community disapproval in the Northeast, as may the use on the roads of acetate produced from industrial wastes such as cheese whey).

Technical background for fermentation of residue biomass does exist but research is needed on both pure cultures and on combined cultures. Of additional importance is the establishment of engineering parameters such as kinetics and overall percent conversions and research on efficient means for recovery of the acetate.

In the following is a brief summary of the subject project workscope followed by the experimental procedures and results. Based on the experimental results, a process description and preliminary cost analysis are then given. Overall, there appears to be substantial potential for commercialization of calcium magnesium acetate production from regional biomass.

15.4 PROJECT WORKSCOPE

There were a number of specific areas that needed to be investigated in an experimental program to judge overall process feasibility for CMA production from residue biomass. Basically, the proposed process for biomass conversion to calcium magnesium acetate involves three steps:

(1)   bioconversion of biomass to acetic acid,

(2)   liquid ion exchange for extraction and isolation of the acetic acid, and

(3)   production of CMA fom extractant/acetic acid solutions.

Within each of these, particular experimental concerns were identified. Resolution of these provided practical input to a realistic, yet preliminary, process model.

15.4.1 Bioconversion of Residue Biomass to Acetic Acid

In this portion of the research, sewage sludge was subjected to anaerobic fermentation for production of acetic acid. Maximization of acetic acid levels was not the goal at this stage. Rather, these experiments were designed to produce an acetic acid broth of fermentation origin. Preliminary data on reaction rates were of interest, as well as characterizations of the fermentations overall. The subsequent work on extraction and CMA production ultimately required actual fermenter broth acids.

15.4.2 Extraction of Acetic Acid

A second aspect of the work was to obtain information on the liquid ion exchange extraction system for acetic acid. In this system, the fermenter broth, which contains the acetic acid, is extracted with an organic base dissolved in a hydrocarbon solvent. Of interest in this program was the screening of several extractant types for extraction efficiency (as judged by equilibrium values) and the preliminary investigation of solvent/extractant recovery capabilities. This information provided an important link in judging overall feasibility.

15.4.3 Production of CMA

Integral to the process is the actual production of CMA. This portion of the process is based upon the addition of a lime, e.g., dolomite, to the extracted acid. The isolation of CMA was critical for evaluating the product formation and recovery aspects of the general process. This information derived from this experimental work, as well as that from the previous two steps, provided some further advancement toward the solidification of a process design.

15.5 EXPERIMENTAL PROCEDURES AND RESULTS

15.5.1 Bioconversion of Biomass to Acetic Acid

The goal of the fermentation portion of this program was to establish directional objectives for operation of a sewage sludge fermentation for maximum volatile acids output. Initially, rather than directly suppressing methane production, the experimental protocol was geared toward establishing baseline conditions for viability of acid-formers in a sludge fermentation; subsequent work would address culture selection and continuous operation. Thus, batch digesters were used in the assessment of the general "health" of the baseline fermentations.

15.5.2 Experimental Protocols

Sewage sludge was used as the substrate for all the fermentations in this work. Both raw sewage sludge (RSS) and anaerobic digester effluent (ADE) were obtained from the Nut Island Sewage Treatment Plant in Quincy, Massachusetts. (The Nut Island Plant produces primary sludge which is then anaerobically treated in stratified digesters). These samples were collected on a one-time basis, stored in a refrigerator at 4°C, and used within one month. The solution and solids analyses of the raw sludge in the week prior collection are included in Table 15.1; the metals analyses of the sludge and effluent are shown in Table 15.2. Data in each are consistent with expected ranges for raw sludge and effluent in this plant.

TABLE 15.1
Analysis of Raw Sewage Sludge*

Day	pH	Alkalinity, mg/l	Total Solids %	Total Volatile Solids, %
1	5.46	500	5.37	79.73
2	5.62	600	5.31	83.20
3	5.49	600	5.05	82.26
6	5.48	700	6.87	83.85
7	5.50	800	5.47	82.47
8	5.61	500	4.60	83.72

*Courtesy, Nut Island Sewage Treatment Plant, Quincy, MA

TABLE 15.2

Metals Analyses of Raw Sludge and Digester Effluent*

Metal	Raw Sludge, mg/1	Effluent, mg/1
Cr	0.01222	0.0053
Cu	0.0968	0.0559
Cd	0.0052	0.0013
Pb	0.0554	0.0417
Ni	0.0134	0.0158
Hg	0.0116	0.0082
Zn	0.109	0.0784

*Courtesy, Nut Island Sewage Treatment Plant, Quincy, MA

A series of batch reactors were used for the fermentations in this work. An initial set of fermentations were run at 37°C and 60°C using constant RSS/ADE ratios (90/10). A second set were run at a constant thermophilic temperature (60°C) with variable RSS/ADE ratios. A third set were run at 37°C, 50°C, and 60°C at variable RSS/ADE ratios (90/10 and 75/25). A final set were run at a constant mesophilic temperature (37°C) and constant RSS/ADE ratios (75/25). The conditions for each of these sets of fermentations are summarized in Table 15.3. The rationale for each set is as follows: Set (1) baseline feasibility and apparatus check; Set (2) survey of thermophilic conditions; Set (3) verification of microorganism conversion profiles; and Set (4) preparation of "best case" samples.

TABLE 15.3

Batch Fermentation Conditions

Set #	Temp. °C	RSS/ADE by Volume	Number of Digesters
1	60	90/10	2
	37	90/10	2
2	60	90/10	2
	60	60/40	2
3	37	90/10	1
		75/25	2
	50	90/10	1
		75/25	2
	60	90/10	2
		75/25	1
4	37	75/25	2

The batch reactors used in each of the fermentations were 250 ml reagent bottles filled to approximately 75% capacity with RSS/ADE prepared in designated ratios (by volume). After flushing the headspace with nitrogen, each of the reactors was sealed and allowed to digest for a period of 14-28 days (or longer, if maximum anticipated results required so) in an appropriate temperature-controlled incubator. Both the concentration of volatile acids and gas production were monitored regularly, the former using gas chromatographic (GC) methods and the latter using a modified manometric system.

The GC's were run on a Hewlett-Packard 5890 gas chromatograph equipped with a flame ionizator detector. Acids were separated on a HP-FFAP (50m x 0.2mm x 0.33um) column using helium as the carrier gas. The best separation was effected with a program set at an initial oven temperature of 150°C for 12 minutes followed by heating at 50/min to 180°C until completion.

Gas production was checked using a manometer-type measuring device. At the time of measurement, a connecting tube on the "manometer" was attached to a closed port of digestion vessel. The port was opened and the rate of gas production over an hour's sampling time was measured via displacement of water in an attached buret (the manometer).

15.5.3 Fermentation Results

In the fermentations of sewage sludge, batch anaerobic digesters were operated under both thermophilic and mesophilic conditions to establish organic acid formation in the breakdown of the sludge organics. The acid formation is promoted by acid-forming microorganisms (acetogens) intermediate in the overall metabolic pathway to produce the final product gases methane and carbon dioxide. The latter step, conversion of the intermediate acids to the gases is controlled by the methane-formers (methanogens). The general pathway for the conversion of glucose (the hydrolytic breakdown product of cellulose, starches, etc.) is shown in Figure 2. There are several species of bacteria, both thermophilic and mesophilic, that can transform glucose according to this pathway. Sludge digestion is supported by a rich population of microorganisms which are responsible for just this type of conversion.

$$C_6H_{12}O_6 \quad \cdots\cdots \rightarrow \quad 3CH_4 \quad + \quad 3CO_2$$
$$glucose \qquad\qquad\qquad methane$$

$$\downarrow \qquad\qquad\qquad\qquad\qquad\qquad \uparrow$$

$$2H_2 + 2CH_3COCOOH \quad \xrightarrow{2H_2O} \quad 2CH_3COOH \; + \; 4H_2 + 2CO_2$$
$$acetic\ acid$$

Figure 2: Anaerobic Digestion of Glucose

The first set of fermentation experiments (Set #1) was designed to compare conversions at 37°C (mesophilic) and 60°C (thermophilic) conditions. Set #2 was arranged for thermophilic conditions only. In Set #3, variable incubation temperatures were used to distinguish microorganism conversion profiles. The final set (Set #4) was run to collect fermentation broth acids in a "best case" scenario for CMA production and isolation.

The data for the first set of fermentations at 37°C and 60°C, presented in Table 15.4, establish the viability of the goal. As illustrated by these data, acetic acid production was minimal using standard microbial populations at thermophilic temperatures. At 37°C, however, acetic acid production reached the 0.8 - 0.9% level expected using these "standard microbes". While the stoichiometry of the glucose conversion reaction (see Figure 2) would predict a maximum of about 3% acids concentration (measured as acetic acid) for an average 5.5% total solids/80% volatile solids sludge (see Table 15.1), feedback inhibition of the "standard acetogens" by the acids formed is usually observed at approximately 0.9%. Thus, these fermentations (Set #1) served to support the initial suppositions of the fermentation science.

334

TABLE 15.4

Sample Acetic Acid Productions: Fermentation Set #1   (90/10 RSS/ADE)

Temp°C	Day	[Acetic Acid], %*
60	1	0.135
	7	0.075
	8	0.019
	10	0.033
	16	0.052
	18	0.036
60	7	0.118
	8	0.048
37	2	0.72
	5	0.76
	6	0.79
	14	0.84
	20	0.96
37	2	0.78
	5	0.77
	6	0.71
	8	0.75
	20	0.88

*[Acetic Acid] of Initial RSS/ADE at 90/10 was 0.04%

The second set of fermentations (Set #2) were set up at 60°C. These
digesters have variable RSS/ADE ratios (see Table 15.3). Following Set #1, a
survey of the thermophilic condition was required. Again, acetic acid
concentrations were low (0.07 - 0.27% maximum) in all cases. As a check of
general viability gas production rates were regularly monitored. A sample of
these for one case (90/10 RSS/ADE) is shown in Table 15.5. The second set of
digesters, then, showed that, at thermophilic conditions using a standard
"wild type" microbial population, acid pools do not accumulate and conversion
to methane is rapid.

TABLE 15.5

Sample Gas Production Rates: Fermentation Set #2 (90/10 RSS/ADE; 60°C)

Digester #	Day	Gas Production Rate, ml/hr
1	3	12
	4	4
	7	2
	10	8
	11	12

A third set of fermentations (Set #3) was planned to scan a broad matrix of conditions. RSS/ADE ratios, as well as temperature, were varied (see Table 15.3). The data for this set are summarized in Table 15.6. Directly contrasting similar conditions in each of the fermentations emphasizes the trends in these "wild type" sludge microorganisms. The acids pool to the level of feedback inhibition for the acetogens, i.e., 0.9%, at 37°C; while at 60°C conversion of acids is rapid so levels are low, ~ 0.2%. The 50°C fermentations, where "slow thermophiles" and perhaps some "mesophilic remnants" might be expected, acetic acid concentrations are only nominally increased over the thermophilic condition (to ~ 0.3%). The conclusions of these fermentations dictated the conditions for the final set of fermentations.

TABLE 15.6

Sample Acetic Acid Productions: Fermentation Set #3

Temperature, °C	RSS/ADE	DAY	[Acetic Acid],%
60	75/25	8	0.143
		9	0.196
		14	0.133
50	75/25	7	0.154
		8	0.182
		13	0.294
37	75/25	6	0.212
		8	0.437
		12	0.818

In Set #4 of the fermentations, two larger scale (i.e., 500 ml) batch digesters were assembled for incubation at 37°C. The objective of this part of the work was to generate sufficient quantities of fermentation broths for extraction of acetic acid and production of CMA; that is, for an overall feasibility trial. The data for these two digesters are summarized in

Table 15.7. Although these digesters were slower in starting (most probably because of the month-long refrigeration of substrate), both were judged "healthy" by both gas production and eventual acetic acid levels. Enough sample was then available for further work on acetic acid extraction and CMA production using actual fermentation samples.

TABLE 15.7

Gas and Acids Productions: Fermentation Set #4   (75/25 RSS/ADE; 37°C)

Digester #	Day	[Acetic Acid], %	Gas Production Rate, ml/hr
1	1	0.084	–
	2	0.228	4
	3	0.157	1
	4	0.150	2
	7	–	3
	8	0.132	3
	9	0.196	3
	10	0	2
	11	0.432	2
	21	0.749	1
2	1	0.095	–
	2	0.228	4
	3	0.154	3
	4	0.080	2
	7	–	3
	8	0.213	4
	9	0.331	2
	10	–	1
	11	0.439	2
	21	0.842	2

15.5.4 Extraction of Acetic Acid

The second step in the production of CMA from sewage sludge biomass is the extraction of acetic acid from the fermentation broth. The acetic acid produced in the anaerobic fermentation is concentrated and isolated from the broth by use of a liquid ion exchange resin (Ex). The liquid ion exchange medium consists of an organic base, such as trioctylphosphine oxide (TOPO), or an amine, such as Adogen 283-D[tm] (Ashland Chemical), a branched di-tridecylamine, dissolved in a suitable hydrocarbon solvent, such as kerosene. Following the extraction of the acetic acid with the organic base, a CMA is produced in a final step by reaction of the acetic acid-rich organic phase with a calcium and/or a magnesium containing substrate (e.g., dolomite).

Ultimately, acetic acid concentrations of about 3% in the broth are anticipated based on the conversion stoichiometry shown in Figure 2. In the basic fermentation experiments of this work acetic acid concentrations of 0.8%

were obtained as expected. The extraction aspects of this work, then, focused on three different "acid broths" to check out the fundamental aspects of the extraction step. In the first case, standard 3% aqueous acetic acid solutions were used; these provided data on equilibrium distributions of the aqueous acid using various organic extractants. The second case used "spiked" fermentation broths; these provided additional data on the equilibria of extractions, particularly as might be influenced by broth debris. The final case exploited the actual fermentation broths produced in the digestion experiments; these were focused on providing a view of the overall feasibility of the proposed CMA process.

15.5.5 Experimental Protocols

The experimental protocols for the extraction work using any of the sludge broths (either HAc spiked or actual fermentation samples) were derived from preliminary work with standard 3% aqueous acetic acid samples. In this work, various extractants were first surveyed in a general sense. Then, more specific work with select extractants was carried out, the objective being to provide the fundamental data on acid equilibria.

Extractants which were explored in the work using standard acetic acid solutions included long chain aliphatic secondary and tertiary amines with low water solubilities, as well as a basic phosphine oxide. These are identified by trade name, chemical composition and chemical type in Table 15.8.

TABLE 15.8
Extractants Investigated

Extractant	Chemical Identity	Type*
Adogen 283-D[1]	di-tridecyl amine, branched	II° amine
Amberlite LA2[2]	branched	II° amine
Adogen 381[1]	highly branched II° amine	III° amine
Alamine 336[3]	tri-osooctyl amine	III° amine
TOPO[4]	tri-C_8, C_{10} straight chain	Basic Phosphine Oxide
	tri-n-octyl phosphine oxide	

1. Ashland Chemical	*Symbol: II° = Secondary
2. Rohm and Haas	III° = Tertiary
3. Henkel	
4. American Cyanamid	

338

Work with acetic acid focused on measuring equilibrium constants for the extraction process, a more satisfactory measure than the conventionally reported distribution coefficient. A series of liquid ion exchange materials (extractant) dissolved in kerosene was used to extract 3% aqueous acetic acid solutions.

The process for the formation of the extractant-acetic acid adduct is described by the following reaction scheme:

$$HAc(aq) + Ex_{(org)} \rightarrow Ex \cdot HAc_{(org)}$$

where

HAc =   acetic acid

Ex   =   extractant

The equilibrium constant, K_{eq}, for the above is defined as:

$$K_{eq} = \frac{[Ex \cdot HAc]_{org}}{[HAc]_{aq}[Ex]}$$

where [ ] indicate molar quantities at equilibrium. Each phase (aqueous and organic) was titrated with sodium hydroxide solution to determine the efficiency of extraction and from these data equilibrium constants for the extraction was determined.

The equilibrium constant for the extraction of acetic acid (HAc) by triocylphosphine oxide (TOPO) was investigated over a four-fold variation in the initial [HAc]/[TOPO] ratio. In that range a standard deviation from the norm of 21.2% was observed with no discernable trend in results. Results for this series of experiments is reported in Table 15.9.

TABLE 15.9

K_{eq} for Extraction of HAc by TOPO

Ratio of Initial [HAc] to Initial [TOPO]	K_{eq}
0.60	6.19
1.20	5.00
1.20	7.05
1.25	8.78
2.00	5.13
2.40	8.17
Mean 6.72 ± 1.43 (σ)	

K_{eq} values for the other extractants were also determined and are reported in Table 15.10. The two most promising candidates were TOPO (K_{eq} = 6.72) and Adogen 283-D (K_{eq} = 10.9). Note that the greater the values of K, the greater is the fraction of HAc removed from the aqueous phase. Thus Adogen 283-D and TOPO were the extractants considered for further work; TOPO was used in the broth extractions.

TABLE 15.10

K_{eq} for Extraction of HAc by TOPO

Extractant	K_{eq}
TOPO	6.72 $\pm$ 1.43*
Adogen 283	10.9 $\pm$ 0.1**
Adogen 381	0.15
Alamame 336	0.42
Amberlite LA-2	2.62

*6 measurements
** measurements; ratio of initial [HAc] to [TOPO] varied from 1.0 to 2.5

Following the results obtained from the standardized solutions, a protocol was established for extraction of the broth-derived samples. In general, a known volume of a fermentation sample was centrifuged and the broth was recovered by decantation. The broth was titrated against 0.851 N sodium hydroxide solution for total acids concentration. (A check of the actual acetic acid concentration was done using GC.) A known volume of this broth was then extracted with twice the volume of a TOPO/kerosene solution which was prepared at a TOPO concentration of stoichiometric equivalence to the broth acids. After thoroughly mixing the broth and TOPO/kerosene mixture, the layers (i.e, the aqueous with residual acid and the organic with TOPO·acid adduct) were separated, and each layer titrated against the sodium hydroxide for acids concentrations.

15.5.6 Extraction Results

The results of three of the broth extractions are presented in Table 15.11. Equilibrium constants, calculated from titration data (spiked sample and sample 5-C) or from material balance closure (Sample 1-D) show excellent agreement with the mean value for K_{eq} reported in Table 15.9 for extraction of acetic acid solutions. (The mean value of K_{eq} for extraction from broth was 6.46 $\pm$ 0.93 as compared with 6.72 $\pm$ 1.43 for acetic acid solutions.)

This close correlation between sludge fermentation extractions and model system extractions suggest that preliminary protocols for TOPO extractions are valid and may with confidence be applied to future work.

TABLE 15.11

Equilibrium Constants for Extractions of Fermentation Broths

Sample	Initial [Acids]* in Broth, M	[Acids] in Aqueous Layer after Extraction, M	[Acids] in TOPO/Kerosene Layer, M	[TOPO] in Kerosene, M	K_{eq}**
1. RSS/ADE Broth 3% HAc Spike	0.578	0.306	0.170	0.250	6.94
2. Fermentation Broth 37°C (Sample 1-D)	0.136	0.085	0.026	0.068	7.28
3. Fermentation Broth 37°C (Sample 5-C)	0.145	0.050	0.017	0.083	5.15

Mean Value for K_{eq} = 6.46 ± 0.93(σ)

*[Acids] determined by titration

$$**K_{eq} = \frac{[TOPO\text{-}Acids]org}{[Acids]aq[TOPO]}$$

15.5.7 Production of CMA

The final step in the CMA proce: he CMA
from the extracted broth. Following the   ell as
other acids and debris) from the fermenta   anic
phases are separated. The organic phase,
TOPO·HAc, is neutralized by direct contac   ium
containing substrate such as limestone (c
(calcium magnesium carbonate). This reac... ....,
which is recycled, and simultaneously produces an acetate salt (calcium
acetate from limestone; calcium magnesium acetate from dolomite). The reaction
for this stage of the process is:

$$4Ex\cdot HAc + CaMg(CO_3)_2 \rightarrow 4Ex + CaMgAc_4 + 2CO_2 + 2H_2O$$

15.5.8 Experimental Protocols

As with the extraction stage of the process, the experimental protocols
for the production of CMA from extracted fermenter broths were derived from
preliminary work using standardized acetic acid solutions with, in some cases,
calcium hydroxide or calcium carbonate. In this work, the stoichiometry of
reaction was investigated and the overall reaction was proven.

The reaction product obtained on reaction of Ex HAc with calcium
hydroxide and calcium carbonate was contained in an aqueous layer which was

separated from the organic. The salt was isolated after evaporation of the water. The white crystals were soluble in water, decomposed on heating without melting, and when heated with ethanol and concentrated sulfuric acid gave the characteristic odor of ethyl acetate. Although not taken as conclusive evidence, the product was identified with a high degree of assurance as calcium acetate.

Native dolomite (calcium magnesium carbonate) will react directly with an organic phase containing TOPO·HAc. A solution (25ml) of 0.50M HAc was shaken with 50ml of TOPO dissolved in kerosene (0.25M). The clear organic phase containing the TOPO·HAc adduct was removed and to this was added crushed dolomite in excess of the stoichiometric quantity required. The organic layer was shaken with the dolomite for varying periods of time. The unreacted dolomite was recovered, washed, dried and weighed and from this was calculated the percent of the theoretical amount of dolomite which reacted. In Table 15.12 are summarized the results for three of these trials. In Trial 1, conducted at room temperature, agitation was continued for only 5 minutes; nevertheless 37% of the theoretical weight of dolomite reacted. This suggested that contact times of less than an hour may be sufficient for essentially complete reaction. Mechanical agitation was employed in Trials 2 and 3, the former at room temperature and the latter at 60°C. In both cases, the theoretically calculated weights of dolomite reacted. It should be pointed out that the shaking times indicated for Trials 2 and 3 in Table 15.12 were much greater than required, but clearly demonstrate that quantitative conversion to acetate is quite feasible.

TABLE 15.12

Reaction of TOPO·HAc with Dolomite

Aqueous phase: 25 ml of 0.50 M HAc
Organic phase: 50 ml of 0.25 M TOPO
K_{eq} = 6.72

Theoretical weight of dolomite required to react with TOPO·HAc = 0.336 g

Trial	Temp.	Contact Time	Weight Initial	Dolomite Recovered	Grams Reacted	Percent of Theoretical
1	Room Temp.	5 min.	2.574	2.450	0.124	36.5
2	Room Temp.	18 hr.	2.512	2.132	0.380	111
3	60°C	60 hr.	2.571	2.226	0.345	101.6

Recovery of the insoluble material from the aqueous layer (before evaporation to give the CMA) followed by analysis indicated that the solid was unreacted dolomite (93.4% $CaMg(CO_3)_2$). Thus, the protocol established that the designated conditions would result in a "clean" process as judged by the reaction products (i.e., regenerated extractant, CMA, water, and carbon dioxide).

The results of this work with the standardized acetic acid solutions produced a protocol for the production of CMA from the fermentation extractions. Basically, an aqueous slurry of dolomite (minimally, a four-fold stoichiometric excess relative to organic acids concentrations) was combined with the kerosene mixture. This was shaken at room temperature for several hours. The aqueous fraction was separated, filtered, and evaporated to dryness for recovery of the salt. Additionally, the overall recovery of acetate from the original broth acetic acid was measured via GC of an acidified aliquot of the aqueous layer.

15.5.9 CMA Production Results

The results of the CMA isolations for fermentation-related samples are shown in Table 15.13.

Based on these three samples, the yields of CMA ranged from 34-100% calculated on the basis of the TOPO·HAc adduct in the organic phase. These yields are quite encouraging in this "one-pass" feasibility test.

One pertinent comment concerns the range of yields represented in the table of results. Actually, it is not a range of yields but two different yields -- nominally 35% for the first two cases and 100% for the third case -- which are represented. The explanation for this lies in the experimental protocol. In the two 35% cases, the contact time between the dolomite slurry and the TOPO·HAc mixture was on the order of three hours; in the 100% case, the contact time was significantly longer (approximately 16 hours). In this feasibility survey, it is important to highlight this contrast in the yields. As with the preliminary work with the standards, the CMA production step of the process is quite noticeably dependent upon the contact between the dolomite and the organic phase. The yield, in these cases, was improved by extending the reaction times. In a process sense, reaction time might be balanced with a change in the particle size of the dolomite or in the slurry preparation.

Having produced the supposed CMA from a fermentation sample, additional work was focused on characterization of the solid as CMA. The most definitive work used GC analysis to analyze the solid in a series of "reconstitution" experiments. In these, a known portion of the CMA was dissolved in a measured amount of water. The sample was acidified and a portion of this was chromatographed. In the first experiment where RSS/ADE was spiked with acetic

343

TABLE 15.13

Production of CMA from Fermentation Samples

Sample	[TOPO·HAc], M	Vol. Organic Phase, ml	Recovered CMA, gm.	Theoretical CMA, gm.	Percent Yield
1. RSS/ADE 3% HAc Spike	0.170	25	0.109[1]	0.319[2]	34.2
2. Fermentation Broth 37°C(1-D)	0.026	37	0.0256[3]	0.0723[4]	35.4
3.Fermentation Broth 37°C (5-C)	0.017	172	0.221[5]	0.220[6]	100.5

Reaction Stoichiometry:

HAc + TOPO → TOPO · HAc

$4TOPO \cdot HAc + CaMg(CO_3)_2 \rightarrow 4\ TOPO + CaMg(Ac)_4 + 2CO_2 + 2H_2O$

[1] Total volume aq. phase recovered from reaction with dolomite = 25 ml;
CMA recovered from 20 ml aliquot = 0.087 gm;
Total CMA = 0.087 (25/20) = 0.109 gm.

[2] Theoretical CMA = (300.57 gm/mole CMA) (1 mole CMA/4 moles HAc) (0.170 moles HAc/liter) (0.025 liters) = 0.319 gm CMA

[3] Total volume aq. phase recovered from reaction with dolomite = 16 ml;
CMA recovered from 5 ml aliquot = 0.008 gm.; Total CMA = 0.008 (16/5) = 0.0256 gm.

[4] Theoretical CMA = (300.57)(1/4)(0.026)(0.037) = 0.0723 gm

[5] Total volume aq. phase recovered from reaction with dolomite = 41.5 ml;
CMA recovered from 12.2 ml aliquot = 0.0649 gm.; Total CMA = 0.0649
(41.5/12.5) = 0.221 gm.

[6] Theoretical CMA = (300.57)(1/4)(0.057)(0.172) = 0.220gm.

acid, the acidification of the CMA solution showed the presence of acetic acid. This is not surprising, but what was of more interest was the reconstitution of actual fermentation-derived CMA samples. The GC's of these samples showed not only the presence of acetic acid from the acetate salt, but also propionic acid and some butyric acid also. The acetate portion of the salt appears to be roughly equal to the propionate portion, and the butyrate portion considerably less than the others (about 20% of the total). Thus, all of the lower volatile acids in a fermentation broth seem to play a role in the CMA process surviving not only the extraction step, but also the actual CMA production step. The choices for this process are then to select for maximum acetic acid yields in the fermentation step, modify the extraction for selective removal of the acetic acid, or use a calcium magnesium mixed acid salt. If the latter can effectively do the deicing job, as would most likely be the case, this is probably the best of all choices, although on a weight basis a mixed acid salt would not as efficiently use the organics.

While the GC data provided supportive evidence for actual CMA production, additional interest was centered on the analysis of the product for contaminants. Because the CMA sample was derived from a sewage sludge fermentation, there can be some concern about the carry-over of sludge contaminants into the CMA. Of particular environmental concern is the contamination of CMA with heavy metals. In the proposed process scheme, cases can be made for either heavy metal elimination or heavy metal concentration based on the fermentation conditions (e.g., pH), the extraction protocol (e.g., use of organic solvents), or the CMA production procedures (e.g., the select exchange using dolomite). However, analysis of the product (as well as, ultimately, the product streams) is the only way to resolve the question of heavy metal contamination.

One such analysis is presented in Table 15.14 for the case where CMA was produced from a fermentation broth (5-C). Interestingly, in all but two cases, concentrations of heavy metals above the threshold levels were not detected in an aqueous sample of an aqua regia digest of the CMA. Either samples were too dilute for analysis or the overall processing scheme selects for metal elimination. The latter is a likely conclusion when examining the results for chromium or lead. Concentrations of chromium were higher than initial, but close to the sensitivity limits suggesting some reason to watch for chromium. Concentrations of lead were very much higher than the initial raising issue with potential lead problems. While the process seemed to select out several of the heavy metals, it also seemed to ''select in" lead, if not chromium also. Further work would have to address if this is a reproducible result and, if so, identify the contaminating source. The carry-over may be from the sludge or from the dolomite.

TABLE 15.14
Metal Analysis of CMA*

Metal	Concentration, mg/l	Sensitivity Limits, mg/l
Cr	0.066	0.06
Cu	0.21	0.03
Cd	-0.10	0.01
Pb	0.236	0.10
Ni	-0.107	0.06
Ag	0.000	0.03
Zn	-0.048	0.008

*Courtesy, Nut Island Sewage Treatment Plant, Quincy, MA

15.6 OVERALL PROCESS DESCRIPTION

In conjunction with the experimental work, a preliminary engineering
evaluation of a competitive process for the conversion of residue biomass,
such as sewage sludge and MSW, to calcium magnesium acetate was done. In the
following is a brief summary of the overall process for bioconversion of
residue biomass to CMA, concluding with a preliminary cost analysis.

15.6.1 Overall Process Concept

The conversion process described makes use of residue biomass, such as
sewage sludge and MSW. In the process, the residue biomass is first
pretreated using available equipment.   The pretreated residue biomass is then
anaerobically fermented to acetic acid in a "suppressed methane" fermenter.

It is assumed that the fermentation will be carried out under
thermophilic conditions using non-sterile cultures of microorganisms derived
from sewage sludge digester effluent, such as Clostridium thermoaceticum. The
acetic acid is recovered from the fermenter broth using continuous liquid ion
exchange extraction. The acetic acid is separated into aqueous medium and
then contacted with dolomite (a limestone of 50:50 atomic ratio
calcium/magnesium).   With some final drying of the spray dried calcium
magnesium acetate, the process is complete. Conventional processes and
equipment are used throughout.

15.6.2 Process Description of Calcium Magnesium Acetate Production from Residue Biomass

The process to produce calcium magnesium acetate deicing salt from residue biomass will require the following basic unit operations.

1. Collection of residue biomass at a receiving/storage facility.

2. Pretreatment of the residue biomass feedstock by a suitable process to render the biodegradable components accessible and fermentable.

3. Fermentation of residue biomass in a "suppressed methane" fermenter. Acetic acid formed in fermentation is continuously removed (see 4 directly following) and the aqueous phase is recycled to the fermenter. Spent biomass feedstock is removed from the fermenter, and after a wash step to remove residual acetic acid, is burned in a boiler (similar to the type used to process spent bagasse in sugar mills) to supply process heat as steam and hot air.

4. Removal of acetic acid in an external column by use of appropriate liquid ion exchange (LIEx).

5. Back extraction of acetic acid from the LIEx by reaction with a dolomite (limestone) slurry. This step takes place in a series of mixers such that the process is essentially concurrent, and so that a highly concentrated (30+%) calcium magnesium acetate solution and lean ion exchanger result (e.g. no solids remain). The lean ion exchanger is returned to the acetic acid extraction column.

6. Spray drying of calcium magnesium acetate solution to a dry powder, and packaging for shipping.

15.6.3 Process Choices

Before process unit operations are discussed in detail certain critical process choices deserve discussion here. One process choice has been to produce calcium magnesium acetate from dolomite, rather than from other materials (such as calcium acetate, magnesium acetate, or other atomic ratios of calcium and magnesium). In the experimental work, it was found that the dolomite reacted at an acceptable rate to permit back extraction of acetic acid from the LIEx as described below. Liquid ion exchange has been selected for separation because the relatively dilute acetic acid stream produced by residue biomass fermentation can, with back extraction by limestone, result in a highly concentrated calcium magnesium acetate solution at relatively high pH, and problems of other approaches (membrane or multiple effect evaporation) are avoided. The product solution is suitable for spray drying directly. The justification for liquid ion exchange is presented in detail later. Other process choices are straightforward and discussed below.

15.6.4 Material Flows, Process Performance and Yield
        Major process aspects will be discussed in turn. Summary of capital
and operating cost estimates are given later in Tables 15 and 16.

15.6.5 Feedstock Consumption and Yield
        As the basis for evaluating the performance of the facility for the
production of calcium magnesium acetate deicing salt, it is assumed that
calcium magnesium acetate production capacity will be 454 metric tons/day (500
U.S. tons/day). (Plant sizes ranging between 91 and 910 metric tons/day (100
and 1000 U.S. tons/day) have been specified for consideration by the Federal
Highway Administration.) In evaluating the process, the overall weight yield
of acetic acid on residue biomass is assumed to be 50%; this is the net yield
after allowing for process losses in wash steps and elsewhere. This
assumption should be reasonable; thermophilic (60°C) operation has been
established in earlier work to yield acetic acid quantitatively from material
converted, e.g., no higher product acids such as propionic or butyric are
formed. Furthermore an acetic acid product yield of 50% is conservative; it
would represent a net yield of 60% of stoichiometric yield based on
cellulosics (which are themselves 75% by dry weight fraction of the biomass).
There should be no obstacle to such yields, which have been equalled or
exceeded by various workers with numerous cellulosic substrates [10]. With
this yield, production of 454 metric tons/day (500 U.S. tons/day) calcium
magnesium acetate deicing salt will require a 40% moisture residue biomass
feed rate of 1148 metric tons/day (1265 U.S. tons/day).

15.6.6 Feedstock Pretreatment
        Prior to fermentation, a residue biomass feedstock such as MSW may
require pretreatment. An example pretreatment for MSW may be the "Iotech
process", which has been developed by the Iogen Corporation of Gloucester,
Ontario, Canada [29]. In this process biomass is "exploded" through a
sequence of exposure to high pressure steam, followed by decompression. This
renders the biomass cellulosic component accessible and fermentable. While
Iogen, the developer, keeps many process details proprietary, they estimate
the complete pretreatment cost to be $22/metric ton ($20/U.S. ton) which comes
to $36/dry metric ton ($33/dry U.S. ton) on the process scale being
considered. Steam consumption can be estimated on sensible heat requirements
at 0.25 kg steam/kg dry biomass. The pretreated MSW material is delivered to
the fermenter by means of a screw conveyor.
        There may also be utility in the pretreatment of raw sewage sludge to
facilitate "suppressed methane" fermentation to acetic acid. For example,

work by Leuschner et al. (preliminary report from NYSERDA, Albany, NY) indicates that acid, base, and neutral pH pretreatment of anaeroic digester effluent at temperatures of approximately 125 - 150°C for short retention times of from 20 - 100 seconds (via steam injection) results in anaerobic methane fermentation of over 85% total volatile solids. Moreover, the material from the pretreatment unit is "solubilized" enabling a high-rate anaerobic filter type fermenter to be used.

In addition, work by Wise et al. (see list of references) has shown that MSW, combined with sewage sludge and anaerobic digester effluent, results in a significant decrease in mixing torque versus time. It was found that such a mixture could not only be readily mixed, but pumped, a practically significant process factor. This earlier work points to a simple MSW/sewage sludge/digester effluent holding tank as a potentially useful "pretreatment".

15.6.7 Estimate of Fermenter Size

A critical factor in the overall economics of calcium magnesium acetate production will be the cost of reactor needed to carry out the fermentation step. Reactor size is fixed by the volumetric productivity and the desired output rate. Note that in this particular fermenter, the feedstock is undergoing "suppressed methane" anaerobic fermentation. Application of this fermenter has been to municipal solid waste [30, 31] to aquatic biomass [32] and even to pretreated fossil fuels such as peat and lignite [33]. Closely related work using whey has been investigated by Wise et al. [34]. The formation of acetate under these conditions is also described by Ljungdahl [35]. More recently, several Japanese workers have investigated acetic acid production that is supportive of this fermentation system [36].

For purposes of this evaluation, it will be assumed that the value for productivity which may be attained is that implicit from kinetic information presented by Ashare et al. [37]. A correction factor is applied for inhibition of product formation by presence of acid product in the reactor. Rate constants near 0.65 day^{-1} have been observed [37] for anaerobic digestion of a variety of substrates to methane. It is assumed that product inhibition in the acid generation step will reduce rates and this rate constant by 75%, to 0.163 day^{-1}. (High product acid levels and low pH in the fermentation broth will be necessary to facilitate the acetic acid extraction by liquid ion exchange.) With this rate constant, the time for 65% conversion of biodegradable solids will be 6.5 days. A further assumption is that the initial solids loading in the fermenter will be 15% by weight. At this initial solids loading, with a dry feed rate of 689 metric tons/day, (759 U.S. tons/day) the reactor volume is computed, with allowance for effect of

consumption of biodegradables, to be 4.5×10^7 liters (1.6×10^6 ft^3).
Because this fermenter may be of simple construction, that is, a set of large
tanks through which liquid circulates, cost per unit volume is low, estimated
at 7 cents/liter (27 cents/gal) for a total of $3,200,000. This cost estimate
-- and those to follow -- includes piping and auxiliaries, but not
engineering, design, and other factors which are itemized separately in the
overall cost summary.

15.6.8 Liquid Ion Exchange Extraction Systems for Acetic Acid Separation

The conventional method employed for separation of acetic acid from
aqueous solutions in concentrations up to 35% acetic acid w/w involve
extraction followed by regeneration of the solvent and recovery of the acetic
acid from the solvent [38]. Since the solvents used tend to have substantial
water solubility themselves, there is also usually some means for removal of
residual solvent from the raffinate. These processes involve use of acetates
or ethers as the extractant with equilibrium distribution coefficients, K_D, in
the range of 0.89 - 0.14 (expressed in terms of weight fraction of acetic acid
in the solvent phase divided by weight fraction of acetic acid in the aqueous
phase at equilibrium). These low values of K_D lead to high solvent to feed
flow ratios and large costs in dewatering and solvent recovery by azeotropic
distillation [39-41].

The best opportunity for reducing the cost of extraction of acetic acid,
or for reducing the lower limit feed concentration for economic recovery, lie
in the development of solvents giving substantially higher K_D and therefore
reducing the large solvent-circulation rates required.

Higher values of K_D can be obtained with chemically complexing
extractants -- notably strong, organic Lewis bases, which can undergo
acid-base complexing with acetic acid. Because of the complexing, such
extractants can also be more selective for acetic acid over water. While more
costly than conventional solvents, a number of strong Lewis-base extractants
have been developed and used commercially for some years in hydrometallurgical
processing (see Tavlarides et al. [42]).

The principal complexing extractants explored to date have been
phosphoryl compounds and amines. Tri-n-butyl phosphate gives K_D of about 2.3.
The phosphoryl oxygen becomes a stronger electron donor with removal of
oxygens from the linkages to the phosphorous atom. Trioctylphosphine oxide
(TOPO) has been studied extensively but has not been used on a large scale.
Tertiary amines are as strong, or stronger, extractants for acetic acid than
are phosphine oxides and are less costly and are available in the Alamine and
Adogen series.

Both TOPO and amines should be used in solvent mixtures with one or more diluents in order to (a) reduce viscosity and control density differences and surface tension, (b) reduce the reboiler temperature for the regeneration column, and (c) increase K_D since these pure extractants are poor solvents for the acid-base complexes formed with the acetic acid. Efforts to define systems lead to di-isobutyl ketone (DIBK) as a diluent having favorable properties of yielding a relatively high K_D and low volatility relative to acetic acid. This latter property is important for solvent regeneration by distillation in the presence of the extractant. For example, the relative volatility of acetic acid to DIBK is about 315 for a solvent containing 40% Alamine 336 in DIBK. Values of K_D for acetic acid extraction are about 2.5 for both 50% Alamine 336 and TOPO in 2-heptanone. These systems show promise for the extraction of acetic acid from dilute aqueous solutions.

15.6.9 Separation Considerations

Separation, which in this case comprises extraction of dilute acetic acid from the fermentation broth and the ultimate production of a concentrated calcium magnesium acetate solution, could be accomplished by several approaches. Two possibilities which can be considered start with neutralizing the fermenter output with lime. Resulting dilute calcium magnesium acetate could then be concentrated by either multiple effect evaporation or membrane (reverse osmosis) approaches. However, with multiple effect evaporation, the decreasing solubility of calcium magnesium acetate with increasing temperature, as well as low concentration of starting material and presence of a variety of organics, would lead to high steam consumption and potentially severe fouling problems. Other cations and organics in the fermenter broth would carry through into the final product as well, with unknown effect. Fouling by fermenter organics in membrane concentration approaches -- in combination with difficulty with membranes, of achieving the osmotic strength represented by the 30+% calcium magnesium acetate desired -- argued against this approach as well. The best process device appears to be generation of concentrated calcium magnesium acetate from dilute fermenter broth through application of liquid ion exchange (LIEx).

15.6.10 Spray Drying

Upon separation, the 35% calcium magnesium acetate solution will be spray dried. Sufficient capacity will be afforded by ten 5.5 meters (18 ft) diameter units along with an eleventh which is spare [43]. At an installed cost of $700,000/unit, the spray dryer costs are estimated to total $7,700,000.

15.6.11 Boiler

The approximately 50% by weight (dry basis) of the feedstock remaining as residue from the fermentation will be available as boiler fuel, at a relatively high moisture level. (Indeed, using this material as fuel avoids a further disposal problem.) The boiler will supply process steam for the pretreatment, as well as heated air (either flue gas or more likely by indirect exchange) for the spray drying step. In costing this boiler, it is assumed that the cost per unit capacity is similar to costs for boilers adapted to burning bagasse in sugar mills. Cost for a boiler capable of burning 1084 metric tons/day at 35% solids (380 dry U. S. tons/day of residue) are estimated at $5,000,000 as derived from data presented by the F. C. Schaefer Company [44].

It is of interest that combustion of the process residue can provide the 0.25 kg steam/kg process residue for the pretreatment as well as 1.16 kwh/kg (1800 BTU/lb) of water to be evaporated in the spray drying step. This is assuming a 75% boiler efficiency based on net heating value *after* compensating for the 1.2 kwh/kg (1860 BTU/lb) biomass lost to evaporating residue moisture. Thus the plant should be very near energy self-sufficiency and the situation improves if residue moisture can be reduced. Electrical power (1000 kw average) is assumed to be provided by outside utilities, but other supplemental energy and fuel requirements will be minimal.

15.6.12 Other Capital Costs

An office and building complex will be required in addition to process equipment. This is estimated to cost $1,000,000. Another significant cost item is the liquid ion exchanger, because of the large holding in back extraction. For the analysis it is computed that the plant inventory is 550 liters which at $10/liter ($38/gal) has a total cost of $5,500,000. Other miscellaneous items such as the screw conveyor, dewatering, and wash steps must be included [45].

15.6.13 Summary of Cost Items: Capital and Operating

The capital costs breakdown for the facility (summarized in Table 15.15) is estimated to be $46,200,000. Operating costs (summarized in Table 15.16) are estimated to be +$24,730,000 for Case A in which residue biomass must be purchased at an estimated cost of $20/U.S. ton and -$4,320,000 for Case B where a credit is taken for receiving residue biomass of $50/U.S. ton. A capital recovery factor of 0.3, exclusive of operating costs (reflecting 50% debt/50% equity, a 10% interest on debt, and an ROI of 15+% at tax rates below 48%), gives a capital related cost component of +9.3 cent/kg (4.22 cents/lb).

The sum of 8.3 cents/kg (4.2 cents/lb) capital related costs and Case A: +16.5 cents/kg (+7.5 cents/lb); Case B: – 2.9 cents/kg (-1.3 cents/lb) operating costs gives a total production cost for calcium magnesium acetate deicing salt of: Case A: 25.8 cents/kg (11.7 cents/lb); Case B: 5.4 cents/kg (2.9 cents/lb). In summary, the preliminary cost estimate indicates that CMA may be produced for $234/U.S. ton if residue biomass must be purchased for $20/U.S. ton. On the other hand, if a credit of $50/U.S. ton may be taken for disposal of the feedstock residue biomass, then CMA may be produced for an estimated $58/U.S. ton. A closely related cost analysis dealing with woody biomass was presented at the American Solar Energy Society meeting June 1988 [46] and an extended paper was published in *Solar Energy*, the archival journal of the *International Solar Energy Society* [21].

TABLE 15.15

Estimate of Capital Costs for Facility to Produce 454 Metric Tons/Day (550 U.S. Tons/day) of Calcium Magnesium Acetate Deicing Salt

Item	Cost ($U.S.)
Receiving Area	$ 250,000
Screw Conveyor	300,000
Fermenter	3,200,000
Acetic Acid Extraction Column $(500M^3; 20M^2 C.S.A.$	2,000,000
Back Extraction (5 x 150,000 liter tanks)	1,400,000
Settling/Separation Tank	700,000
Wash Tanᴸs	2,500,000
Final Dewatering	2,500,000
Spray Dryers	7,700,000
Boiler	5,000,000
Product Storage	200,000
Building	1,000,000
Total Equipment Cost	26,750,000
Contractors' Overhead and Profit	2,675,000
Engineering and Design	1,338,000
Subtotal Plant Investment	30,763,000
Contingency	3,076,000
	33,839,000
Interest During Construction	4,061,000
Start-up (Includes LIEx Inventory Purchase)	7,500,000
Working Capital	800,000
TOTAL Capital Requirements	$46,200,000

TABLE 15.16

Estimate Annual Operating Costs for Facility to Produce 454 Metric Tons/Day
(500 U.S. Tons/Day) Calcium Magnesium Acetate Deicing Salt (90% Plant Service
Factor; $1.49 times 10^5 Metric Tons (164,250 U.S. Tons/Year) Output)

Item	Cost/Year ($U.S.)
Feedstock:	
Case A (Penalty): Purchase Residue Biomass @ $22/metric ton ($20/U.S. ton)	+$8,300,000
Case B (Credit): Receive Residue Biomass @ $55/metric ton ($50/U.S. ton)	-$20,750,000
Pretreatment $22/metric ton ($20 U.S. ton)	8,300,000
Calcium Oxide (powder)	2,330,00
Labor:	
Operating	960,000
Maintenance	300,000
Supervisory	300,000
Administration and Overhead	1,200,000
Utilities:	
Electirc (1000 kw) at 8 cents/kwh	700,000
Supplemental Fuel	200,000
Supplies:	
Operating (Includes LIEx makeup)	600,000
Maintenance	200,000
Local Taxes and Insurance	1,340,000
TOTAL Operating Cost/Year Case A:	$+24,730,000
Case B:	$-4,320,000

354

15.7 OVERALL SUMMARY OF RESULTS

The work specifically detailed in this chapter focused on evaluating the feasibility of producing CMA from sewage sludge. The evaluation was based on preliminary experimental work in each of the following three areas: (1) fermentation of sewage sludge for production of volatile acids, most importantly acetic acid; (2) liquid ion exchange extraction of acetic acid from the fermentation broth, and (3) production of CMA from extracted acetic acid. Overall, it was demonstrated that CMA can be produced from the fermentation of sewage sludge; thus feasibility has been demonstrated from a scientific perspective. Further, a preliminary engineering cost analysis indicates that CMA may produce at competitive costs, provided a creidt is taken for receiving this type of feedstock.

It is evident that additional research work is required to address optimization of a process. This research is within the realm of applied science and is directed toward each of the process areas, i.e., the fermentation, the extraction, and the production of CMA. In the fermentation area subsequent work would focus both on raising the acetic acid production levels to the theoretical 3% maximum and on more clearly defining the kinetic parameters of this production in a continuous fermentation mode. Further work on the extraction part of the process would target the determination of efficiencies in a continuous system also with special attention given the role of impurities, solvent recovery, and solvent/extractant ratios. Finally, additional work on the CMA production step would address more specifically the issues of product formation and recovery. With further attention to these research areas, the potential for residue biomass bioconversion to CMA appears substantial, and the production of this alternative deicing salt seems promising.

ACKNOWLEDGEMENT

This material is based upon work supported by the National Science Foundation under award number ISI-8861087. Any opinions, findings and conclusions or recommendations expressed in this publication are those of the authors and do not necessarily reflect the view of the National Science Foundation.

REFERENCES

1   H.H. Bungay and L.W. Hudson, Presentation at the Bioconversion
     Symposium, University of Waterloo, Canada, 1984, as well as Personal
     Communication (1987). This presentation has been published as the
     chapter "Calcium Magnesium Acetate from Biomass" in the text Biomass
     Conversion Technology: Principles and Practice, M. Moo-Young, J.C.
     Lamptey, B.R. Glick, and H.R. Bungay, Editors, Pergamon Press (1987).
2   Engineering News Record, 212, No. 1, pp. 36-39, January 5, 1984.
3   L.W. Hudson, New York State Energy Research and Development Authority,
     Albany, N.Y., Personal Communication (1987).
4   Report No. FHWA-RD-108, "Alternative Highway Deicing Chemicals", S.A.,
     Dunn R.U. Shenk (October 1979).
5   Report No. FHWA-RD-77-52, "Survey of Alternatives to the Use of
     Chlorides for Highway Deicing", J.A. Zenewitz (May 1977).
6   D.D. Ernst and T. Wieman, "CMA Research Project, Washington Department
     of Transportation", (undated).
7   J.C. Wambold, "Evaluation of Wet Skid Resistance Using Four Deicing
     Salts Prepared for PA DoT, Task Order 12", (July 1983).
8   "Preliminary Evaluation of CMA for Use as a Highway Deicer in
     California", Prepared for CA DoT, 56 Gidley (September 1986).
9   R.L. Schultz, "Report on Wet Skid Resistance Using Various Concentration
     of Two Deicing Chemicals", Prepared for Washington State DoT.
10   D.G. Manning and L.W. Crowder, "A Comparative Field Study of CMA and
     Rock Salt", Prepared for Ontario Ministry of Transportation and
     Communications, (September 1987).
11   J.H. Defoe, "Evaluation of CMA as an Ice Control Agent, Final Report",
     Prepared for Michigan Transportation Commission, (June 1984).
12   B.H. Chollar, "Field Evaluation of CMA During the Winter of 1986-1987",
     Public Roads, 52, 13 (1988).
13   M. Sheeler, "Experimental Use of CMA", Final Report, Iowa Highway
     Research Project HR-253, (September 1984).
14   C.W. Marynowski, J. Jones, D. Tuse, and R.L Boughton, "Fermentation as
     an Advantageous Rate Route for the Production of an Acetate Salt for
     Roadway Deicing", (SRI International), Paper presented at AICHE Summer
     National Meeting, Philadelphia, PA, August 19-22, 1984.
15   C.M. Marynowski, J.L Jones, and E.C. Gunderson, Report No.
     FHWA-RD-83-062 (Phase II) to USDT, FHA, "Production of CMA for Field
     Trials", (SRI International, April 1984).
16   C.M. Marynowski, J.L. Jones, R. Boughton, and D. Tuse, Report No. FHWA
     RD-82-145 (Phase I) to USDT FHA "Process Development for Production of
     CMA".
17   A.B. Gancy, Report No. FHWA RD-86-006 to USDT, FHA, "Preparation of High
     Quality CMA Using a Pilot Plant Process", (Gancy Chemical Corp.),
     (January 1986).
18   J. Solash, Report No. FHWA/RD-87/045 to USDT, FHA, "Preferred Drying
     Method of CMA Solutions", (Energy Minerals Research Co.), (November,
     1986).
19   R.U., Schenk, Report No. FHWA/RD-86/005 to USDT, PHA, "Ice-Melting
     Characteristics of CMA", (Bjorksten Research Laboratories), (January
     1986).
20   E.C. Clausen and J.L. Gaddy, "Organic Acids from Biomass by Continuous
     Characteristics of CMA", (Bjorksten Research Laboratories), (January
     1986).
21   D.L. Wise and D. Augenstein, "An Evaluation of the Bioconversion of
     Woody Biomass to Calcium Magnesium Acetate Deicing Salt", Solar Energy,
     41, 453 (1988).

356

22   D.L. Wise, A.P. Leuschner, and P.F. Levy, Suppressed Methane
      Fermentation of Selected Industrial Wastes: A Biologically Medicated
      Process for Conversion of Whey to Liquid Fuel, Chapt. 11. in Vol. 26 of
      Developments in Industrial Microbiology, Published by Society for
      Industrial Microbiology, 1985.
23   D.L. Wise, Editor, Fuel Gas Systems, CRC Press, Inc. 1983.
24   D.L. Wise, Paper presented at the Intersociety Energy Conservation
      Symposium, Am. Soc. Aeron. and Astro., Philadelphia, Pa., August 10,
      1987.
25   L.G. Ljungdahl, L.H. Carriera, R.J. Garrison, N.E. Rabek, L.F. Gunter,
      L. F. and J. Wiegel, "CMA Manufacture (II): Improved Bacterial Strain
      for Acetate Production", Federal Highway Administration, U. S. Dept. of
      Transportation Report. No. FHWA/RD-86/117.
26   L.G. Ljungdahl, L.H. Carreira, R.J. Garrison, N. Rabek, and J. Wiegel,
      "Comparison of Three Thermophilic Acetogenic Bacteria for Production of
      Calcium-Magnesium Acetate", Biotechnol. Bioeng. Symposium 15, 207-223,
      1985.
27   L.G. Ljungdahl, "The Autotrophic Pathway of Acetate Synthesis in
      Acetogenic Bacteria", Ann. Rev. Microbiol. 40, 415-450, 1986.
28   Meyer-J, Wolin, Research Director, Wadsworth Center for Laboratories and
      Research, New York State Dept. of Health, 1988, personal communication.
29   B. Foody, President, Iogen Corporation, Gloucester, Ontario, Canada,
      Personal Communication, 1988.
30   D.C. Augenstein, D.L. Wise, C.L. Cooney, "Packed Bed Digestion of Solid
      Wastes", Resource Recovery and Conservation, 2, 1976/1977, 257-262.
31   D.L. Wise, R.L. Wentworth, D.C. Augenstein, and C.L. Cooney,
      "Multi-State Digestion of Municipal Solid Waste to Fuel Gas" in Resource
      Recovery and Conservation, 3, 1978, 41-59.
32   P.F. Levy, J.E. Sanderson, E. Ashare, and S.R. de Riel, "Alkane Liquid
      Fuels Production from Biomass", in Liquid Fuel Developments, D.L. Wise,
      Editor, CRC Press, Inc., 1983, 159-188.
33   P.F. Levy, J.E. Sanderson, and S.R. de Riel, "Development of a
      Biochemical Process for Production of Olefins from Peat, with Subsequent
      Conversion to Alcohols", in Organic Chemicals from Biomass, D.L. Wise,
      Editor, Benjamin/Cummings Publishing Company, 1983, 173-218.
34   D.L. Wise, A.P. Leuschner, and P.F. Levy, "Suppressed Methane
      Fermentation of Selected Industrial Wastes: A Biologically Mediated
      Process for Conversion of Whey to Liquid Fuel", Vol. 26, Developments in
      Industrial Microbiology, 1985, 197-207.
35   L.G. Ljungdahl, Formation of Acetate Using Homoacetate Fermenting
      Anaerobic Bacteria", in Organic Chemicals from Biomass, D. L. Wise,
      Editor, Benjamin/Cummings Publishing Co., 1983, 219-248.
36   Y. Nomura, M. Iwahara, and M. Hongo, "Acetic Acid Production by an
      Electrodialysis Fermentation Method with A Computerized Control System",
      Applied and Environmental Microbiology, 54, 1988, 137-142.
37   E. Ashare, R.L. Wentworth, D.L. Wise, "Fuel Gas Production from Animal
      Residue: Part II: An Economic Assessment", Resource Recovery and
      Conservation, 3, 1979, 359-386.
38   C.J. King, "Acetic Acid Extraction", in T.C. Lo., M.H.I. Baird, and C.
      Hanson, Ed., Handbook of Solvent Extraction, John Wiley and Sons, NY,
      1983.
39   P. Eaglesfield, B.K. Kelly, and J.F. Short, Ind. Chem., 29, 147-243
      (1953).
40   J.M. Wardell, and C.J. King, J. Chem. Eng. Data, 23, 144 (1978).
41   N.L. Ricker, J.N. Michaels, and C.J. King, J. Separation Process
      Technol. 1(1), 36 (1979).
42   L.L. Tavlarides, J.H. Bae, and C.K. Lee, Separation Sci. and Tech., 22,
      (2 3), p.581 (1987).

43   Peters and Timmerhaus, "Cost Estimation for Chemical Engineers", 1980
      Edition.
44   See F.C. Schaefer Company analysis of a "Sugar Cane to Ethanol Plant
      Cost" in Lipinsky, E., et. al., U.S. DoE Report TID-29400/2, Batelle
      Columbus Laboratories, 1978.
45   Perry's Handbook for Chemical Engineers, McGraw-Hill.
46   D.L. Wise, and D.C. Augenstein, "An Evaluation of Biomass Conversion to
      Calcium Acetate, an Organic Deicing Salt", Symposium Paper presented at
      Am. Solar Energy Society Meeting, MIT, June, 1988.

Chapter 16

CALCIUM MAGNESIUM ACETATE (CMA) MANUFACTURE FROM GLUCOSE BY FERMENTATION WITH THERMOPHILIC HOMOACETOGENIC BACTERIA

Juergen Wiegel, Laura H. Carreira, Ron J. Garrison, Nancy E. Rabek and Lars G. Ljungdahl
Center for Biological Resource Recovery, Department of Biochemistry, Department of Microbiology, University of Georgia, Athens, GA 30602

16.1 INTRODUCTION

Calcium magnesium acetate (CMA) is considered for use as an organic biodegradable non-corrosive deicing salt and as an additive to coal-fired combustion units to control sulfur emision. These uses are discussed in several chapters of this book. In this chapter we wish to present our work to produce CMA by fermentation from glucose obtained from corn grain and with dolime as a source of calcium and magnesium. The project was initiated and supported by U.S. Department of Transportation, Federal Highway Adminstration. It consisted of the following tasks:

Task A - Establish baseline data: This involved: (1) Surveying the literature for available strains of homoacetate fermenting bacteria, including *C. thermoaceticum*, and obtaining these strains; (2) Establishing baseline data for these strains such as pH and temperature optima, ratio of acetate formed to glucose consumed, tolerance to glucose, acetate, calcium, magnesium and other metals present in dolime and/or needed for acetate production, doubling time, level of other ingredients in culture medium, and optimizing the fermentation conditions; (3) Carrying out continuous culture experiments with and without cell recycling.

Task B - Develop an improved bacterial strain using standard microbiological procedures and techniques. The strain should have greater than 2.5 acetate/glucose ratio; ferment below pH 5.5 in the presence of dolime; have a production rate of 5 g/liter per hour in the presence of 50 g/liter of acetate ions; utilize 90 percent of the substrate; be stable and thus keep selected properties superior to the wild type strain. The mutant strains should be compared with wild type strains under different fermentation conditions.

Task C - Characterize CMA product. The product shall be liquid broth containing CMA having a pH from 8.5 to 9.0 as adjusted with dolime. Analyses include the content of acetic acid, Ca, Mg and other metals, total carbon, proteins and phosphate.

Task D - Refine economic case studies. This study should be based on an earlier report by Marynowski et al. [1] and should take into account the results obtained in Tasks A, B and C.

16.1.1 The Homoacetate Fermentation

Before presenting methods, experimental details and results we feel a limited review of the physiology of *C. thermoaceticum* and other homoacetate fermenting bacteria is warranted. It will help in the understanding of the fermentation process and of some of our experimental approaches. For detailed reviews of homoacetogenic bacteria including their biochemistry, see [2-5].

The homoacetogenic bacteria are unique in that they are able to ferment a variety of substrate to acetate and in addition have the ability to reduce carbon dioxide and other one-carbon compounds to acetate. It is very likely that the acetogenic bacteria are important in natural anaerobic habitats, such as in soils, muds, sediments in fresh and sea water, in intestinal tracts of animals, and in manmade anaerobic environments like sewage sludge digestors and composts. In these environments, degradation of organic compounds occurs with methane and carbon dioxide as products. The degradation is brought about by consortia of anaerobic bacteria. Acetate is a key intermediate in this process [6] and it has been estimated that about 10×10^{12} kg per year of acetate is metabolized world-wide in anaerobic habitats [2]. Thus, a large quantity of acetate is formed in nature and also catabolized. Clearly, acetate is a compound compatible with the natural environment.

The homoacetogenic bacteria have the ability to convert glucose, xylose and some other hexoses and pentoses almost quantitatively to acetate according to the following reactions:

$$C_6H_{12}O_6 \longrightarrow 3CH_3COOH \qquad (1)$$
$$2C_5H_{10}O_5 \longrightarrow 5CH_3COOH \qquad (2)$$

Although research on acetogenic bacteria has been conducted for more than 50 years, general interest and appreciation of, acetogenic bacteria are recent [2-5]. This interest probably developed as a consequence of the energy crisis of 1972 to 1973 and of new anaerobic techniques. These were recently reviewed [7]. Before 1970 the only recognized homoacetogenic bacterium available in a pure culture was *Clostridium thermoaceticum*. It was isolated by Fontaine et al. [8]. Today, at least 18 homoacetogenic bacteria have been described, and of them 14 have been discovered within the last six years [5].

At the start of the research project discussed here there were three thermophilic species among the homoacetogenic bacteria: *Acetogenium kivui* [9], *C. thermoaceticum* [8] and *Clostridium thermoautotrophicum* [10]. There seem to be several advantages using thermophilic bacteria in industrial fermentations [11]. These bacteria carry out fermentations from 50°C to over 100°C. At these temperatures, the temperature control of exothermic processes appears easier, and cooling of sterilized media may be eliminated. Density, viscosity, and surface tension of media are decreased, whereas ionization, solubility and diffusion of substrates are increased. Furthermore, as far as is known there are no pathogenic bacteria among the thermophiles. Consequently, the investigation presented in this report is an evaluation of the three thermophilic acetogenic bacteria for the production of CMA.

Recently, it has been realized that the acetogenic bacteria, in addition to fermenting sugars to acetate, also have the ability to synthesize acetate from several one-carbon compounds, including a gas mixture of hydrogen and carbon dioxide [10,12], carbon monoxide [13], formate and methanol [14], as shown in the following reactions.

$$2CO_2 + 4H_2 \longrightarrow CH_3COOH + 2H_2O \qquad (3)$$
$$4CO + 2H_2O \longrightarrow CH_3COOH + 2CO_2 \qquad (4)$$
$$4HCOOH \longrightarrow CH_3COOH + 2CO_2 + 2H_2O \qquad (5)$$
$$4CH_3OH + 2CO_2 \longrightarrow 3CH_3COOH + 2H_2O \qquad (6)$$
$$2CO + 2H_2 \longrightarrow CH_3COOH \qquad (7)$$

The fact that the acetogens can grow on H_2/CO_2 and CO as sole sources of carbon and energy, classifies these bacteria as autotrophs [2-5]. This ability extends their possible use in industrial production of acetate. An additional metabolic potential of acetogens was the discovery by Bache and Pfennig [15] that *Acetobacterium woodii* in the presence of CO_2 uses the O-methyl groups of a number of methylated aromatic acids including vanillic, syringic, 3,4,5-trimethoxybenzoic, ferulic, sinapic, 3,4,5-trimethoxycinnamic, 2,4-dimethoxybenzoic and 3-methoxycinnamic for the synthesis of acetate. These investigators found also that *A. woodii* reduces the double bond of the acrylic acid side chain of ferulic, caffeic, sinapic and 3,4,5-trimethoxycinnamic acids to a propionic acid side chain. The ability to use O-methyl groups of phenylmethylethers for acetate synthesis has now been found for many acetogenic bacteria including *C. thermoaceticum* [16]. Wu et al. [17]

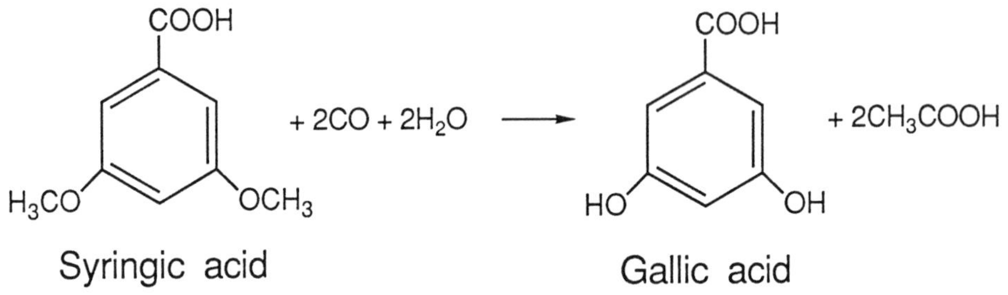

Syringic acid + 2CO + 2H₂O → Gallic acid + 2CH₃COOH

REACTION 8

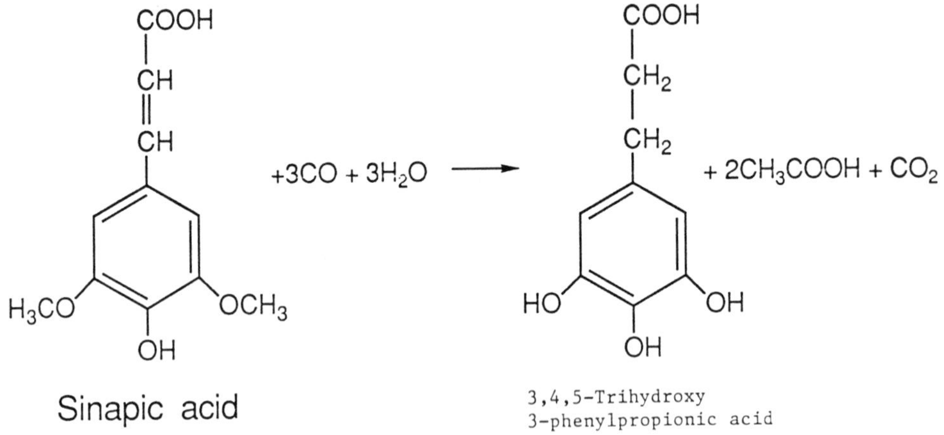

Sinapic acid + 3CO + 3H₂O → 3,4,5-Trihydroxy 3-phenylpropionic acid + 2CH₃COOH + CO₂

REACTION 9

have presented results that indicate that the conversions of syringic and sinapic acids to respective phenols and acetate in the presence of CO may occur according to reactions (8) and (9).

   We will now very briefly consider the complete conversion of glucose to acetate. The details of the fermentation has been established largely in work with *C. thermoaceticum* [2,4,5]. The fermentation can be visualized as a two-step process that occurs via two separate pathways. The first step represented by reaction (10) is the fermentation of glucose to acetate and CO_2; the section step (reaction 11) is the synthesis of acetate by fixation of CO_2.

$$C_6H_{12}O_6 + 2H_2O ---> 2CH_3COOH + 2CO_2 + 8H^+ + 8e \qquad (10)$$

$$2CO_2 + 8H^+ + 8e ---> CH_3COOH + 2H_2O \qquad (11)$$

Sum: $C_6H_{12}O_6 ---> 3CH_3COOH$ $\qquad$ (12)

   Reactions (10) and (11) constitute sums of many enzyme catalyzed reactions; reaction (10) is the common glycolytic pathway leading to the formation of pyruvate from glucose and the subsequent conversion of the pyruvate to acetate and CO_2. Reaction (11) is the newly discovered pathway of autotrophic synthesis of acetyl-coenzyme A from CO_2 [2-5]. A detailed scheme of the complete conversion of glucose to acetate via the two pathways is shown in Fig. 1. Clearly, the homoacetate fermentation is a very complicated biochemical process involving many enzymes and cofactors, including vitamins and trace metals. Of the latter, iron is a constituent of many of the enzymes, whereas tungsten, selenium and iron are required for formate dehydrogenase, nickel, zinc, and iron are needed for carbon monoxide dehydrogenase/acetyl-CoA synthase, and cobalt is required for the corrinoid protein. That the acetate fermentation depends on these metals must be considered in developing an industrial process for acetate production using acetogenic bacteria. The most remarkable of the enzymes in the acetyl-CoA pathway is carbon monoxide dehydrogenase/acetyl-CoA synthase. It catalyzes not only the reduction of CO_2 to CO, but also the formation of acetyl-CoA by condensing a methyl group coming from the methyl corrinoid-enzyme, CO and Coenzyme-A.

16.2 EXPERIMENTAL METHODS
16.2.1 Bacterial Strains

   All work discussed in this article has been done with the bacteria *Clostridium thermoaceticum* [8], *Clostridium thermoautotrophicum* [10] and

Acetogenium kivui [9]. Two strains of each bacterium were obtained. The strains of *C. thermoaceticum* were designed Ljd and Wood. These strains originate from a single strain used in 1952 in the laboratory of Harland G. Wood at Case Western Reserve University [18]. Separate broth cultures of *C. thermoaceticum* have been maintained for many years in our laboratory, strain Ljd (ATCC 39073), and in H. G. Wood's laboratory, strain Wood. *C. thermoautotrophicum*, strains JW701/5 and JW701/3 (DSM 1974) were isolated by Wiegel et al. [10] and maintained by us. *A. kivui* was isolated by Leigh et al. [9] and deposited in the German Collection of Microorganisms (DSM), Göttingen. Strain DSM 2030 was obtained from the German collection. The second strain, labelled R.W., was kindly supplied by Ralph S. Wolfe of the University of Illinois, Urbana.

16.2.2 Media and Microbiological Technique

The compositions of basic media for *C. thermoaceticum*, *C. thermoautotrophicum* and *A. kivui* are given in Table 16.1. Trace mineral, vitamin and reducing solutions have the compositions and were prepared as outlined in Tables 16.2 and 16.3, respectively. Resazurin solution was prepared by dissolving 100 mg of resazurin in 100 ml of water. Resazurin is a redox indicator, that is added to the medium to ascertain that it is anaerobic with an E'_o below -100 mV. During the experimental work the compositions of the media were varied somewhat to study effects on fermentations. Carbon dioxide was used as a gas phase over bacterial cultures throughout this investigation. Most chemicals were of analytical grades. Dolime (High Magnesium Carson Lime) was from Carson Lime Company, Plymouth Meeting, Pennsylvania, and hydrolyzed corn starch, was from Corn Products Corporation, Summit-Argo, Illinois. Deionized water was used in all media.

The acetogenic bacteria are anaerobes. The successful handling and cultivation of them requires techniques to effectively remove oxygen (air) from the medium and to prevent the introduction of air during handling. Current methods for anaerobic microbial work are based on techniques originally described by Hungate [22] and these were recently reviewed and updated by Ljungdahl and Wiegel [7]. Continuous fermentations and cell-recycling experiments were performed in the set-up as described in Figure 2.

Continuous fermentations have also been performed in a new type of fermentor that we call "rotary fermentor". The first one we constructed consists of a glass tube in which a stack of needle pinched synthetic fiber pads was rotated. The pads are of a fibrous polyester material, Reemay 2033, made by DuPont and commercially available from Filtration Sciences, Mount Holly Springs, PA. The pads are spaced by small teflon washers in a stack of

75 and are held together by a central teflon rod mounted on stirring bar.
Each pad is 4.5 cm in diameter. The height of the stack is about 6.5 cm. The
stack is rotated at 80 rpm. The purpose of the stack of fiber pads is to
provide a large area for cell attachment and cell-to-substrate interaction.
The rotation movement should prevent the build up of a concentration gradient
of acetate around the bacterial cells. Such build up may prevent efficient
fermentation. We have also set up a second rotary fermentor that is about
double the size of the first. It is equipped with portholes to allow sampling
at different levels of the fermentor. The fermentor set-up is shown in Figure
3, and the design of the fermentor vessel with the synthetic fiber pads is
outlined in Figure 4. Although these fermentors for anaerobic fermentations
are of our design, the basic idea is that of Clyde [24]. A similar type of
fermentor has also been described by Del Borghi et al.[25].

The rotary fermentor seems to combine the features of a fixed-film
fermentor with the bacterial cells immobilized and of the continuous fermentor
with cell recycling. Thus, the discs serve as a support of the bacterial
cells and also as a filtration device keeping the cells in the fermentor and
preventing washout. The cell concentration in the fermentor will be high,
which allows fast and efficient fermentation with shorter medium retention
time than in a regular continuous fermentor. Furthermore, the design allows a
pH gradient to be formed inside the fermentor. Thus, medium at optimum pH is
introduced at the bottom of the fermentor. At this place a rapid fermentation
occurs. As the medium moves up the fermentor, the pH decreases due to acetic
acid production. Cells in the upper part of the fermentor will automatically
be selected to survive and ferment at the lower pH. Consequently, the final
fermentation liquid, when leaving the fermentor at the uppermost point, will
contain the highest concentration of acetic acid at the lowest possible pH.

16.2.3 Assay Methods

Bacterial growth was followed by determining the optical density at 554
nm with a Hitachi-Perkin-Elmer double-beam spectrophotometer adapted to hold
two Hungate tubes (medium blank and sample) or at 600 nm with an LKB Ultrospec
4050 spectrophotometer. Temperature studies were performed using a
temperature gradient incubator (Scientific Industries, Bohemia, NY) modified
by us. Glucose and acetic acid were determined with a Varian 5060 high-
pressure liquid chromatograph equipped with column heater, variable-wavelength
UV detector (which was set at 210 nm), and a refractive index detector. A
BioRad 87 H organic acids column was used. It was maintained at 70°C and was
preceded by a guard column with material similar to that of the BioRad column.
The eluent solution was originally 0.013N H_2SO_4 (Baker reagent grade) in HPLC-

grade water. However, when analyses were done of media containing high amounts of Ca^{2+} or Mg^{2+}, the elution solution was 0.026N H_2SO_4 containing 5 percent acetonitrile (Fisher Scientific Co., HPLC grade). The elution rate was 0.4 ml/min. Metal analyses were performed using plasma emission spectroscopy [26]. Protein was determined using the method by Goa [27].

16.3 ESTABLISHING BASELINE DATA OF WILD TYPE STRAINS
16.3.1 Batch Cultures in Anaerobic (Hungate) Tubes

The comparison of the three homoacetogenic bacteria has been published [23]. A summary is given below. Although the results do not indicate which of the bacteria would be best for industrial production, they nevertheless show that the two *Clostridia* are more tolerant to dolime, calcium and magnesium than *A. kivui*. The latter grows faster than the *Clostridia*, however, it does not produce acetate to a higher concentration than that corresponding to the fermentation of 1 percent glucose in the medium. The six stains (two of each bacterial species) were grown in the media given in Table 16.1.

(i) pH Range and Optimum. The relationship between initial pH of the medium and growth rate (doubling time) as well as maximum growth (maximum absorbance or O.D.) was established by inoculating 10 ml batch cultures containing media with pH adjusted between 4.5 to 8.5. The pH was not further adjusted during the fermentation. The results are shown in Figure 5. The pH range for the *C. thermoaceticum* strains is between 5.7 and 7.8, with strain Ljd being somewhat more tolerant on the basic side. The *A. kivui* strains have an almost identical response to pH, the range being from below 5 to 7.6, and with a broad pH optimum between 6.2 and 6.8. These results are similar to those previously reported [9]. The strains of *C. thermoautotrophicum* also have similar pH ranges and they grow from below 5.4 to 7.3. However, the strains differ in that the pH optimum for strain 701/5 is rather sharp, around 6.6, whereas for strain 701/3 it is broader. It should be noted that the pH optimum for *C. thermoautotrophicum* growing on glucose is more at the alkaline side than when using glycerate or methanol as substrate [10].

It should be noted that at extreme pH values the growth started only after an initial lag period. In all cultures, the pH decreased during the culturing due to acetic acid production. For instance, with *C. thermoaceticum* inoculated at pH 6.3 the final pH was 4.7 and with the culture inoculated at pH 7 the final pH was 5.4. It should further be noted that for *C. thermoaceticum* and *C. thermoautotrophicum*, growth rates at different pH conditions corresponded well with the maximal optical density values. However, high cell yields of *A. kivui* was obtained only at pH above 6.2. At more acid pH the maximum growth dropped, although growth was observed at a pH as low as 5.

The decrease of growth rates and maximum cell productions at initial pH values of 5.8 and lower have been interpreted to mean that undissociated acetic acid is more inhibitory than the acetate ion [28]. In agreement with this interpretation are the findings by Baronofsky et al. [29] who noticed that when pH falls below 5 the ionized acetic acid moves passively into the cells and abolishes the pH gradient between the cytoplasm inside the cell and the exterior environment. Such a pH gradient appears to be necessary for energy production by the cell [30]. However, with the wild type strains we observed some acetic acid production in a batch culture but no cell growth at pH values below 5. Using adapted strains in the rotary fermenter allowed growth and acetic acid production below pH 4.5. This and fermentations using pH-control will be discussed below in the sections describing "continuous fermentations", respectively.

(ii) Effect of Temperature. Figure 6 is a comparison of growth rates at pH optima of respective bacterial strains. The two *C. thermoaceticum* strains grow between 48 and 65°C. Growth was here defined as a significant increase in OD in less than 4 days. Strain Ljd has a faster growth rate compared with strain Wood at temperatures about 55°C. The temperature range for *A. kivui* is from 45 to 66°C. Strain RW appears to grow somewhat faster than strain DSM at the lower temperatures. The *C. thermoautotrophicum* strains differ somewhat from each other. Thus, strain 701/5 grows between 45 and 67°C, whereas the range for 701/3 is between 47 and 64°C.

(iii) Effect of Initial Glucose Concentration and Acetate Tolerance. Media was prepared for optimum growth, but with glucose in concentrations from 0.5 to 20 percent. Good growth was obtained with all bacterial strains up to a glucose concentration of 10 percent. Growth was observed, but at lower rates at higher concentrations, e.g., 16 percent for *C. thermoaceticum*, strain Ljd, 20 percent for *C. thermoautotrophicum*, strain 701/3 and 14 percent for *A. kivui*, strain DSM 2030. However, although good growth was observed at these high concentrations of glucose, in media without pH control only 2 percent of glucose was converted to acetate by *C. thermoaceticum* and 1 percent or less for *C. thermoautotrophicum* and *A. kivui*. The conversion of 2 percent glucose completely to acetate would yield 2 percent of acetate or about 330 mM in the culture fluid. As will be shown below, this concentration of acetate is about the maximum for allowing growth of the wild-type strains. It can be concluded that growth of the three acetogens is not inhibited by glucose at concentrations of 10 percent or less.

Table 16.4 gives the concentrations of sodium and potassium acetate at which growth was observed within 120 h after inoculation. Acetate was added before inoculation to media otherwise having optimal concentrations of glucose, yeast extract, and phosphate (see Table 16.5). Incubations were at optimal temperatures and pH values. Initially, inocula were from regular cultures without any acclimation to high concentrations of acetate. Thereafter, cultures acclimated to grow at the highest initial concentration of acetate were used as inocula. The results of these experiments indicate that the clostridial acetogens have higher tolerance for sodium ions than for potassium ions. *A. kivui* seems to tolerate both ions almost equally well. It is of considerable interest that Yang and Drake [33] demonstrated that growth of *C. thermoaceticum* is independent of sodium ions, whereas growth and acetate synthesis from CO_2 via the acetyl-CoA pathway by *A. kivui* is sodium dependent. These observations suggest that in *A. kivui* energy generation may be a Na^+-dependent system, whereas in the clostridia ATP is generated by a proton-dependent ATPase shown to be present [30]. It should be noted that *C. thermoaceticum* contains also an Na^+/H^+ antiporter [34] which could be involved in either energy generation or more likely in a system controlling the cytoplasmic pH.

The results shown in Table 16.4 also demonstrate that the three acetogens, when cultivated at a high sodium or potassium acetate concentration, acclimate and are then able to grow at higher concentrations of acetate. Our findings agree with those of Wang and Wang [28] who thoroughly investigated the effect of sodium, potassium and ammonium acetate on the fermentation of glucose by *C. thermoaceticum*. In fermentations controlled with sodium hydroxide at pH 6.9, a maximum concentration of 56 g of acetate per liter (about 930 mM) was reached. Our results show that *C. thermoaceticum* can grow with simple acclimation in an acetate concentration of 850 mM.

(iv) Tolerance to Calcium, Magnesium, and Dolime. Since the object of this investigation was to evaluate acetogenic bacteria for the production of CMA with dolime as the source of calcium and magnesium, the effects of Ca^{2+}, Mg^{2+}, and dolime on the fermentations of glucose were investigated. $MgCl_2$ or $CaCl_2$ from 10 to 400 mM, or dolime up to 2.5 percent, were added to regular media, which were inoculated with a 1 percent inoculum and then incubated at the optimal temperatures. Dolime is not completely soluble in the media. As a result, the concentrations of Ca^{2+} and Mg^{2+} were not known. Furthermore, dolime contains many additional elements in trace amounts that may effect the fermentations. Precipitates were also formed with $MgCl_2$ or $CaCl_2$ presumably consisting of carbonates or phosphates of Mg and Ca. Thus, with these compounds the concentrations of Mg^{2+} and Ca^{2+} were also not precise. However, during the fermentations most of the precipitates dissolved.

The highest concentrations of Ca^{2+}, Mg^{2+}, or dolime at which growth was obtained within 120 hours are given in the summary (Table 16.5). Of the three bacterial species, *C. thermoaceticum* apparently best tolerates both Ca^{2+} and Mg^{2+}, whereas *A. kivui* is the least tolerant. *C. thermoautotrophicum* 701/5 seems best to endure dolime. The dolime used contained 153 mg of Mg and 378 mg of Ca per g (Table 16.6). Table 16.6 also gives amounts of other metals in the dolime. These analyses were done because the acetogens are dependent on several trace metals to grow. On the other hand, too high concentrations of trace metals are inhibitory. We have not optimized the concentration of each trace method or element in the media. However, by adding trace metals five times the amounts normally used (Table 16.1) we observed decreased cell growth and fermentation rate, whereas lesser amounts of the metals caused a decrease in the acetate/glucose ratio with a lower yield of acetate. Thus, the amounts of metals previously established (Table 16.1) seem to be about optimum. The amount of trace metals in dolime are not inhibitory, however, the use of dolime in media will add to the concentrations of the trace metals. We feel that this has to be considered if dolime is going to be used to control pH of media during fermentations.

The solubilization of dolime in 0.3 M acetic acid was determined. The 0.3 M concentration was chosen since this corresponded to the concentration of acetic acid obtained with 2 percent glucose in the medium used with *C. thermoaceticum*. The dolime dissolved rather slowly but within 2 h 12 g of dolime was solubilized by the acetic acid. Additional dolime was not dissolved. The rate of dolime solubilization is faster than the rate of acetic acid production in fermentations and we feel that dolime can be used as a pH-controlling substance during fermentation.

(v) Requirements of Yeast Extract and Phosphate. The concentration of phosphate in the medium was evaluated as to the range within which good growth occurred and the optimal concentration as well as minimal concentration. The use of a low phosphate concentration is desirable due to the eutrophication of surface wastes caused by phosphate. Moreover, the minimal and optimal yeast extract concentrations were determined. The values are given in Table 16.5 together with the comparison of the three species.

16.3.2 pH-Controlled and Continuous Cultures

(i) pH-Controlled Cultures. The optimum pH-values established with batch cultures using Hungate tubes with only the initial pH adjusted may not be the optimum pH values in continuous fermentations. Therefore, the pH optimum values were redetermined in pH-controlled batch cultures. These were performed in 100 ml glass fermentors designed by us and made in the glass shop at the University of Georgia. Alternatively, we used a New Brunswick C30 Benchtop Fermentor with a working volume of 375 ml and modified by us for strict anaerobic bacteria. The pH was controlled by adding 8N NaOH as needed using the equipment labelled 10, 11, 12, and 13 in the cell-recycling setup as described in Figure 2. In later experiments the pH was controlled by adding a 10 percent (w/v) suspension of dolime in water. This slurry has a pH between 11.8 and 12.5 (depending on the batch of dolime). To pump it, we used a Harvard Apparatus Peristaltic Pump, Model 1203 with black butyl-rubber tubing. Other peristaltic pumps were not able to handle the thick slurry or the tubing. In the beginning of fermentations at pH values below 6, it was necessary to control the pH by using a gas phase of CO_2 at a pressure of up to 5 psi (35 kPa).

In general, optimum values for temperature and pH established with batch cultures in Hungate tubes were found also in pH-controlled fermentations. Table 16.7 summarizes the results of the fermentations with *C. thermoaceticum*, strain Ljd. With the medium of Table 16.1 the optimum pH is 6.8 and the doubling time (t_d) 4.9 hours. At pH values 7.0 and 6.6 the doubling times obtained were 7.3 and 7 hours, respectively. The results with *C. thermoautotrophicum* 701/5, were almost identical to those of *C. thermoaceticum*, whereas for *A. kivui* DSM, the pH-optimum with pH-controlled cultures was slightly lower, and found to be 6.2. The doubling time for *A. kivui* at pH optimum and at 63°C was 2.5 hours. In fermentations with pH controlled at 6.8 and 5, the doubling times were 8.3 and 7.0 hours, respectively, for *A. kivui*. The acetate/glucose ratios obtained in fermentations were usually between 2.5 and 2.8 with the three acetogens.

It should be stressed that for efficient fermentations of glucose and starch hydrolysates to acetate, the acetogenic bacteria should be kept vigorously growing in continuous cultures or when using batch cultures, with transfers into new medium about every third day using 10 percent inocula. Only after the establishment of such vigorously growing cultures should they be used as inocula for experiments to determine growth parameters.

As shown in Table 16.7 we used both glucose and corn starch hydrolysate as substrates. For the latter we used first acid-hydrolyzed corn starch and later enzyme-hydrolyzed corn starch. Both hydrolysates were from CPC International (according to the present information the costs of the two hydrolysates per glucose unit are about the same). The acid hydrolyzed corn starch is a syrup containing about 65 percent (w/w) of glucose and contains in parts per million: Cl, 175; K, 12; Na, 260; and traces of Al, Fe, Cu and Pb. The metal contamination is low and has no significant effect on the fermentation and composition of media. The enzyme-hydrolyzed corn starch is a

white powder with about 95 percent (w/w) of glucose. We encountered no
significant changes in the bacterial growth rate or acetate production rate in
batch or continuous cultures when substituting this corn starch hydrolysates
for glucose.

(ii) <u>Continuous Fermentations Without Cell-Recycling</u>. Continuous
fermentations without cell recycling were performed in the apparatus outlined
in Figure 2 but without the cell-recycling train (labelled 3 to 6). The
fermentations were carried out first under the optimal conditions listed in
Table 16.5, and in media essentially as listed in Tables 1, 2 and 3 without
limiting the glucose concentration. In these experiments we were not
concerned with the fact that in some fermentations the glucose was not
completely utilized.

The influence of the volumetric retention time in continuous
fermentations using *C. thermoaceticum*, *C. thermoautotrophicum*, and *A. kivui* on
acetate production rates and final acetate concentrations, optical densities,
and glucose utilizations is demonstrated in Figure 7. The glucose
concentrations in these fermentations were initially 2 percent for the two
Clostridia and 1 percent for *A. kivui*. The same maximum acetate production
rate of 2 g/liter per h was obtained with the three bacteria. However, with
A. kivui, which has a growth rate (see Table 16.5) about 3 times faster than
the growth rates of the *Clostridia*, this rate was obtained at a volumetric
retention time of less than 2 hours. The best production rate with *C.
thermoautotrophicum* was obtained with a volumetric retention time of about 4
hours and with *C. thermoaceticum* 6 hours. In the continuous fermentations,
the best retention time was somewhat lower than the shortest doubling time
obtained in batch cultures with Hungate tubes. This seems to indicate faster
growth rates in continuous fermentations than in the batch cultures, which may
possibly be due to selection of faster growing bacteria in the continuous
fermentations.

In addition to the above described basic continuous fermentations, some
were performed under other conditions. One of these involved reducing agents.
The "classic" medium for *C. thermoaceticum* contains 0.05 percent sodium
thioglycolate as a reducing agent [35]. In our assays of acetate in culture
fluid using HPLC, we found that the presence of thioglycolate in the fluid
interfered with the assay. Therefore, we replaced thioglycolate with 40 ml
per liter of the reducing solution preparing according to Table 16.3, and
containing sodium sulfide and cysteine. This solution was found to be a good
replacement for thioglycolate, actually, the growth rate and glucose to
acetate conversion by *C. thermoaceticum* improved. For the latter, the
acetate/glucose ratio was 2.45 with thioglycolate and 2.85 with sodium

sulfide-cysteine. Our experience is now that in larger fermentors much lower
amounts of reducing solution, as low as 4 ml per liter, can be used. The
reducing solution is needed only at the beginning of the fermentation to lower
the redox potential of the medium to allow anaerobic growth. Once growth has
started the bacteria maintain a suitable redox potential, provided anaerobic
conditions prevail.

Yeast extract is an expensive ingredient of media. It can generally be
replaced by cheaper materials such as technical grade yeast extract, yeast
autolysates and corn-steep liquor. In preliminary experiments, these
materials were tested with *C. thermoaceticum* and good growth was obtained.
However, they were not used further since an optimization has to be done
individually for each industrial preparation in the scaled up fermentation.
In media for *C. thermoaceticum*, the normal amount of yeast extract is 0.5
percent. We tested media with 0.3 percent of yeast extract during continuous
fermentation conditions and found that the production rate of acetate with a
retention time of 7.4 hours was only 54 percent of that with 0.5 percent yeast
extract. Similarly, with 5 hours retention time the production rate dropped
to 35 percent of that with 0.5 percent.

Additional experiments with *C. thermoautotrophicum* revealed that good
acetate production was obtained only in media containing bicarbonate and with
carbon dioxide in the gas phase. The latter could not be replaced by nitrogen
maintaining anaerobic conditions. Thus, carbon dioxide must be considered a
necessary substrate for the acetogenic bacteria. This is of course in
agreement with the metabolic pathways shown in Figure 1.

It can be concluded from the experiments with pH-controlled batch
cultures and continuous cultures that it would not be difficult to culture the
acetogens in either way. The pH-controlled cultures demonstrated that
essentially all available glucose is fermented to acetate, even concentrations
as low as 0.02 percent w/v. During both types of fermentations, the acetate/
glucose ratios, with few exceptions, were 2.5 or higher. We did not try to
utilize all the glucose in the continuous fermentations to avoid limitations
in the adaptation process. However, we have shown in later experiments that
it is relatively easy to maintain a quasi-steady state at a low glucose
concentration with all glucose fermented. Under such conditions glucose is
added to the fermentation separately from the rest of the medium. It was
mentioned previously that for efficient fermentation and acetate production,
the acetogenic bacteria should be kept vigorously growing. In the continuous
fermentation vigorous cell growth was accompanied by efficient fermentation.
This demonstrates that acetate formation and cell growth are coupled.

(iii) <u>Fermentations With Cell-Recycling</u>. The maximum production rate for acetate with the continuous culture was about 2 g/liter per hour. This rate is low. However, an increase in production can usually be obtained by increasing the cell mass (number of living cells) in the fermentation vessel. This can be accomplished with cell-recycling. The process involves removal of the cells from the fermentation fluid leaving the fermentor and then their return to the fermentor, in which the cells will again actively ferment.

The cell-recycling setup is that of Figure 2. In this setup, number 4 is a filter system for removal of the cells from the culture fluid to have them returned to the fermentor. We tested two filter systems, the Pellicon Cassette System (Millipore Corporation, Bedford, MA) and the Hollowfiber Cartridge System (Amicon Corporation, Danvers, MA). In our experience, the hollow-fiber system was superior. The clostridial cells, *C. thermoaceticum* more so than *C. thermoautotrophicum*, showed a tendency to undergo lysis in contact with the Pellicon Cassette system, and this system clogged. With the hollow-fiber system, less lysis and clogging occurred. The latter was prevented by reversing the flow through the system every 12th hour. Hollowfiber Cartridge Systems are available in sizes large enough for industrial fermentations.

Lysis was not a big problem with *A. kivui* but this bacterium developed into a flocculating and clumping strain during the continuous fermentations. This was caused by an accumulation of filamentous bacterial cells. The phenomenon of flocculation may be of considerable advantage in the industrial production of acetate, since flocculation would make it easier to separate the cells from the cultural fluid.

Results of cell-recycling fermentations are presented in Figure 8 for *C. thermoaceticum* and for *C. thermoautotrophicum* and *A. kivui* in Tables 8 and 9, respectively. These tables also give comparisons with results from continuous fermenations without cell-recycling. The highest production rate obtained for *C. thermoaceticum* was 4 g/liter per hour with a concentration of 18.2 g/liter of acetic acid at a volumetric retention time of 4.9 h and cell recycling of about 55 percent. The corresponding values for *C. thermoautotrophicum* were 3.8 g/liter per hour, 4 hours, and 47 percent; for *A. kivui*, the values were 4.76 g/liter per hour, 0.9 h, and 72 percent. The major finding in these experiments is that under our conditions the best results were obtained between 47 and 72 percent recycling, which indicates that stationary cells significantly lose their fermentative activity. It can be pointed out that although *A. kivui* exhibited the fastest fermentation rate, the concentration reached in the fermentation fluid was much lower than with the *Clostridia*.

A summary of fermentations using batch, continuous and cell-recycling fermentations is presented in Table 16.10. This table also includes results of fermentations with hydrolyzed corn starch as substrate. Although there are small differences in yields and production rates with the corn starch in comparison with glucose, we consider the hydrolyzed corn starch to be as good a substrate as purified glucose.

16.4 IMPROVED BACTERIAL STRAINS
16.4.1 Desired Properties of Improved Strain

The properties desired of an improved acetogenic bacterial strain were discussed in the "Introduction". In summary, such a strain should produce 2.5 mol or more of acetate per mol of glucose fermented. The fermentations should be at a low pH value to efficiently dissolve dolime; the production rate should be 5 g of acetate per liter per hour and the final concentration of acetate in the "beer" should be 50 g/liter or higher.

Efforts have already been made to obtain strains of *C. thermoaceticum* that produce acetic acid at pH 5.5 [37,38]. One strain was isolated that grew and produced acetic acid at pH 4.5. It had a doubling time of 36 h, and the highest acetic acid concentration reached was 4.5 g/liter. These studies also demonstrated that at pH 7.0 a redox potential in the medium as low as -300 mV is required by *C. thermoaceticum* to grow.

The fact that Schwartz and Keller [37, 38] were able to isolate the pH 4.5-tolerant strain gives room for optimism for the possibility of isolating still better strains. On the other hand, Baronowsky et al. [29] concluded that it is unlikely that mutations at one or a few loci would result in *C. thermoaceticum* strains with higher acetic acid tolerance than their parental strains. This statement was based on the concept that unionized acetic acid diffuses passively across the cytoplasmic membrane and destroys the ability of the cells to generate energy for growth. As pointed out previously (section 3, b 3), acetate production appears tightly coupled to cell growth.

The production rate shall approach or exceed 5 g/liter per h and the final concentration of acetate in the culture fluid shall be 50 g/liter or higher. Although studies of acetogenic bacteria (*C. thermoaceticum*) have been performed to evaluate them for industrial production of acetate from glucose by fermentation, virtually nothing has been published regarding the rate of production of acetate. Most studies have been with batch cultures and the ability to obtain a high concentration of acetate in the fermentation fluid. Thus, in batch cultures at pH 6.9 Wang and Wang obtained 56 g/liter with *C. thermoaceticum* [28]. Schwartz and Keller obtained 20 g/liter at pH 7 [37]. We obtained with w.t. cells at pH 6.8 an acetate concentration of 34.4 g/liter (572 mM) (Table 16.7) and by using acclimated cells as inoculum 51 g/liter (Table 16.4). Thus, *C. thermoaceticum* wild-type strain is able to produce a concentration of 50 g/liter. The rate of 5 g/liter per h is apparently also within reach, *C. thermoaceticum* and *C. thermoautotrophicum* in cell-recycling experiments (Table 16.10) produced 4 g/liter per h and *A. kivui* 4.8 g/liter per h of acetate.

It would be desirable to use a glucose concentration of 6% or higher in media and to ferment at least 90% of the glucose to acetate. The rationale to use media containing 6 percent of glucose is that in batch culture, if all the glucose is fermented, the fermentation broth would contain about 55 g/liter of acetate. As was discussed in section 3 a) 3), the acetogens grow well in 10 percent of glucose and also in 50 g/liter of acetate. However, with high glucose concentration they do not utilize more glucose than what corresponds to a 2 percent concentration in the medium. Thus, to circumvent this problem, the fermentation is started with 2 percent glucose and, as the fermentation proceeds, concentrated glucose solution is added to the medium so that its concentration is maintained at or close to 2 percent. This procedure was followed by Wang and Wang to produce 56 g of acetate per liter [28].

If dolime is going to be used to control pH of fermentations, the bacteria should have high tolerance for calcium and magnesium. In titrations of dolime with acetic acid, we found that 18 g of acetic acid neutralized 12 g of dolime. Our goal is at least 50 g of acetic acid per liter, which means that 50 x 12/18 = 33 g of dolime will be added to the fermentation per liter. As shown in Table 16.5, *C. thermoaceticum* and *C. thermoautotrophicum* grow in the presence of 20 g and 30 g of dolime, respectively and we feel that this requirement can be met.

From the above discussion it is clear that the wild type strains almost possess all of the properties of the desired bacterial strain. However, these properties are not found together under the same fermentation conditions, e.g., a high acetate production rate is not achieved at a high concentration of glucose or acetate and not at a low pH. The main goal is to develop a strain that ferments at a low pH with a high acetate production rate and concomitantly produces the high acetate concentration desired. We will now discuss our attempts to reach this goal by adaption, selection of mutants and by using a rotary fermentor design.

16.4.2 Developing Improved Bacterial Strains

It is not possible to describe the isolations of all strains that were obtained and investigated as a result of either adaptation or mutation procedures. Thus, we will concentrate on the procedures used and describe some results of the most promising strains. Many strains were isolated that showed promising properties, but as the evaluation of them proceeded, it became clear that they did not fulfill the expectations. This was especially evident with *A. kivui*. According to our findings, this bacterium apparently is very resistant to mutation agents. We isolated mutants that were able to grow at high concentrations of pyruvate in the medium (for rationale of this selection procedure, see below) and also of acetate. However, these mutants did not produce high CMA in the culture fluid, and they were more sensitive to dolime than the clostridial species. Eventually, we felt that it was not possible to continue mutant work involving *A. kivui* and that it was better to concentrate our effort on the clostridial species. Consequently, the work (essentially negative results) with *A. kivui* will not be discussed.

(i) <u>Treatments with Mutagenic Agents</u>. As mutagenic agent we used ultraviolet (UV) irradiation (main wavelength 254 nm), nitrous acid (NA), methane sulfonic acid (EMS) and nitrosoguanidine (NTG).

Ultraviolet light affects especially pyrimidines, and causes formation of thymine dimers covalently linked. It causes breaks in the nucleotide chain. Some UV induced lesion can be repaired by photoreactivation using visible light. Bacterial cells to be UV-treated were of early log phase in regular media. The irradiation was done with the bacterial suspension in a quartz-flow cell of 1 cm light path polished on three sides. The bacterial sample could either be pumped directly from a fermentor or injected and withdrawn manually. The irradiation at 254 nm was with two germicidal lamps placed 7 cm from the flow cell. Before introducing a bacterial suspension into the flow cell, it was sterilized with 70 percent ethanol.

The UV-treatment consisted of irradiation for different lengths of time, e.g., 0, 1, 2, 3, 4 and 10 min. The suspension was then serially diluted in regular medium with 2 percent agar (before the agar solidified) according to the procedure known as the agar shake method [7]. After inoculation, the agar was allowed to solidify. Cultures were left to grow at $58^{\circ}C$. From these cultures single colonies could be picked using anaerobic techniques and transferred into selective media. The technique also allowed determination of surviving cells. With *C. thermoaceticum* strain Ljd, a reduction of viable cells of 10^3 (1 min irradiation), 10^5 (2 min) and 10^9 (3, 4 and 10 min) was obtained. An example of a "mutant" is strain Wood 6-2-1-P derived from *C. thermoaceticum*, strain Wood. The designation implies first an irradiation of 6 min and that it was the 2nd colony found to grow in 500 mM sodium acetate, pH 6.1 medium, and subsequently was grown in a fermentor at pH 5.5. The cells of this culture were irradiated a second time for 1 min and cells growing in the presence of 30 mM pyruvate at pH 5.3 were selected.

Nitrous acid (NA) is a mutagen that causes the transition of guanine-cytosine to adenine-thymine. The reaction involves deamination. The treatment of NA was generally as follows: 5 ml of a bacterial culture in mid-log phase was centrifuged. The bacteria were washed with 0.1 M sodium acetate, pH 4.6, and after recentrifugation suspended in 1/5 vol of freshly prepared nitrous acid solution (0.05 M sodium nitrite in 0.1 M sodium acetate, pH 4.6). The suspension was incubated for 10, 15 or 20 min at 37°C. At the end of the mutation period, 10 ml of the minimal medium (without substrate) was added to stop the action of the nitrous acid. The bacterial cells were recovered by centrifugation and resuspended in 0.5 ml minimal medium and from this suspension a serial dilution (10^8 times) was made as described above using the agar shake method. Isolated mutants were designated NA and with the number of minutes of the NA treatment, e.g., NA 10.

Nitrosoguanidine (NTG) (N-methyl-N^1-nitro-N-nitrosoguanidine) induces transition mutations similar to nitrous acid involving guanine-cystosine convertion to adenine-thymine. It also causes frame shifts at low frequency. The treatment with NTG was similar to that described for nitrous acid. Cells were grown to mid-log phase. After centrifugation and washing, the cells of a 5 ml culture were suspended in 1 ml minimal medium containing 100 µg of NTG. Cells were exposed to the NTG at 37°C for 10, 15, and 20 min and the reaction was stopped by the addition of 10 ml minimal medium. After centrifugation the cells were again resuspended in 0.5 ml minimal medium and used for a dilution series.

Methane sulfonic acid (EMS) action is primarily to alkylate guanine. Cultures for EMS treatment were grown to late log phase in regular media. For the exposure to EMS, 10 ml of a culture was centrifuged and the cells were washed with minimal medium. After the wash, the cells were resuspended in 1 ml minimal medium. EMS, 0.1 ml in 2.5 ml minimal medium, was then added. Incubation followed at 37°C for 45, 60, and 90 min. The reaction was stopped by diluting the cell mixture to 100-fold with minimal medium. This suspension was used to inoculate a dilution series involving the agar-shake method.

A number of "mutant" strains were isolated after treatments described above. They were selected by procedures outlined in the next section. The word "mutant" is for these strains since they had properties apparently different from the present strains; however, genetic analyses have not been done and we do not know if the altered strains are true mutants.

(ii) <u>Selection Conditions</u>. "Mutant" strains and also wild type strains (to look for adaptation) were exposed to several growth conditions designed to select bacteria growing with desirable properties. The conditions initially involved growth at low pH, or in the presence of high acetate, and a combination of the two, e.g., pH 6.6 + 500 mM sodium acetate, pH 6.1 + 500 mM sodium acetate, pH 5.3 + 100 mM sodium acetate, pH 5.3 without acetate, pH 6.8

+ 850 mM sodium acetate and pH 6.8 + 850 mM acetic acid neutralized by dolime.

In another selection procedure, cells were tested for growth in: (a) formate, 50 mM, alone; (b) pyruvate, 50 mM, alone; (c) glucose, 2 percent, alone; (d) glucose, 2 percent, and formate, 50 mM, combined, and (e) glucose, 2 percent, plus either 11, 30 or 50 mM pyruvate at pH values from 5.1 to 6.8. The rationale for using a pyruvate containing medium is based on the fact that pyruvate plays a fundamental role in the metabolism of acetogenic bacteria. Thus, it is formed from glucose and is the precursor of acetate in the pyruvate-ferredoxin oxidoreductase reaction (Figure 1) and it is a supplier of the "CO" unit of [CO-Ni-E]. This unit is the precursor of the carboxyl group of acetate in the pathway of autotrophic synthesis of acetate from CO_2 and other one-carbon compounds. In preliminary experiments we found that pyruvate strongly inhibited growth of the acetogens, especially at pH 5.5 and below. Formate, similar to pyruvate, is an important metabolite in acetogens. It is the precursor of the methyl group of the acetate synthesized via the autotrophic pathway. It is conceivable that the inhibition observed by us and several other workers at low pH of the acetate fermentation is not directly related to acetic acid, but instead is due to accumulation of formic acid or pyruvic acid inside the cell, when it is exposed to low external pH.

To select for fast-growing strains with high rates of acetate production fermentations were performed in chemostats (Figure 2) under conditions similar to those described above. The concept also involved performing continuous fermentations at decreasing retention times. The latter was not successful; we were unable during the course of this investigation to obtain a strain that grew faster than the wild-type strains. However, selection of bacterial strains fermenting fast at low pH may have been achieved using the new rotary fermentor. The result with this fermentor will be described later in this report.

(iii) "Improved" Bacterial Strains. Table 16.11 lists examples of "Improved" bacterial strains derived from C. thermoaceticum and C. thermoautotrophicum. It is evident that it is relatively easy to obtain strains that grow and produce acetate at very high concentrations of CMA. Striking is the property of strains Ljd EMS and 701/5 P-NA 20 to produce CMA at a concentration of over 2 M (about 150 g of CMA per liter). The following is an account of work with the "mutant" strains.

C. thermoautotrophicum, strain 701/5 wild-type, was exposed to 50 mM pyruvate with pH controlled at 5.5. After some growth, 200 mM pyruvate was added and the culture was left for 5 days. Surviving cells (culture 701/5 P) were exposed to nitrous acid for 10 and 20 min. From the 20 min NA-treatment, the strain 701/5 P-NA 20 was selected. It grew overnight in 11 mM pyruvate.

It was used to inoculate a 100 ml fermentor. The medium in the fermentor was controlled to contain glucose between 20 to 100 mM and a pH between 5.4 to 7.2 by additions of dolime. After 9 days and with 6 additions of glucose the acetate concentration was 800 mM. The culture was diluted to contain about 400 mM of acetate and additional glucose was added; after 7 days the acetate was 900 mM. After two additional dilution cycles with addition of glucose, an acetate concentration of 1240 mM was obtained. The acetate/glucose ratio was 2.2. Figure 9 shows the progress of the fermentation during the first 24 days through the initial growth period and acetate production and two following cycles. The fermentation was carried through additional cycles and the results are shown in Table 16.12. As shown during cycle 4, the concentration of CMA was well above 2 M. The highest rate of acetate formation was 12 g/liter per 24 h. This is slow in comparison with fermentations using cell recycling and media containing no acetate initially. Under these conditions we obtained with the wild type strains over 4 g/liter per h (Table 16.10).

Although the fermentation rate is low, strain 701/5 P-NA 20 and strains of similar properties could be useful for industrial production of CMA. This is because they produce a very high concentration of CMA in the "beer". An example is the following calculation using strain 701/5 P-NA 20:

1. Acetate after 24 h   1050 mM
    Initial <u>acetate</u>   <u>887 mM</u>
    Acetate produced   163 mM $= 9.78$ g$\bullet$l$^{-1}\bullet$24 h^{-1}
2. Acetate after 74 h   1351 mM
    Initial <u>acetate</u>   <u>917 mM</u>
    Acetate produced   434 mM $= 8.68$ g$\bullet$l$^{-1}\bullet$24 h^{-1}

In example 2, if we start with a 300 liter culture and allow the fermentation to proceed for 3 days, we will have 300 liters of fluid containing 24518 g of acetate ions. If we then harvest 100 liter from this culture and add 100 liter of new medium, the culture will again reach over 1300 mM of acetate within 72 h. Thus, we can harvest 100 liter every 72 h containing between 8 to 10 kg of acetate ions neutralized with dolime (about 12 kg Ca, Mg acetate).

The results with *C. thermoaceticum*, strain Ljd-EMS were similar to those of *C. thermoautotrophicum* strain 701/5 P-NA 20. With strain Ljd-EMS growing in a medium originally containing 800 mM acetate/dolime complex, we reached concentrations of 2030 mM (=122 g/liter of acetic acid). The production rate at this concentration was about 14 g/liter per 24 h.

The mutant strains were compared with respective parent wild-type strains as to ranges and optima for pH and temperature, tolerances to glucose, acetate and dolime, and acetate/glucose ratio. The tests were run in media containing 20 mM phosphate compared to 80 mM in earlier experiments, 0.5 percent yeast extract, 0.75 percent sodium bicarbonate, 2 percent glucose from corn starch and trace minerals as used regularly.

The pH study carried out in pH-controlled 100 and 750 ml fermenter units revealed that all strains, including the wild-types, grow at a faster doubling time than we initially found (Table 16.5), e.g., Wood w.t. had a doubling time of 3 h compared with 5 h as found earlier in the growth experiment. The growth rates were similar between the w.t. and the mutant strains. The pH optima for the w.t. were those reported earlier, or slightly lower. For strains Wood EMS 90, Wood 6-2-1-P and Ljd EMS, the pH optima were 6 to 6.5, 6 to 7, and 6 to 6.6, respectively. The pH optima for the 701/5 strains were around 6.7.

Temperature optima were not determined for the mutant strains except for Ljd EMS. The strains were selected at 60°C, the optimum temperature of the wild type strains, and it was assumed that this temperature was optimum also for the mutants. The Ljd EMS strain was selected for temperature study because its pH optimum was lower than that of the w.t. strain. The optimum temperature range at pH 6.5 was found to be between 51 to 65°C with a doubling time between 1.2 to 1.8 h; at pH 7.2 the temperature range was between 55 to 66°C with a doubling time from 1.3 to 2 h.

The tolerance of the mutant strains for glucose was similar to that of the wild types as described earlier, section 3 A) 3) and Table 16.5. All strains grew best at 2 percent glucose with the Wood series having doubling times of less than 2 hrs, ratios of acetate produced/glucose consumed were between 2.2 to 2.8. The Ljd series had doubling times from 2.3 to 2.7, with ratios of 1.9 to 2.5. The 701/5 series had doubling times of 2.6 to 3.9 and ratios of 2.4 to 2.9. None of the strains grew in media containing 15% glucose, or in media with less than 1 mM phosphate.

The tolerance to sodium acetate was tested using media with concentrations from 200 to 1200 mM. Other conditions were optimal. All strains grew at 200 mM and produced an additional 200 mM of acetate with acetate/glucose ratios of 2.9 or better except Wood w.t. (ratio 1.71). Except for Wood 6-2-1-P and 701/5 p-NA 20, all strains grew at 400 mM of acetate and produced an additional 150 mM with ratios of 2.0 or better. An exception was 701/5 P-NA 20 NTG 50 which yielded an acetate/glucose ratio of 1.6. Only Wood w.t. strain grew at 650 mM sodium acetate. Cells grown at 400 mM sodium acetate were used as inocula to media containing higher concentration of sodium acetate to see if the cells would acclimate. None of the cells did except those of Ljd EMS.

The results of the tolerance tests using sodium acetate contrast to those obtained using acetate neutralized with dolime. Several of the mutant strains grew in media containing 1 M CMA and as described above strains Ljd EMS and 701/5 P-NA 20 grew in 2 M CMA at pH 6.9. Growth at these high concentrations of CMA was obtained only after many growth cycles in media containing increasing amounts of CMA. Cells obtained from these fermentations when transferred to "normal" media did not grow well. We have interpreted the above results to indicate that high dolime concentrations are needed for good growth of these "mutants". Also, the acetate concentration may have to be high.

Continuous cultures with and without cell recycling, similar to those performed for the wild type strains, section 3 B) 2) and 3), were performed with some of the "mutant" strains. The results were disappointing. Thus, strain 701/5 P-NA 20 was tested using the normal medium with 0.75 percent Na_2HCO_3, 20 mM phosphate, 0.5 percent yeast extract and corn starch hydrolysate corresponding to 2.5 percent of glucose. The pH was controlled with 5 N sodium hydroxide and, as before, the temperature was 60°C and CO_2 was bubbled through the medium. Unlike the wild-type strain, the "mutant" strain did not respond well to the recycling system. At best, the acetate concentration in the culture fluid reached 120 mM and the production rate of 1.5 g/liter per h. This was obtained using a 63 percent recycling and a pH of 6.6. Under similar conditions but with 16 percent cell recycling, the production rate was 0.9 g/liter per h, with an acetate concentration of 80 mM. Corresponding data with *C. thermoaceticum* Ljd wild type, were 4 g/liter per h and 303 mM. Clearly, the "mutants" selected to grow at 850 mM CMA or higher, in respect to production rates, are less efficient than the wild type strains when tested under "normal" growth conditions.

16.4.3   Rotary Fermentor Experiments

The theory behind and the design of the "rotary fermentor" was described in Section 16.3.2 and Figures 3 and 4.

The first experiments with the rotary fermentor were with *C. thermoautotrophicum*, wild type strain. This strain was selected because of its lower tendency to lyse in comparison with *C. thermoaceticum* w.t. strains. After the fermentor was inoculated, the culture was allowed to grow up in batch-like fashion. A series of continuous culture/batch cycles was then investigated, decreasing the retention time during continuous cycles each time from a long 7 h to a short 3 h.

Frankly, the results were impressive. First, the retention time of 3 h is short. Secondly, the pH of the culture fluid leaving the fermentor was about 5.2.

3 h retention time   ----- 2.6 g acetic acid per liter/h

3.5 h retention time ----- 2.7 g acetic acid per liter/h

3.75 h retention time ----- 2.8 g acetic acid per liter/h

These results are good, especially considering the low pH of the final fermentation fluid. The medium used in the fermentations contained 20 mM phosphate initially at pH 6.8 and 0.75 percent sodium bicarbonate. It should also be emphasized that the results were obtained without optimizing the medium composition and the fermentor design. The results demonstrate that by changing the fermentation design, rates of fermentation can be vastly improved.

Interestingly, the bacterial cells, especially those at the top of the fermentor, seem to have adapted to the low pH and appeared different than those of the original wild-type strain. The adapted cells are referred to as 701/5-RF1; RF1 standing for rotary fermentor, run 1.

Even more favorable results were obtained with *C. thermoaceticum* Ljd-EMS in the rotary fermentor. As with *C. thermoautotrophicum*, the cells selected may have changed. The resulting culture will be referred to as Ljd-EMS-RFB. The bacterium was tested in our second rotary fermentor that has a volume of 500 ml. The medium contained 5 mM phosphate pH 6.8, 0.5 percent yeast extract, 0.5 percent $NaHCO_3$. Corn starch hydrolysate solution containing an equivalent of 25 percent of glucose was added separately from the rest of the medium. This has worked quite well and the amount of glucose lost in the effluent can be reduced to virtually none after stabilization of the culture growth conditions and adjustment of the glucose pumping rate.

Initial results were similar to, or slightly less impressive than those obtained with the smaller rotary fermentor. But after various growth conditions were tested and the cells had been adapted to the system, an acetate production rate of 4.3 g/liter per h was obtained at a retention time of 1.3 h. The effluent culture fluid contained 91 mM acetate and had a pH of 5.04. This is a high production rate, especially considering the low pH of the effluent culture fluid. It appears that the colonization/ adaptation process of the bacterial cells is accelerated in the rotary fermentors by brief (30 min) periods of batch growth from time to time. These short breaks during continuous culture growth resulted in an increase of the cell mass and helped to keep the pH below 5.3.

The fermentor was continuously run for over 2 months. During this time, several experiments were performed, including a study of the effect of medium retention time. The results are given in Table 16.13 and Figure 10. The

rotary fermentor could be run with a retention time as low as 0.32 h. This
time gave an acetate production rate of 9.5 g/liter per h, about 11 g/l was
the highest production rate obtained by us with this fermentation system.
Unfortunately, the concentration of acetate in the culture fluid was only 50
mM at a pH of 6.6. The best results were obtained with a retention time of
about 1 h, at which, during a 14-day run, the acetate production rate was
constantly around 5 g/liter per h at a concentration of 100 mM but with a pH
of 5 in the effluent culture fluid. These values compare favorable with those
obtained by Schwartz and Keller (36, 37) obtained for their low pH mutant.

Additional preliminary experiments indicated that the phosphate and
yeast extract concentrations of the medium can be lowered when used in the
rotary fermentor. Thus, good fermentations were obtained with 1 mM phosphate
and 0.1 percent yeast extract. The use of low concentrations of phosphate and
yeast extract would lower considerably the medium cost. In addition, the use
of less phosphate is beneficial for the environment.

Our experience with the rotary fermentor is positive. The use of
support material within the fermentor for cell colonization seems to work
better than cell recycling. The slow stirring seems to prevent product build
up around the cells and allows a fast fermentation. In addition, there
appears to be some selection of cells adapted to ferment at low pH values. As
with any new system, many variables must still be researched and fine-tuned
for maximum efficiency. These include control of substrate addition and
utilization, medium composition, effect of active pH control with possible use
of dolime, further design modifications allowing higher rates of formation of
acetate as well as a higher concentration in the "beer". The latter may
perhaps be accomplished by using a train of fermentors coupled together.

16.5 PRODUCTION OF 150 lb (67 kg) OF CMA

A part of the CMA-project was to produce a minimum of 150 lb of CMA by
fermentation to be used for environmental studies. This was done in a New
Brunswick stainless steel fermentor with a capacity of 100 gallons. Eight
runs were performed as listed in Table 16.14. The fermentations were at
59/60°C using essentially the medium for *C. thermoaceticum* given in Table
16.1. The preparation of the medium was done by dissolving all ingredients
except the hydrolyzed corn starch in about 370 liter of water. The medium was
then sterilized for 2 h at 121°C. At the end of the sterilization period,
sterile carbon dioxide was bubbled through the medium which was allowed to
cool to 60°C, the temperature of the fermentation. The hydrolyzed corn
starch, dissolved in about 30 liters of water and sterilized separately, was
added. The fermentor was inoculated with 30 liters of a vigorously growing

culture of *C. thermoaceticum*. The fermentation was done with bubbling of CO_2 through the medium. The duration was for 5 to 6 days. During the fermentation, analyses of acetate and glucose concentrations were performed and pH was followed. At onset the pH was 6.8, but as acetic acid was produced it rapidly decreased to below 6. Several portions of a slurry of dolime (sterilized) were then added to keep the pH above 6. From the beginning, the glucose (starch hydrolysate) concentration was about 2 percent, and it fell off relatively rapidly. It was maintained at about 0.5 percent during the main part of the fermentation by addition of more sterile starch hydrolysate. At the end of the fermentation (the acetic production rate was slow) the level of glucose was allowed to decrease to a value as low as possible. The fermentation was ended by addition of dolime to obtain a pH of about 8.

Results of the analyses of the eight runs are given in Table 16.14. The acetate/glucose ratios were somewhat low. Several explanations are available, one is that a low bicarbonate concentration was used in the medium. This was done to promote solvation of dolime. In addition, it was difficult to keep a record of the volume, since several additions were made of dolime and substrate and medium was withdrawn for samples and also to prepare seed cultures for the next run. One of the most interesting observations is that magnesium of the dolime was preferentially solubilized over calcium. Thus, the CMA produced was more a magnesium acetate than calcium acetate. The ratio between the Mg/Ca was from 1.8 to as high as 3.

The CMA solution of the third run of Table 16.14 was analyzed for metals using plasma emission spectroscopy. The results are given in Table 16.15. Both the total broth containing undissolved dolime and precipitates and the broth after removal of the solids were analyzed. As noted above, most of the magnesium was soluble, whereas a large portion of the calcium remained undissolved. Iron and phosphorus were also mostly in the solids.

16.6 FINANCIAL ANALYSES

Financial analyses were performed for the various approaches tested for of CMA-production. These analyses were for a facility producing 1000 ton of CMA per day. The basis for the analyses were the laboratory results presented in this chapter. Four processes were considered: Batch fermentation producing 11.24% acetate solution and harvesting one third of the fermentor volume every third day; Continuous fermenation without cell recycling producing 0.75% acetate solution with a fermentor retention time of 3 h; Continuous fermentation with cell recycling producing 2.24% acetate solution with a fermentor retention time of 4.5 h; Rotary fermenation producing 1% acetate solution with a fermentor retention time of 1 h. The analyses were performed by L. F. Gunter at the Department of Agricultural Economics, University of Georgia. We wish to express our thanks to Dr. Gunter for carrying out the analyses. Details of the analyses will not be given here, they are available in a DOT-Report [38].

The financial analyses were based on the cost relationships and assumptions given by Maryknowski et al. [1], but involved modifications for continuous processes and for the low level of CMA in the fermentation broth. One important difference from Maryknowski et al. [1] was that they assumed the use of an 8 to 9% glucose concentration in the media, whereas we found that only 2% were fermented.

The results obtained by Dr. Gunter indicated that the CMA prices for the different processes were as follows: Batch fermentation, $0.53/kg; continuous without cell recycling, $0.86/kg; continous with cell recycling, $0.60/kg; and rotary fermenation, $0.76/kg. These calculations are based on results obtained using laboratory test tubes and 100 ml fermentors and on 1985 tax structure and credit conditions. Clearly, they have to be reexamined as the CMA-process is researched, most importantly by performing pilot plant studies. We feel that CMA can be produced by fermentation at a considerably lower price than given here. In support of this is the work by Wise and Augenstein [39] who evaluated the production of calcium acetate from woody biomass. They arrived at a cost of $0.258/kg.

The evaluation carried out by Dr. Gunter gave some insights into the economics of CMA production. Thus, an inverse relation exists between the concentration of CMA in the fermentation broth and the plant gate price of CMA. It favors processes which yield a more concentrated broth. The production of more dilute solutions requires increases in both the capital investment, for additional drying capacity, and in annual operating costs for fuel, utilities, water and labor. The relative advantages of a more concentrated broth may be reduced somewhat, however, by whole plant design optimization for more dilute solution processes, or by improved water removal techniques. The development of an economical process to recover acetic acid or CMA from dilute solutions is fundamental for an industrial production using homoacetogenic bacteria. Examples of possible processes are those described by Busche et al. [40] and Kuo and Gregor [41].

The CMA plant gate price is more sensitive to changes in annual operating costs than to changes in capital investment. Design, equipment and process modifications to increase the economic efficiency of the production process should therefore concentrate on achieving savings in annual operating costs. A fifty percent increase in plant facility investment increased CMA prices by three to four cents per pound. This was due partly to the

investment tax credit and tax savings from the accelerated depreciation
regulations in place in 1985. In contrast, an increase in annual operating
costs of only ten percent increased the CMA prices required by two to three
cents per pound.

The affect of glucose syrup prices on the CMA plant gate price is
approximately one to one, with a three cent increase in the price of a pound
of glucose requiring a three cent increase in the price of CMA. The prices
used in the calculation referred to here were for glucose $0.26/kg and for
dolime $90.00/ton.

16.7 CONCLUSIONS AND RECOMMENDATIONS

Homoacetogenic bacteria ferment glucose and some other sugars to
acetate, which is the only product. It is a very efficient fermentation and
the theoretical yield of fermenting one mol of glucose is three mol of acetate
(acetate/glucose ratio = 3) or when expressed in weight, 100 g of glucose
yields 100 g of acetic (equivalent to 127 g of CMA). Since acetate is the
only product, it is also a very clean fermentation that would not need a
method to separate the acetate from other fermentation products. The
homoacetogenic fermentation seems ideal for an industrial production of
acetate from a renewable resource such as glucose.

In this report we have described properties of three thermophilic
homoacetogenic bacteria. *C. thermoaceticum*, *C. thermoautotrophicum*, and *A.
kivui* and outlined possible fermentation procedures. We have also
demonstrated that it is possible to develop improved bacterial strains through
selection and with the use of mutagens.

The basic studies performed did not single out one of the bacteria as
being superior. However, during the course of this investigation, work with
A. kivui was discontinued for the following reasons: It was less tolerant to
high concentrations of glucose and dolime than the clostridia, and we were
unable to obtain improved or mutant strains from it that were superior to the
original wild-type strain. In contrast to *A. kivui* from the clostridia,
strains were obtained having much higher tolerance for CMA than the wild-type
strains. Strains were also obtained that produced acetic acid at pH 5 and
slightly lower, at ratio of up to 9.5 g/liter per hour in rotary fermentors.

The three bacteria investigated and the improved clostridial strains
ferment purified glucose and hydrolyzed corn starch equally well. They
utilize 98 percent of the substrate and the yield of acetate is from 85 to 98
percent of the theoretical. The fermentation must be under anaerobic
conditions with a gas phase of carbon dioxide over the fermentation broth.
The best fermentation occurs with the substrate concentration in the medium

between 1 to 2 percent, and it is recommended that substrate be added in increments during the fermentation in such a manner that its concentration does not exceed 2 percent. The pH, an important factor of the fermentation, can be controlled by the addition of dolime in combination with bubbling of gaseous carbon dioxide through the fermentation broth.

Four fermentation processes were evaluated: continuous fermentation, continuous fermentation with cell-recycling, continuous rotary fermentation and batch fermentation with CMA-tolerant improved bacterial strains. The required plant-gate CMA prices for these processes were calculated to be in $/lb 0.391, 0.271, 0.344, and 0.240, respectively. Clearly, the batch fermentation appears the most economical, however, the continuous rotary fermentation, when developed, may be the preferred process.

The price of $0.24/lb obtained in the batch procedure is above that of $0.2158/lb calculated for the chemical production of CMA with glacial acetic acid and dolime as feed-stock [42]. Glacial acetic acid constitutes 75 percent of the cost of CMA production by the chemical method. A slight increase in the price of glacial acetic acid and a slight improvement of the fermentation process would bring the cost of CMA by the two processes within parity. Thus, we may conclude that an economical process involving fermentation of hydrolyzed corn starch to CMA can be developed.

To improve the fermentation process for CMA, we recommend that future work should involve pilot plant studies in fermentor volumes from 100 to 1000 liter. These studies would be necessary to corroborate the results obtained with bench-top fermentors. The pilot plant studies should involve attempts to simplify the medium of which yeast extract and the reducing solution are the most expensive ingredients. It is possible that these can be eliminated or drastically reduced. Furthermore, phosphate and trace metal solution should not be needed since these ingredients would be provided by the addition of dolime.

The continuous rotary fermentor was developed by us in the last month of this study. With it a higher fermentation rate is obtained (almost 10 g of acetate per liter per h) than in any other fermentor. However, the concentration of acetic acid is low (about 1 percent) in the harvested fermentation broth. Consequently, the cost to concentrate the CMA is very high and is reflected in the price of $0.76/kg of CMA with this process. We firmly believe that if the rotary fermentation process is developed to its full potential, it will be economical. Cost calculations indicate that if the harvested fermentation broth contains 2 percent of acetate which can be achieved, the price for CMA would decrease to $0.60/kg. The possibility should also be considered of developing a fermentation process involving two

388

stages, perhaps by combining the rotary fermentor process with the batch process.

The study presented here demonstrates that *C. thermoaceticum* and *C. thermoautotrophicum* can be used in an industrial process to ferment hydrolyzed corn starch to CMA. It should be kept in mind, however, that during the last 5 years 10 new homoacetogenic bacteria have been discovered. It is possible that one of them can be superior to the two clostridial bacteria. Furthermore, the demonstration that improved strains can be selected from the *Clostridia* after treatment with mutagens indicates the possibility of isolating even better improved strains. Finally, this study considered hydrolyzed corn starch as substrate. The homoacetogenic bacteria are able to utilize xylose and some other sugars. Xylose and glucose can be obtained from wood and other plant materials that can be considered a future source of substrate.

ACKNOWLEDGEMENTS

The authors would like to express their sincere thanks and appreciation for gifts, services and support as follows:

Professor Harland G. Wood, Case Western Reserve University, Cleveland, Ohio for his laboratory culture of *Clostridium thermoaceticum*.

Professor Ralph S. Wolfe, University of Illinois, Urbana, Illinois for a culture of *Acetogenium kivui*.

Carson Lime Company, Plymouth Meeting, Pennsylvania for a free supply of dolime (high magnesium carson lime).

Corn Products Corporation, Summit-Argo, Illinois for a free supply of hydrolyzed corn starch.

Dr. Ronald A. Makula, Director of the Fermentation Plant, University of Georgia, for assistance to produce 150 lb of CMA in the 400-liter fermentor.

Dr. L. F. Gunter, Department of Agricultural Economics, University of Georgia, for economical evaluation of CMA production.

U.S. Department of Transportation, Federal Highway Administration, for financial support, Contract RFD# DTFH-61-83-R-00124.

REFERENCES

1   C.W. Maryknowski, J.L. Jones, R.L. Boughton, D. Tuse, J.H. Corlopassi, and J.E. Gwinn, 1983. Process Development for Production of Calcium Magnesium Acetate (CMA), FHWA/RD-82/145, Washington, DC: Federal Highway Administration.
2   L.G. Ljungdahl, 1986. The autotrophic pathway of acetate synthesis in acetogenic bacteria. Annual Reviews of Microbiology, 40:415-450.
3   G. Fuchs, 1986. CO_2 fixation in acetogenic bacteria; variations on a theme. FEMS Microbiol. Rev. 39:181-213.

4   H.G. Wood, S.W. Ragsdale, and E. Pezacka, 1986. The acetyl-CoA pathway: a newly discovered pathway of autotrophic growth. <u>Trends Biochem. Sci.</u> 11:14-18.

5   H.G. Wood, and L.G. Ljungdahl, 1990. Autotrophic character of the acetogenic bacteria. In <u>Variations on Autotrophic Life</u> (J. M. Shively and L. R. Barton, eds.), Academic Press (in press).

6   R.H. Mah, R.E. Hungate, and K. Ohwaki, 1976. Acetate, a key intermediate in methanogenesis. In <u>Microbial Production and Utilization of Gases</u> (H. G. Schlegel, G. Gottschalk, N. Pfenning, eds.). Göttingen: E. Goltze, K. G., pp. 97-106.

7   L.G. Ljungdahl, and J. Wiegel, 1986. Working with anaerobic bacteria. In <u>Manual of Industrial Microbiology and Biotechnology</u> (A. L. Demain and N. A. Solomon, eds.). Washington, DC: American Society of Microbiology, pp. 84-96.

8   F.E. Fontaine, W.H. Peterson, E. McCoy, M.J. Johnson, and G.J. Ritter, 1942. A new type of glucose fermentation by *Clostridium thermoaceticum* n. sp. <u>J. Bacteriol.</u> 43:701-715.

9   J.A. Leigh, F. Mayer, and R.S. Wolfe, 1981. *Acetogenium kivui* - A new thermophilic hydrogen-oxidizing, acetogenic bacterium. <u>Arch. Microbiol.</u> 129: 275-280.

10   J. Wiegel, M. Braun, and G. Gottschalk, 1981. *Clostridium thermoautotrophicum* species novum, a thermophile producing acetate from molecular hydrogen and carbon dioxide. <u>Curr. Microbiol.</u> 5:255-260.

11   J. Wiegel, and L.G. Ljungdahl, 1986. The importance of thermophilic bacteria in biotechnology. <u>CRC Crit. Rev. Biotechnol.</u> 3:39-107.

12   R. Kerby, and J.G. Zeikus, 1983. Growth of *Clostridium thermoaceticum* on H_2/CO_2 or CO as energy source. <u>Curr. Microbiol.</u> 8:27-30.

13   D.R. Martin, A. Misra, and H.L. Drake, 1985. Dissimilation of carbon monoxide to acetic acid by glucose-limited cultures of *Clostridium thermoaceticum*. <u>Appl. Environ. Microbiol.</u> 49:1412-1417.

14   J. Wiegel, and R. Garrison, 1985. Utilization of methanol by *Clostridium thermoaceticum*. <u>Am. Soc. Microbiol. Ann. Meeting</u>, p. 165, Abstract I 115.

15   R. Bache, and N. Pfenning, 1981. Selective isolation of *Acetobacterium woodii* on methoxylated aromatic acids and determination of growth yields. <u>Arch. Microbiol.</u> 130:255-261.

16   S.L. Daniel, and H.L. Drake, 1988. Acetogenesis from methoxylated aromatic acids by *Clostridium thermoaceticum*. <u>Am. Soc. Microbiol. Ann. Meeting</u>, p. 198, Abstract I-105.

17   Z. Wu, S.L. Daniel, and H.L Drake, 1988. Characterization of a CO-dependent O-demethylating enzyme system from the acetogen *Clostridium thermoaceticum*. <u>J. Bacteriol.</u> 170:5705-5708.

18   H.G. Wood, 1952. A study of carbon dioxide fixation by mass determination of the types of C^{13}-acetate. <u>J. Biol. Chem.</u> 194:905-931.

19   J.R. Andreesen, A. Schaupp, C. Neurauter, A. Brown, and L.G. Ljungdahl, 1973. Fermentation of glucose, fructose and xylose by *Clostridium thermoaceticum*: effect of metals on growth yield, enzymes, and the synthesis of acetate from CO_2. <u>J. Bacteriol.</u> 114:743-751.

20   W.E. Balch, G.E. Fox, L.J. Magrum, C.R. Woese, and R.S. Wolfe, 1979. Methanogens: reevaluation of a unique biological group. <u>Microbiol. Rev.</u> 43:260-296.

22   R.E. Hungate, 1969. A roll tube method for cultivation of strict anaerobes. <u>Methods Microbiol.</u>, Vol. 33 (J.R. Norris and D.W. Ribbons, eds.) New York: Academic Press, Inc., pp. 117-132.

23   L.G. Ljungdahl, L.H. Carreira, R.J. Garrison, N.E. Rabek, and J. Wiegel, Comparison of three thermophilic acetogenic bacteria for production of calcium-magnesium acetate. <u>Biotechnol. Bioeng. Symp.</u> 15: 207-223.

24   R. Clyde, 1983. Horizontal stainless fermentor. U.S. Patent No. 4 407 954.

25   M. Del Borghi, A. Converti, F. Parisi, and G. Ferraiolo, 1985. Continuous alcohol fermentation in an immobilized cell rotating disk reactor. <u>Biotechnol. Bioeng.</u> 27:761-768.

390

26   J.B. Jones, Jr. 1977. Elemental analysis of soil extracts and plant tissue by plasma emission spectroscopy. Commun. Soil Sci. Plant Anal. 8:349-365.

27   J. Goa, 1953. A microbiuret method for protein determination-determination of total protein in cerebrospinal fluid. Scand. J. Clin. Lab. Invest. 5:218-222.

28   G. Wang, and D.I.C. Wang, 1989. Elucidation of growth inhibtion and acetic acid production by Clostridium thermoaceticum. Appl. Environ. Microbiol. 47:294-298.

29   J.J. Baronofsky, W.J.A. Schreurs, and E.R. Kashket, 1984. Uncoupling by acetic acid limits growth of acetogenesis by Clostridium thermoaceticum. Appl. Environ. Microbiol. 48:1134-1139.

30   D.M. Ivey, and L.G. Ljungdahl, 1986. Purification and characterization of the F_1-ATPase from Clostridium thermoaceticum. J. Bacteriol. 165:252-257.

31   L.L. Lundie, Jr. and H.L. Drake, 1984. Development of a minimally defined medium for the acetogen Clostridium thermoaceticum. J. Bacteriol. 159:700-703.

32   M.D. Savage, and H.L. Drake, 1986. Adaptation of the acetogen Clostridium thermoautotrophicum to minimal medium. J. Bacteriol. 165: 315-318.

33   H. Yang, and H.L. Drake, 1990. Differential effects of sodium on the hydrogen- and glucose-dependent growth of the thermophilic acetogen, Acetogenium kivui. Appl. Environ. Microbiol. 56:81-86.

34   J. Terracciano, W.J.A. Schreurs, and E.R. Kashket, 1987. Membrane H^+ conductance of Clostridium thermoaceticum and Clostridium acetobutylicum: Evidence for electronic Na^+/H^+ antiport in Clostridium thermoaceticum. Appl. Environ. Microbiol. 53:782-786.

35   L.G. Ljungdahl, 1983. Formation of acetate using homoacetate fermenting anaerobic bacteria. In Organic Chemicals from Biomass (D. L. Wise, ed.). Menlo Park, CA: Benjamin Cummings Publ. Co., pp. 219-248.

36   R.D. Schwartz and F.A. Keller, Jr. 1982 a. Isolation of a strain of Clostridium thermoaceticum capable of growth and acetic acid production at pH 4.5. Applied and Environmental Microbiology 43:117-123.

37   R.D. Schwartz, and F.A. Keller, Jr. 1982. Acetic acid production by Clostridum thermoaceticum in pH-controlled batch fermentations at acidic pH. Appl. Environ. Microbiol. 43:1385-1392.

38   L.G. Ljungdahl, L.H. Carreira, R.J. Garrison, N.E. Rabek, L.F. Gunter, and J. Wiegel, 1986. CMA manufacture (II): Improved bacterial strains for acetate production RFD# DTFH-61-83-R-00124. U.S. Department of Transportation, Federal Highway Administration.

39   D.L. Wise, and D. Augenstein, 1988. An evaluation of the bioconversion of woody biomass to calcium acetate deicing salt. Solar Energy 41:453-463.

40   R.M. Busche, E.J. Shimshick, and R.A. Yates, 1982. Recovery of acetic acid from dilute acetate solutions. Biotech. Bioeng. Symp. 12: 249-262.

41   Y. Kuo, and H.P. Gregor, 1983. Acetic acid extraction by solvent membranes. Separation Sci. Tech. 18:421-440.

42   A.B. Gancy, 1986. Preparation of high quality calcium magnesium acetate using a pilot plant process, FHWA/RD-86/006. Washington, DC: Federal Highway Administration, pp. 1-26.

TABLE 16.1

Basic Media for *Clostridium thermoaceticum*, *Clostridium thermoautotrophicum* and *Acetogenium kivui*

Ingredient	*C. thermoaceticum*[1] Amount/liter	*C. thermoautotrophicum*[2] Amount/liter	*A. kivui*[3] Amount/liter
Glucose	20.0g	10.0g	10.0g[3]
Yeast Extract	5.0g	2.0G	-
Yeast Extract, Difco	-	-	2.0g[3]
$NaH_2PO_4 \cdot H_2O$	-	-	4.5g
$NaHCO_3$	7.5g	-	-
K_2HPO_4	7.0g	-	220.0mg
KH_2PO_4	5.5g	-	220.0mg
NH_4Cl	-	500.0mg	310.0mg
$(NH_4)_2SO_4$	1.0g	-	220.0mg
$MgSO_4 \cdot 7H_2O$	250.0mg	-	90.0mg
$MgCl_2 \cdot 6H_2O$	-	180.0mg	-
$CaCl_2 \cdot 2H_2O$	-	-	6.0mg
$Fe(NH_4)_2(SO_4)_2 \cdot 6H_2O$	39.0mg	-	-
$FeSO_4$	-	5.0mg	-
$FeSO_4 \cdot 7H_2O$	-	-	2.0mg
$Co(NO_3)_2 \cdot 6H_2O$	30.0mg	-	-

TABLE 16.1 (concluded)

Ingredient	C. thermoaceticum[1] Amount/liter	C. thermoautotrophicum[2] Amount/liter	A. kivui[3] Amount/liter
$Na_2WO_4 \cdot 2H_2O$	3.3mg	–	–
$Na_2MoO_4 \cdot 2H_2O$	2.4mg	–	–
$Na_2HPO_4 \cdot 7H_2O$	–	3.1g	–
$Na_2HPO_4 \cdot 12H_2O$	–	–	6.1g
Na_2SO_4	–	1.0g	–
NaCl	–	–	450.0mg
$NiCl_2$	240.0µg	–	–
$ZnSO_4$	290.0µg	–	–
Na_2SeO_3	17.0µg	–	–
Trace mineral solution[4]	–	5.0ml	10.0ml
Vitamin solution[5]	–	0.5ml	–
Resazurin solution[6]	–	1.0ml	1.0ml
Reducing solution[7]	20.0ml	40.0ml	10.0ml[3]

[1]The medium is a modification of that used by Andreesen et al. [19].
[2]From Wiegel et al. [10].
[3]Based on medium published by Leigh et al. [9] but modified.
[4]For composition see Table 16.2.
[5]Composition according to Balch et al. [20].
[6]For composition see the text.
[7]For composition see Table 16.3.

TABLE 16.2

Trace Mineral Solution[1]

Ingredient	mg/liter
Nitrilo acetic acid	1500
$MgSO_4 \cdot 7H_2O$	3000
$MnSO_4 \cdot H_2O$	500
NaCl	1000
$FeSO_4 \cdot 7H_2O$	100
$Co(NO_3)_2 \cdot 6H_2P$	100
$CaCl_2$(anhydrous)	100
$ZnSO_4 \cdot 7H_2O$	100
$CuSO_4 \cdot 5H_2O$	10
$AlK_2(SO_4)_3$(anhydrous)	10
Boric acid	10
$NiCl_2$	50
$Na_2MoO_4 \cdot 2H_2O$	10
Na_2SeO_3(anhydrous)	1
Na_2WO_4	10

[1] The nitriloacetic acid is suspended in about 500 ml of water. It is dissolved by titrating with 2-3 N KOH until the pH is stabilized at 6.5. The rest of the ingredients are then added and dissolved in the order they are listed. Finally, the volume is adjusted to 1000 ml. The trace mineral solution is a modification of one previously composed by Wolin et al.[20]

TABLE 16.3

Reducing Solution[1]

Ingredients	Amount
NaOH 0.2 N	200 ml
Na_2S $9H_2O$	2.6 g
Cysteine HCl H_2O	2.5 g

[1] The sodium hydroxide solution is brought to boiling and made anaerobic by bubbling with N_2/H_2 (95/5 percent) gas mixture. The solution is allowed to cool somewhat and the sodium sulfide and the cysteine are then added. Using the anaerobic Hungate technique, 8 ml aliquots are transferred to Belco tubes which have been preflushed with the N_2/H_2 gas mixture. The tubes are stoppered with butyl rubber stoppers, wired, and autoclaved for 15 min at 15 psi.

TABLE 16.4

Sodium and Potassium Acetate concentrations Allowing Growth and Acetate Production by Acetogens from Ljungdahl et al. [23]

Bacterial Strains	Initial[1]		Acclimated[2]	
	NaAc (mM)	KAc (mM)	NaAc (mM)	KAc (mM)
A. kivui				
DSM	400	350	700	400
RW	400	350	700	350
C. thermoautotrophicum				
701/3	200	50	300	150
701/5	300	250	400	350
C. thermoaceticum				
Wood	500	nd	850(930)	(305)[3]
Ljd	550	nd	850	350

[1] Highest concentration of acetate in optimum medium before inoculation in which growth and acetate production was obtained within 120 h. As inocula, 1 percent, were used cultures previously not grown at high acetate concentrations. A. kivui was grown at 63°C, pH 6.7, and with 1 percent glucose. The Clostridia at 60°C, pH 6.8, and with 2 percent glucose.

[2] Highest concentration of acetate in medium before inoculation with 10 percent inocula of cultures previously grown on media containing the initial highest acetate concentration in which growth was obtained.

[3] Data from Wang and Wang with pH controlled culture [28].

TABLE 16.5

Baseline Data for Strains of *C. thermoaceticum*, *C. thermoautotrophicum*, and *A. kivui* in Batch Cultures. From Ljungdahl et al. [23]

Property	*C. thermoaceticum*		*C. thermoautotrophicum*			*A. kivui* DSM 2030
	Wood	Ljd	701/3	701/5	RW	
Temperature, °C						
Range	47-65	51-65	48-65	42-66	35-73	35-73
Optimum	56-58	60	62	60	60-65	60-64
pH						
Optimum	6.6	6.8	6.5	6.6	6.2-6.8	6.3-6.8
Range	5.7-6.8	5.7-7.65	5-7.2	4.8-7.3	5-7.6	5-7.6
Doubling time, h	4-6	6-8	5	5	2-3	2-3
Glucose, %						
Range	0.5-16	0.5-10	0.5-20	0.5-11	0.5-10	0.5-14
Optimal[1]	2	2	1	1	1	1
Yeast extract %[2]						
Minimal	0.005	0.005	None	None	None	None
Optimal	0.5	0.5	0.5	0.4-0.5	0.05	0.2
Phosphate, mM						
Range	2-80	2-80	2-80	2-80	2.9-60	2.9-60
Optimal	40	40	20-40	20-40	10	2.9
$CaCl_2$, tolerance, mM	250	250	25	25	25	25
$MgCl_2$, tolerance, mM	250	250	200	350	50	200
Dolime tolerance, %	2	1.25	0.5	3	0.75	0.5

TABLE 16.5 (concluded)

1 Optimal concentration of glucose was that with fastest growth and complete conversion to acetate in the batch culture.

2 Yeast extract was found to be required for both strains of *C. thermoaceticum*; however, a minimal medium without yeast extract has been developed for this bacterium [31] and recently also for *C. thermoautotrophicum* [30]. The latter publication also describes vitamin requirements for many acetogens, e.g., *C. thermoautotrophicum* and *C. thermoaceticum* require nicotinic acid, whereas *A. kivui* has no requirement. In the absence of yeast extract, *C. thermoautotrophicum* JW 701/3 and 701/5 required at least 1 mg/ml nicotinic acid.

TABLE 16.6

Concentrations of Metals in CMA

Element	Concentrations mg/1 g Dolime	µmol/g Dolime
Silver, Ag	0.013	0.001
Aluminum, Al	n.d.	n.d.
Indium, In	n.d.	n.d.
Arsenic, As	0.04	0.5
Gold, Au	0.015	0.007
Boron, B	0.015	1.4
Barium, Ba	0.01	0.07
Beryllium, Be	0.001	0.1
Bismuth, Bi	0.036	0.17
Calcium, Ca	380.0	9400.0
Cadmium, Cd	0.003	0.03
Cobalt, Co	0.01	0.16
Chromium, Cr	0.04	0.7
Copper, Cu	0.003	0.05
Iron, Fe	2.2	39.0
Gallium, Ga	n.d.	n.d.
Potassium, K	0.55	14.0
Lithium, Li	0.0016	0.2
Magnesium, Mg	153.0	6300.0
Manganese, Mn	0.1	1.9
Molybdenum, Mo	0.013	0.14
Sodium, Na	0.07	3.0
Nickel, Ni	0.014	0.24
Phosphorus, P	0.2	6.8
Lead, Pb	0.06	0.28
Rhodium, Rh	0.06	0.55

TABLE 16.6 (concluded)

Concentrations of Metals in CMA

Element	Concentrations mg/l g Dolime	µmol/g Dolime
Antimony, Sb	0.03	0.26
Scandium, Sc	0.0004	0.008
Selenium, Se	0.12	1.5
Silicon, Si	8.5	300.0
Tin, Sn	0.045	0.38
Strontium, Sr	0.045	0.5
Titanium, Ti	0.05	1.1
Thallium, Tl	0.016	0.8
Uranium, U	0.42	1.8
Vanadium, V	0.013	0.09
Yttrium, Y	0.005	0.05
Tungsten, W	n.d.	n.d.
Zinc, Zn	0.01	0.15
Zirconium, Zr	0.004	0.045

TABLE 16.7

Fermentations with *C. thermoaceticum*, Strain Ljd, Using pH-Controlled Batch Cultures

Condition and Property Measured	Substrate		
	Glucose		Hydrolyzed Starch
	Best exp.	Range[a]	
ph-optimum	6.8	6.6-7.3	6.8
Doubling time, h: Substrate 2%	4.9	4.4-7.3	9.1[b]
10%	7.8		
Final O.D. (A_{554}): Substrate 2%	5.5	4.4-5.6	5.8
10%	19		
Acetate produced, mM: Substrate 2%	238	204-238	277
10%	572		
Glucose used, mM:[c] Substrate 2%	84 (77%)	76-91 (65-77%)	102 (100%)
10%	266 (50%)		
Acetate/glucose: Substrate 2%	2.9	2.5-2.9	2.7
10%	2.2		

[a] Results given in ranges from several experiments. Fermentations were at 60°C and pH 6.8.

[b] This value is high, generally the doubling time for starch hydrolyzate is similar to fermentations with glucose as substrate.

[c] The values in parentheses are percent of added glucose that has fermented.

TABLE 16.8

Results of Continuous Culture (A) and Cell-Recycling Culture (B) with *C. thermoautotrophicum* strain 701/5[1]

Medium Retention Time	Optical Density	Acetate conc.	Glucose conc.	$\frac{Acetate}{glucose}$	Acetate Production rate
A. Hours	O.D.	mM^3	mM^3	ratio	g/l/h
3.4	4.0	75.2	82.6	2.74	1.33
4.0	4.8	122.0	62.0	2.55	1.84
5.0	6.1	131.0	58.8	2.60	1.57
5.6	6.3	142.0	55.0	2.58	1.52
6.6	6.4	154.0	52.0	2.65	1.40
11.0	5.9	181.0	40.5	2.60	0.99
22.0	4.3	250.0	17.8	2.70	0.68
B.					
4.0	5.3(47)[2]	251.0	20.7	2.82	3.80
4.8	4.8(60)	227.0	32.6	2.93	2.86
4.9	8.4(100)	180.0	26.5	2.15	2.20
5.3	4.6(56)	241.0	28.8	2.90	2.72
5.4	4.3(60)	215.0	32.2	2.69	2.37

[1]   Conditions were pH 6.8, 60°C and with 2 percent glucose.

[2]   Values within parentheses are percent cell-recycling

[3]   The values represent acetate concentration obtained and glucose left in the effluent from the fermentor.

TABLE 16.9

Results of Continuous Culture (A) and Cell-Recycling Culture (B) with *A. kivui* strain 701/5[1]

Medium Retention Time	Optical Density	Acetate conc.	Glucose conc.	Acetate glucose	Acetate Production rate
A.[2] Hours	O.D.	mM[4]	mM[4]	ratio	g/l/h
1.0	1.7	32.8	40.0	1.95	1.84
1.6	2.0	52.0	39.1	2.75	1.95
3.8	2.6	59.2	34.7	2.70	0.93
5.5	2.5	62.4	33.7	2.80	0.74
8.0	2.5	91.6	26.5	2.90	0.69
11.0	2.3	104.0	22.4	2.90	0.57
B.[2]					
0.9	n.d.[3]	72.2	23.0	2.58	4.76
1.0	"	62.5	30.5	2.66	3.75
1.1	"	64.2	31.1	2.93	3.50
1.4	"	71.7	24.6	2.72	3.07
2.1	"	98.6	15.6	2.64	2.82

[1] Conditions were pH 6.2 and 63°C.

[2] In A the medium contained 1% glucose and in B hydrolyzed starch 1.35% corresponding to 9% glucose.

[3] The optical density was not possible to accurately determine since cells tended to flocculate.

[4] The values are the concentrations of acetate and glucose in the effluent fluid from the fermentor and represent the concentration of acetate reached and glucose remaining after the fermentation.

403

TABLE 16.10

Comparison of *C. thermoaceticum*, *C. thermoautotrophicum* and *A. kivui* in Batch and Continuous Cultures with and without Cell-Recycling

	Retention Time (h)	Glucose left (mM)	Glucose used (mM)	HAc/glucose ratio	HAc produced mM	HAc produced g/l/h
C. thermoaceticum, strain Ljd						
2% glucose batch culture	–	2.0	108.0	2.74	296.0	–
2% glucose continuous culture	5.22	47.3	62.7	2.63	165.0	1.90
3% glucose cell-recycling	4.20	2.0	155.0	2.00	303.0	4.00
Corn starch-2% glucose batch culture	–	0.0	102.0	2.72	277.0	–
Corn starch-2% glucose continuous culture	7.60	56.0	54.0	2.57	139.0	1.10
Corn starch-3% glucose cell-recycling	4.20	91.0	69.0	2.85	197.0	3.00
C. thermoautotrophicum, strain 701/5						
2% glucose batch culture	–	3.0	106.0	2.83	291.0	–
2% glucose continuous culture	4.00	62.0	48.0	2.55	122.1	1.84
3% glucose cell-recycling	4.00	20.7	89.3	2.82	252.0	3.80
Corn starch-2% glucose batch culture	–	3.0	103.0	2.77	285.0	–
A. kivui, strain DSM 171						
1% glucose batch culture	–	2.0	52.0	2.83	147.0	–
1% glucose continuous culture	1.60	39.1	18.9	2.75	52.0	1.95
Corn starch-1% glucose batch culture	–	3.0	53.0	2.87	152.0	–
Corn starch-1% glucose continuous culture[1]	0.91	23.0	28.0	2.58	72.2	4.76
Corn starch-1% glucose cell-recycling[1]	0.93	30.0	23.0	2.68	61.6	3.97

[1] Flocculation occurred.

TABLE 16.11

Improved Bacterial Strains of *C. Thermoaceticum* and *C. thermoautotrophicum*

Strain	Selection Conditions		Comments
	pyruvate	850 mM CMA	
C. thermoaceticum			
Wood w.t.	none	none	produces 450 mM acetate at pH 6.9
Ljd w.t.	none	none	produces 450 mM acetate at pH 6.9
Ljd EMS	none	yes	produces over 2 M CMA of pH 6.9, no growth on xylose
Wood EMS 90	55 mM	yes	produces over 1.3 M CMA
Wood 6-2-1-P	55 mM	yes	produces 92 mM CMA at pH 6.8, and 400 mM CMA at pH 6.2
C. thermoautotrophicum			
701/5 w.t.	none	none	produces 500 mM acetate at pH 6.9
701/5 P-NA 20	55 mM	yes	produces over 2M CMA at pH 6.9, slow growth on xylose
701/5 P-NA 20	55 mM	yes	grows rapidly at first; produces 1300 mM CMA in batch cultures. No growth on xylose.
NTG 50			

TABLE 16.12

Fermentation with *C. thermoaceticum*, strain 701/5 P-NA 20 in a 100 ml Fermentor[1]

Dilution Cycle	Acetate in Culture Fluid
#	mM
1	850
2	930
3	1680
4	2173
5	1525
6	1885
7	1800

[1] See Figure 9 for explanation of "Dilution Cycle" and fermentation conditions.

TABLE 16.13

Effect of Retention Time in Rotary Fermentor during Continuous Fermentation with *C. thermoaceticum*, strain Ljd-EMS-RFB[1]

Retention Time	Acetate Production Rate	Acetate Concentration in culture fluid	pH of Culture Fluid
h	$gl^{-1} \cdot h^{-1}$	mM	
0.32	9.5	50	6.6
0.58	6.1	59	6.2
0.65	7.0	76	5.6
0.96	5.8	94	5.2
1.10	5.4	102	5.0
2.20	3.9	135	4.8

[1] Medium used contained 0.5 percent yeast extract, 0.5 percent sodium bicarbonate and 5 mM phosphate, pH 6.8.

TABLE 16.14

Analyses of 400-liter Fermentations for the Production of CMA from Corn Starch Hydrolysate

Run[1]	Glucose added[2]	Glucose left	Acetate produced	Acetate[3]/glucose ratio	Dolime added[2]	Soluble[4] Ca	Soluble[4] Mg	Total[4] Ca	Total[4] Mg	Soluble Mg/Ca ratio
	mM	mM	mM	ratio	g	mM	mM	mM	mM	ratio
1	222	41	412	2.28(2.54)	7261	48.3	146	107	145	3.02
2	250	5	393	1.60(1.80)	8864	71.0	129	135	222	1.82
3	278	5	502	1.84(2.04)	10586	58.8	168	142	179	2.86
4	278	4.3	432	1.54(1.78)	9425	54.3	144			2.65
5	250	0	403	1.61(1.81)	8275					
6	385	4	625	1.64(1.85)	11500					
7	374	25	565	1.62(1.82)	16615					
8	379	36	476	1.31(1.48)	13025					

[1] Runs 1 to 5 were with *C. thermoaceticum* w.t. strain, runs 6 and 7 with *C. thermoautotrophicum*, 701/5 P-NA 20, and run 8 with *C. thermoaceticum*, Wood EMS 90.

[2] Total glucose and dolime added throughout the fermentation.

[3] Acetate/glucose ratio, was obtained by Acetate produced/Glucose added-glucose left.

These values are lower than expected. However, the calculations were done assuming a constant volume during the fermentation of 400 liters, whereas by addition of glucose and dolime during the fermentation the actual volume was about 450 liters. Using this volume the acetate/glucose ratios obtained are those given within parentheses.

[4] Total concentrations of Ca and Mg include both undissolved dolime suspension and dolime that has dissolved. Soluble Ca and Mg include the concentration after the undissolved dolime has been removed.

TABLE 16.15

Concentrations of Elements in CMA-Solution Obtained by Fermentation[1]

Element	Total concentrations[2]		Soluble concentrations[2]	
	mg•l⁻¹	mM	mg•l⁻¹	mM
Silver, Ag	1.62	0.015	0	0
Aluminum, Al	51.85	1.922	26.11	0.97
Indium, In	0	0	0	0
Arsenic, As	1.56	0.021	1.27	0.017
Gold, Au	0.34	0.002	0.21	0.001
Boron, B	0.31	0.029	0.32	0.03
Barium, Ba	0.27	0.002	0.04	0.0003
Beryllium, Be	0.16	0.018	0.005	0.0006
Bismuth, Bi	0.86	0.004	0.72	0.003
Calcium, Ca	5695.0	142.09	2356.5	58.8
Cadmium, Cd	0.05	0.0004	0.05	0.0004
Cobalt, Co	3.48	0.059	2.03	0.034
Chromium, Cr	0.50	0.01	0.43	0.008
Copper, Cu	0.22	0.003	0.36	0.006
Iron, Fe	55.23	0.99	0.71	0.013
Gallium, Ga	2.75	0.039	1.89	0.027
Potassium, K	1037.0	26.52	1049.0	26.83
Lithium, Li	0.02	0.003	0.02	0.003
Magnesium, Mg	4357.5	179.25	4084.0	168.0
Manganese, Mn	4.59	0.084	0.50	0.009
Molybdenum, Mo	0.67	0.007	0.57	0.006
Tin, Sn	1.49	0.013	1.46	0.012
Sodium, Na	577.7	25.13	593.7	25.82

TABLE 16.15 (concluded)

Concentrations of Metals in CMA-Solution Obtained by Fermentation[1]

Element	Total concentrations[2]		Soluble concentrations[2]	
	ppm	mM	ppm	mM
Nickel, Ni	0.28	0.005	0.22	0.004
Phosphorus, P	505.5	16.32	6.96	0.225
Lead, Pb	0.95	0.005	0.55	0.0003
Rhodium, Rh	0.15	0.001	0	0
Antimony, Sb	1.0	0.008	0.7	0.006
Scandium, Sc	0.01	0.0002	0	0
Selenium, Se	2.10	0.027	1.56	0.020
Silicon, Si	63.2	2.25	52.55	1.87
Strontium, Sr	1.07	0.012	0.54	0.006
Titanium, Ti	0.96	0.02	0.28	0.006
Thallium, Tl	2.60	0.013	0.46	0.002
Uranium, U	4.78	0.02	2.01	0.008
Vanadium, V	0.99	0.019	0.79	0.016
Yttrium, Y	0.13	0.001	0.03	0.0003
Tungsten, W	2.43	0.013	1.54	0.008
Zinc, Zn[3]	2.56	0.039	12.39	0.19
Zirconium, Zr	0.05	0.0005	0.03	0.0003

[1] The CMA-solution is that of fermentation 3 of Table 14.

[2] Total concentrations are the fermentations fluid including the undissolved dolime. For the analyses the dolime was stirred up and nitric acid was added to dissolve the suspended dolime. Soluble concentrations were obtained after that the undissolved dolime had been removed.

[3] The soluble zinc is higher than the total, which obviously cannot be correct. We presently have no explanation for this.

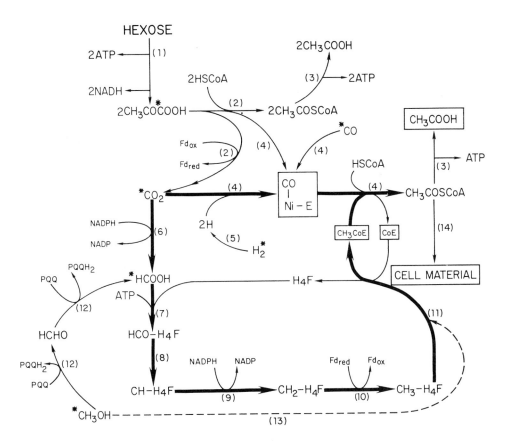

Figure 1:

The autotrophic acetyl-CoA pathway (heavy arrows) and connected metabolism of
hexoses, methanol, and CO; H_4F tetrahydrofolate; CoE, corrinoid enzyme; CO-Ni-
E, carbon monoxide dehydrogenase/acetyl-CoA synthase with CO moiety bound to
nickel; PQQ, pyrroloquinoline quinone; Fd, ferredoxin. Enzymes or reaction
sequences are as follows: 1, glycolysis; 2, pyruvate-ferredoxin
oxidoreductase; 3, phosphotransacetylase and acetate kinase; 4, carbon
monoxide dehydrogenase/acetyl-CoA synthase; 5, hydrogenase; 6, formate
dehydrogenase; 7, formyl-H_4 folate synthetase; 8, methenyl-H_4 folate
cyclohydrolase; 8, methenyl-H_4 folate cyclohydrolase; 9, methylene-H_4 folate
dehydrogenase; 10, methylene-H_4 folate reductase; 11, transmethylase; 12,
methanol dehydrogenase; 13, methanol-cobamide methyltransferase; 14,
anabolism.

410

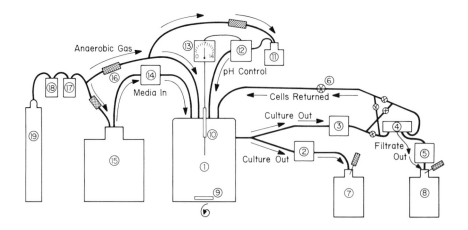

Figure 2:

Set-up for continuous and cell-recylcing studies under anaerobic conditions.
(1) New Brunswick Fermentor Model C-30; New Brunswick Scientific Co., Inc.,
P.O. Box 986, 44 Talmadge Road, Edison, NJ 08818. (2) Manostat Cassette Pump
Junior Model, Catalog No. 72-510-000 (115V) with pumping cassette, Catalog No.
72-550-000; Manostat, 519 Eight Avenue, New York, NY 10018. (3) Harvard
Apparatus Peristaltic Pump, Model 1204; Harvard Apparatus, 22 Pleasant
Street, South Natick, MA 01760. (4) Amicon Hollow Fiber Cartridge Filter,
Type HlMPO1-43 with Hollow Fiber Cartridge Filter, Type HlMPO1-43 with Hollow
Fiber Cartridge Adaptor, Model DH2, No. 54077; Amicon Corporation Scientific
Systems Division, 17 Cherry Hill Drive, Danvers, MA 01923. (5) Same as (2)
above. (6) Regulating valve -- e.g., Nupro part no. B-4JR; Nupro Company,
4800 East 345th Street, Willoughby, OH 44994. (7,8) Outflow container-glass
carboy. (9) Magnetic stir bar. (10) Phoenix pH electrode, Catalog No.
5573705-DL; Phoenix Electrode Co., 6103 Glenmont Street, Houston, TX 77081.
(11) Container for NAOH-glass flask. (12,13) New Brunswick pH Controller,
Model pH-40; New Brunswick Scientific Co., Inc., P.O. Box 986, 44 Talmadge
Road, Edison, NJ 08818. (14) LKB Peristaltic Pump, Type 4912-A; LKB
Instruments, Inc., 9319 Gaither Road, Gaithersburg, MD 20877. (15) Media
Container-glass carboy. (16) Sterile Cotton Gas Filter. (17) Backflow
Trap--glass flask. (18) Sargent-Welch Gas Purifying Furnace, Model S-36517
with copper granules filled gas purifying tube, Model S-36518; Sargent-Welch
Scientific Company, 7300 North Linder Avenue, P.O. Box 1026, Skokie, IL
60077. (19) Gas Cylinder--CO_2. (20) Connection tubing, Butyl rubber tubing,
Viton or Versinik. From Ljungdahl et al. [7] (Reproduced with permission).

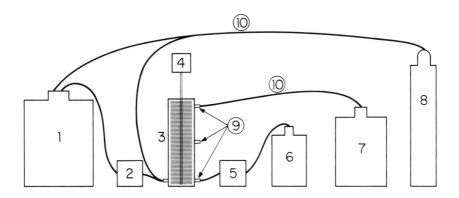

Figure 3:

Rotary fermentor outline. Medium reservoir, Carboy; (2) Peristaltic pump,
Model 1203, Harvard Apparatus, South Natick, MA; (3) Rotary fermentor of
glass, University of Georgia Glass Shop. The fermentor contains a stainless
steel rod to which are attached 75 pads consisting alternatively (1:2) of
DuPont Reemay and 65/35 polyester/wood materials. The fermentor volume is
about 400 ml; (4) Stirring motor, Dayton, Model 22810, W. W. Grainger Co.,
Atlanta, GA; (5) Peristaltic pump, Model TR15-ISCO, Lincoln, NE; (6) Glucose
reservoir, Carboy; (7) Fermentation liquid (product) reservoir, Carboy; (8)
CO_2 gas tank; (9) Portholes for sampling; (10) Butyl rubber tubing.

412

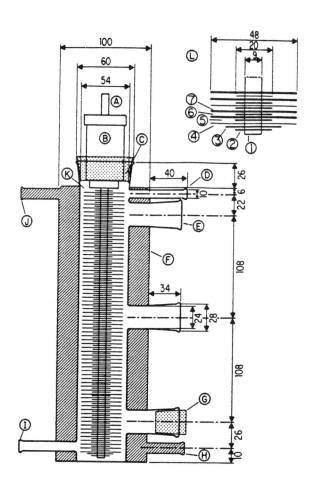

<u>Figure 4</u>:

Design of rotary fermentor vessel. (A) Stirring shaft (0 mm stainless steel
rod) to be connected to stirring motor. (B) Lipseal stirrer assembly (without
coupling ring and motor studs). Cole-Parmer Instrument Co., Chicago, IL. (C)
Rubber stopper - #11. (D) Culture outlet. (E) Sampling port. (F) Water
jacket. (G) Rubber stopper - #4. (H) Water jacket inlet. (I) Media/CO_2
outlet. (J) Water jacket outlet. (K) Fermentor-showing stirring rod with
pads. (L) Expanded section of stirring rod with pads. 1) Stirring rod. 2)
Bottom pin to hold pans on. 3) Bottom washer-stainless steel-2 mm x 30 mm.
4) DuPont Reemay 2033 pad - 0.4 mm x 48 mm. 5) Teflon washer - 0.8 mm x 20
mm. 6) 60:40 pad - 2.5 mm x 48 mm. 7) 60:40 pad.

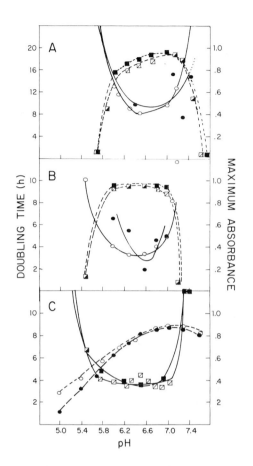

Figure 5:

Effect of pH on growth of acetogenic bacteria. Growth rates (solid lines) and growth maxima (dashed lines) as a function of the initial pH of the medium. (A) *C. thermoaceticum* strain Ljd, solid circles and squares; strain Wood, open circles and squares. (B) *C. thermoautotrophicum* strain JW 701/3, open marks; strain JW 701/5, solid marks. (C) *A. kivui* strain RW, open marks; strain DSM, solid marks. The maximum A_{554} was determined after 120 h. From Ljungdahl et al. [23]. (Reproduced with permission).

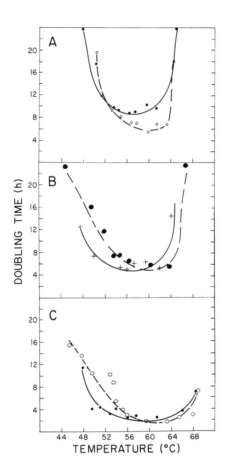

Figure 6:

Growth rates as a function of a temperature of acetogenic bacteria in media
initially at pH-optima. (A) *C. thermoaceticum* strain Ljd, open circles;
strain Wood, solid circles, (B) *C. thermoautotrophicum* strain 701/5, solid
circles; strain 701/3, +; (C) *A. kivui* strain RW, solid circles; strain DSM,
open circles. Cells used for inocula were for each temperature shown at the
listed temperatures to avoid slow growth due to temperature adaptation. The
media for *C. thermoaceticum* contained 2 percent (w/v) glucose and for *C.
thermoautotrophicum* and *A. kivui* 1 percent (w/v) glucose [23]. (Reproduced
with permission).

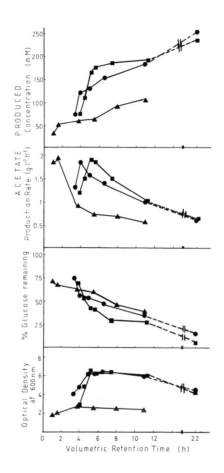

<u>Figure 7</u>:

Continuous fermentations with *C. thermoaceticum*, strain Ljd (-■-); *C. thermoautotrophicum*, strain 701/5 (-•-); and *A. kivui*, strain DSM (-▲-). Acetate concentration reached and production rate, glucose concentration and cell growth (OD) are plotted as a function of volumetric retention time. The media for the clostridial strains contained 2% glucose (w/v) and for *A. kivui* 1 percent. The clostridial fermentations were at 60°C and pH 6.85 and for *A. kivui* at 63°C and pH 6.25 [23]. (Reproduced with permission).

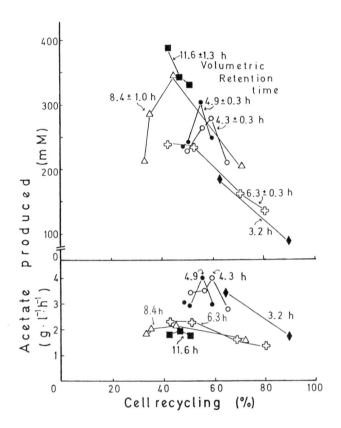

<u>Figure 8</u>:

Fermentations with cell-recycling using *C. thermoaceticum* strain Ljd. Cell recycling in percent and the volumetric retention times are indicated in the figure. Zero percent cell cycling means that no cells were returned to the fermentation unit, and with 100% cell-recycling all cells were returned. The fermentations were at 60°C, pH 6.85, and with 3 percent glucose in the medium reservoir [23]. (Reproduced with permission).

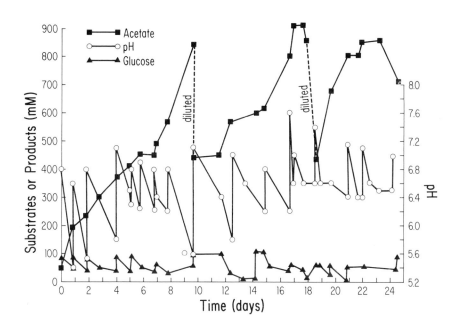

<u>Figure 9</u>:

Fermentation of glucose to CMA using *C. thermoautotrophicum*,
strain 701/5 P-NA 20. The fermentation was performed in a 100 ml fermentor
equipped with pH probe and portholes for sampling, addition of glucose and
dolime. Glucose concentration was kept between 0.5 to 2 percent and pH
between 5.6 to 7.2 by the addition of dolime. At days 9 and 18 culture fluid
was diluted by removal of an aliquot, that was replaced by fresh medium, as
indicated in the text.

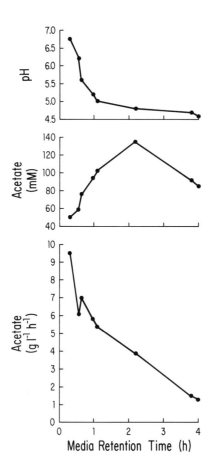

Figure 10:
C. *thermoaceticum*-Ljd-EMS-RFB. Fermentation of hydrolyzed corn starch with C. *thermoaceticum*, strain Ljd-EMS-RFB, in a rotary fermentor. Effect of retention time on pH.

Chapter 17

PROCESS EVALUATION OF CMA PRODUCTION FROM WOODY BIOMASS

Bryan L. DeSouza and Donald L. Wise
Northeastern University, Chemical Engineering Department
342 Snell Engineering Center, Boston, MA   02115

17.1 SUMMARY

CMA, calcium magnesium acetate, is rapidly becoming an important chemical with increased need, primarily as a degradable non-corrosive deicing salt. Another important use already seen is in the combustion of coal. CMA is used to capture the sulphur released during combustion which prevents the formation of sulfur dioxide and thereby reduces acid rain. This chapter features a brief background of the types of feedstocks which are suitable and available with a concentration more on the modelling and economic evaluation of the selected process. A computer model was written using ASPEN PLUS^{TM} a process simulator software package by Aspentech Inc. The entire process was simulated. A cost model was also written in conjunction with the model which evaluated the economics of the process. As seen from the results, the process is economically viable. The preliminary price calculated using a base model of 500 US tons/day (454 metric tons/day) was $ 249/metric ton. The cost of CMA is up to $ 700/ton when produced from acetic acid derived from petroleum or natural gas. Sensitivity analysis performed on the model gave an estimate as to the ranges which certain key variables should have in order for the process to be economically feasible. In addition, this analysis demonstrated the areas in which more research and development may be required. The results show that the process studied here is economically desirable to produce CMA from woody biomass as compared to using acetic acid derived from petroleum or natural gas. As seen from these aspects, there are no factors preventing CMA from being manufactured on an industrial scale.

17.2 INTRODUCTION & BACKGROUND
The Problem
The massive use of chloride salts for roadway deicing has led to the serious problems listed below:
Environmental
- Pollution of aquatic habitats
- Pollution of drinking water as a result of sodium and chloride ions in runoff.

- High sodium ion concentration in soil leading to withering of vegetation.

Corrosion

- Deterioration of Portland concrete bridge decks through chloride ion corrosion of reinforcing steel.
- Corrosion of steel structural members and other highway appurtenances. Corrosion of underground electrical cables and piping.
- Corrosion of vehicle chassis.

Resource waste & maintenance costs

- Gasoline wasted in slow moving traffic caused by repair work on bridges and decks.
- Cost of labor, replacement material, and transport of material to site.

For example, it is estimated that in New York state alone, the overall cost of damage using this kind of deicing salt was more than $ 500 million per year (1).

The obvious advantages of using NaCl and $CaCl_2$ are not only its immediate availability but also its economic desirability. The current prices of NaCl run at approximately 2.5 cents/lb and $CaCl_2$ at 4.5 cents/lb (1).

Approximately 12 million metric tons of NaCl are used per year in the U.S. and about 1.2 million metric tons of $CaCl_2$ in the severe frost belt areas which amount to nearly 20 % of the annual production of NaCl and roughly 35 % for $CaCl_2$ (2).

Even though costs resulting from damages are very high, they must be balanced against the safety and economic benefits derived from keeping the pavements and roads clear during winter (1).

17.2.1 Alternative Approach

Research work in the direction of finding an alternative deicing agent was initiated during the mid-1970's by the FHWA. CMA (calcium magnesium acetate) and methanol were identified as potential candidates primarily because of their non-corrosive and acceptable deicing properties.

The immediate advantage of using CMA as a deicing salt is a substantial reduction, or possible eradication, of corrosion and pollution problems cited earlier in this chapter. But none of the current processes to manufacture CMA can be used on a commercial scale. The main limiting factor is the acetate ion source which comes from acetic acid. Commercial sources available now are derived from petroleum and natural gas but the demand for acetic acid will be so great that a mere 10 % replacement of NaCl (1.2 million tons) will expend one half of the total 1981 production of acetic acid in the US.

The need for CMA is not limited to its use as a deicing salt. CMA has also demonstrated excellent adaptability in techniques to control sulfur dioxide pollution in the combustion of coal. The process follows the addition of CMA to the coal at a specific stage in the combustion process when the calcium and magnesium ions combine with the sulfur in the coal. This combination is known as sulfur binding. The entire process is investigated by Manivannan et al (7), (this text) as a main use of CMA for the control of sulfur pollution.

17.2.2 Prior Art

Inexpensive sources of acetic acid using fermentation routes of biomass were researched. The type of fermentation discussed (U.S. pat No. 4,636,467) is mixed hydrolysis fermentation. The disadvantage in this type of fermentation is high ion concentration. The Ca & Mg ions are reacted in the fermenter itself thereby reducing the fermentation activity of the micro-organisms. Two U.S. patents No. 4,606,836 & No. 4,444,672, deal with specific procedures for making calcium magnesium acetate with raw acetic acid bought from the market; however, these procedures are not economically feasible and worse the supply is scarce. In an interesting U.S. patent, No. 4,488,978, diverse methods are discussed of varying the texture of the final product to increase its deicing efficiency through physical and chemical alterations. This approach merits further study as the product can be made at a lower cost if its effectiveness increases.

17.2.3 Organic Residues

Many organic residues are readily available as cheap sources for the fermentation process wherein the main product is acetic acid (5). They may be categorized by the amounts available, the ease with which they can be converted, and the extent of pretreatment required before fermentation. These residues are:

- Cheese whey: An easy-to-convert dairy by-product which essentially requires no pretreatment.
- Sewage sludge: Large scale amounts available which need little or no pretreatment.
- Corn: An excellent source as far as supply is concerned but requires initial pretreatment.
- Wood residue: An almost unlimited source which exceeds the demand but requires pretreatment prior to fermentation.
- MSW: Municipal solid waste consists of nearly 40-80 % of

cellulosic matter which does not need as severe a pretreatment as corn or wood chips.

17.3 PROCESS DESCRIPTION

As shown in figure 1, the overall process utilizes wood chips as the primary organic feedstock which is pretreated using a steam explosion process. The operation involves exposure to repeating cycles of steam at approximately 2000 psia which breaks up the cellulosic matter from the fiber rendering it more fermentable. This pretreated pulp is fed to a fermenter in a 5-15 % by wt concentration wherein it gets converted to acetic acid (up to 90 %) anaerobically. The acetic acid is then extracted using a liquid ion exchange carrier in an organic phase, contacted with the fermentation broth in a counter-current extractor. The unconverted biomass is washed, dewatered and then burned directly as a boiler fuel. The acetic acid which was carried over is now back-extracted and reacted with hydrated dolime to give a 35 wt % solution of CMA. The solution is spray dried to give the final product. Optionally, the final product can be modified in many different ways to improve it's deicing effiency.   Unit operations involved:

- Collection unit, and buffer stock unit of wood chips.
- Pretreament of wood chips using the steam explosion process.
- Packed bed fermentation of pretreated wood pulp.
- Extraction of acetic acid from fermentation broth by the liquid ion exchanger.
- Back extraction of acetic acid from the liquid ion exchanger by direct contact with a saturated solution of dolime to give a 35 wt % of CMA.
- Separation of the liquid ion exchanger organic phase from the CMA aqueous phase.
- Spray drying of the CMA solution to give the final product CMA.

17.4 PROCESS MODEL AND COMPUTER SIMULATION

A process model was developed based on the quantity of 500 US tons/day of CMA. A black box representation of the process is made and all streams and blocks are identified and labelled.

In the procedures involving computer simulation, stream data and component properties models were set up in order to define the physical properties of the solids being simulated in the process. The solids concerned here were mainly wood pulp and dolomite. None of these solid fractions of the flowsheet streams participated in equilibrium interactions with the corresponding liquid fraction of the streams.

In order to further build the model a great deal of information was acquired by making practical assumptions. Adding information is just a matter of editing the program. The choice of simulation package was ASPEN PLUS™ by Aspentech Inc, Cambridge MA U.S.A. The ASPEN PLUS™ simulator allows good versatility where costing and sensitivity of process variables against each other are concerned, compared to other available simulators which demand another software package like a spreadsheet, to be interfaced manually. The ASPEN PLUS™ simulator was also used to evaluate the costing as much in detail as possible. A few of the preliminary costs were user-provided because the information was readily accessible and the relative error was within appreciably closer limits.

17.5 PROCESS ECONOMICS & EVALUATION

After modelling the process, the next step was to check its economical feasibility. The model now incorporated every minor detail such as pumps, piping, equipment sizing, flow rates, utilities etc. This would lead to additional sensitivity parameters and further analysis of the process. Costing and evaluation of this process then determined the final feasibility which is critical in starting any new project. Results of the sensitivity analysis further fixed a range within which a certain variable could fluctuate and still leave the process economically feasible although not necessarily desirable.

17.6 PROCESS MODEL

17.6.1 Step-By-Step Preliminary Modelling:

1. Clearly state the basis on which to work (Time).
2. Acquire the stoichiometry of all reactions.
    *   Estimate the conversions.
    *   Estimate the kinetics involved for rate of reaction.
3. Set up system boundaries for which a particular block can be defined.
4. Identify the total number of independent components and write a balance equation for each component.
5. Check streams:
    *   Physical streams: no reaction involved.
    *   Chemical streams: reaction involved.
6. Flows & concentrations:
    *   Specify the flow (quantity) of each component.
    *   Specify the total flow and composition.
    *   Specify the flow of one component and the composition.

424

For simplicity, No.1 above has been chosen to represent stream flows and concentrations.

17.6.2 Basis

Plant sizes ranging between 91 and 910 metric tons/day have been specified by the Federal Highway Administration. A calcium magnesium acetate production capacity of 454 metric tons/day (500 US tons/day) is chosen.

17.6.3 Stoichiometry

Stoichiometry of reactions:

Fermenter

Cellulosic fraction of wood to acetic acid:

$$\sim (C_6H_{12}O_6)_n \sim \rightarrow 3CH_3COOH \tag{1}$$

Reactions involving cellulose are in fact a complex set of biological reaction pathways. The reaction shown above represents the overall reaction effect.

Back Extraction

Back extraction of acetic acid by hydrated dolime to give CMA.

$$2CH_3COOH + Ca(OH)_2 \rightarrow Ca(C_2H_3O_2)_2 + 2H_2O \tag{2}$$
$$CH_3COOH + Mg(OH)_2 \rightarrow Mg(C_2H_3O_2)_2 + 2H_2O \tag{3}$$

17.6.4 Kinetics of a Fermenter

The fermenter in this process was modelled as a plug flow reactor. Preliminary modelling is shown below (6).

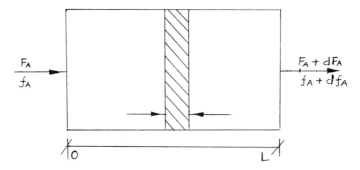

F_A = molal flow rate of reactant A (Cellulosic materials). f_a = fraction conversion of reactant A (acetic acid)

if r_A represents the rate of reaction:

$F_A \rightarrow F_A$

The mass balance around the reactor is:

$$F_A = (F_A + dF_A) + (-r_A)dV_R \qquad (4)$$

$$dF_A = r_A dV_R \qquad (5)$$

also,

$$F_A = F_{AO}(1 - f_A) \qquad (6)$$

where F_{AO} = initial fraction of A at zero conversion at the inlet of fermenter. On differentiation, combining the last two equations we have:

$$(8) \qquad \frac{dV_R}{F_{Ao}} = \frac{df_A}{-r_A}$$

which may be integrated over the entire reactor volume to give:

$$(9) \qquad \frac{V_R}{F_{Ao}} = \int_{fout}^{fin} \frac{df_A}{(-r_A)}$$

Using this equation we either calculate the size of the reactor knowing the rate of reaction OR for a given fixed size of reactor, we can calculate the maximum flow rate.

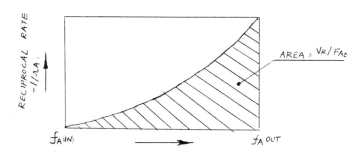

17.6.5 Setting System Boundaries

Identify the various blocks involved. Analysing the flowsheet, the following blocks have been identified:

     1.   Pretreatment

2.   Fermenter (Treated as plug flow reactor)
3.   Extractor (Streams interacting with fermenter)
4.   Back extractor (Reaction with dolomite)
5.   Other auxiliary units:
   •   Washer
   •   Boiler
   •   Settling tank
6.   Spray Dryer

17.6.6 Components Present:

1.   Water
2.   Acetic acid
3.   Dolime
4.   Calcium magnesium acetate
5.   LIX

Balance equations for each block/boundary for each component.

1) Water:	IN:	OUT:
Pretreatment	water (wood chips) water (steam)	water (out)
Fermenter	pulp washer	extractor depleted solids
Washer	depleted solids HAC (lean)	solids to boiler HAC (lean) to fermenter
Mixer 1-2	make up water	water out in CMA slurry
Settling/sep tank	water from back ex-traction	water out in CMA slurry

2) Acetic Acid:	IN:	OUT:
Fermenter	depleted from washer	enriched (from Fermenter)
Washer	lean (from extractor) depleted solids (from fermenter)	wash stream solid biomass
Mixer	from fermenter	to separation Tank
Separation tank	from mixer	to extraction column

3) Dolime:	IN:	OUT:
Mixer 1	unhydrated Dolime	hydrated Dolime
Mixer 2	hydrated Dolime	zero (0)

4) Cellulosics/woodchips/pulp:

	IN:	OUT:
Pretreater	woodchips in	woodchips out (swollen hydrated)
Fermenter	woodchips in (1-x)	unconverted biomass (to boiler via wash)

where x is a preliminary base-line conversion estimated either using reaction kinetics or prior estimated values.

17.7 PRACTICAL PROBLEMS ENVISIONED
17.7.1 Fermenter

The fermenter is modelled as a plug flow reactor. In designing such a fermenter whereby the biomass moves under the influence of gravity, it is important that the downward velocity be controlled in some way and also be held constant. This downward velocity is a constraint for plug flow reactors.

Acetic acid will have to be stripped from this fermenter at a constant rate and in a counter-current manner to keep the pH levels within limits of ± 1 within the desired value. Following the fermenters is a series of washers which are used to strip off residual acetic acid from the unconverted biomass from the fermenter. Also, a centrifuge will have to dewater the biomass which will then be ready for direct burning in the boiler.

17.7.2 Extractor

The extractor contacts an aqueous phase with an organic phase, concentrations of both streams here play an important role in getting the final diffusion coefficients of mass transfer. The extractor should be big enough to ensure clean separation of both phases after contact.
Emulsification of the organic phase is not desired and should be avoided by controlling the temperature or density of the streams and possibly using a de-emulsifying agent. Carryover of extractant ultimately leads to higher make-up amounts which in turn leads to increased product cost, owing to the high cost of the liquid ion exchangers used here.

17.7.3 Back Extractor

Many flow configurations were put forth for the interacting extractor streams, each one of them will have to be simulated for diffusion and mass transfer rates on a lab scale. The idea of staged extraction should also be considered here as shown in figure 2. The main problem seen with flow diagram A involved the use of a slurry of dolomite, whereby some of the slurry was carried over through the reactor which created further separation problems.

The result might lead to reduced product separability and reduced quality.
Flow model B used a saturated solution of calcium magnesium hydroxide. Due to
the inverse solubility of these compounds, the product dilution was far too
much, which led to increased costs by way of concentrating a 0.2 % solution to
some relatively higher concentration like 35 % by wt. This higher
concentration would then be ready for spray drying. Flow model C used a plug
flow reactor where the calcium hydroxide solution was controlled to an amount
slightly sub-stoichiometric than needed so that all of the Ca & Mg is
consumed, and also a relatively high concentration of CMA product is obtained
ready for direct spray drying.

17.4 COST ANALYSIS
17.4.1 Base Model
The process computer model was appended with a cost model. Stand-alone
cost models were also developed using data from the combined process and cost
model. These stand-alone models were used to make numerous runs, varying
parameters of interest as discussed in section 9. The base model costing was
done in 4 steps.
- Equipment sizing and costing.
- Estimation of fixed capital cost.
- Estimation of operating costs.
- Economic evaluation.

Complete output reports were generated for each step mentioned above as shown
in tables 1 to 4 below. ASPEN PLUS™ uses a *factor method* to generate cost
estimates, a series of pre-determined factors are used to multiply the base
cost in order to estimate the total installed equipment costs. Operating
costs are estimated in detail including utilities, raw material, labor, waste
treatment, supplies and running royalties. Other elements of operating costs
are estimated using factors applied to fixed capital costs. Total investment
is the sum of the total capital cost, working capital and startup costs listed
in tables 1, 2 and 3 below.

17.4.2 Profitability Analysis
ASPEN PLUS™ provides cash flow profiles over periods of both the plant
and project operation phases shown in table 4 below. The economic life of the
CMA plant was assumed as 20 years. Profitability was calculated using two
methods. The first method followed an IRR (interest rate of return) at 15 %.
Using this rate of return an initial selling price was calculated as shown in
section 9. Alternatively a second method used was to fix a product selling
price and then determine the IRR. The results of these back-calculations were

used for cost sensitivity analysis. These results can be seen in Table 1.

17.5 COST SENSITIVITY STUDIES

A few parameters of special interest were studied in order to check the effect on cost of product. The higher the sensitivity (shown by steeper changes in cost for the same percentage change in variable) the more attention the variable needed as far as a means of cost reductions were concerned. The main variables studied in this report were:

- % Solid loading in fermenter, as 5, 10, & 15 wt %.
- % Conversion of pulp in fermenter, ranging from 70 % base to 90 %.
- Extraction diffusion coefficients (as percent acetic acid transferred).
- Back extraction coefficients (percent conversion basis).
- Cost of feedstock ranging from $-50 to $ +65 per US ton.
- Plant capacity, of 500, 1000 and 2000 tons per day of CMA.

Cost sensitivity analysis done on the variables mentioned in Section 8 are graphically illustrated in figures 3 to 7. As seen from the sensitivity curves, variables involving operating costs such as feedstock price, pretreatment cost and raw material costs generally fall into one category. This category can be called the less sensitive parameters.

Other parameters such as % solid loading, % acetic acid, directly affect capital orientated costs. Results from the computer costing base model show a break up of 72 % of capital orientated costs and 21 % towards operating costs. This clearly justifies the high sensitivity of parameters working directly towards capital costs.

The costing method followed by the computer package is equivalent to some of the most rigorous methods using hand calculations and other engineering costing routines. This fact was further demonstrated by the relatively low sensitivity of the rate of return with change in the selling price of the product. It would be desirable to reduce the bulk of the streams being handled in the fermenter section as well as in the extractor sections of the process. This can be done by working with higher concentrations of acetic acid in the broth, from 3.5 % as shown by the base computer model up to 13 % percent.

17.6 CONCLUSIONS

As seen from the cost sensitivity studies the most sensitive operating cost variable is the feedstock price. This process can be said to have a feedstock advantage where overall operating costs are concerned. The potential being even greater when the cost of feedstock approaches zero and

430

negative values. Also, the most sensitive capital cost orientated variable is
percent loading of solids in the pretreater and fermentor. The reason being
an increase in equipment size with increased solid loading. It is obvious
from the above fact that more severe pretreatment would lead to reduced solid
loading (5 % & less) and hence a reduction in cost. The variables that merit
further reasearch are percent conversion of pulp to acetic acid (pretreatment
severity) and pretreatment costs. Concluding these studies and modelling, it
would be appropriate to say that this process is viable. Other requirements
that could be factors in the realization of a CMA plant are sites which are
available in and around the frost belt areas and establishing a steady supply
of biomass residues as primary feedstock.

TABLE 17.1

Costing Summary

INVESTMENT		$52,882,000
Physical Plant		1,375,000
Nondepreciable Items		3,133,000
Interest During Construction	(15.00%)	2,123,000
Startup		-6,875,000
Tax Credit		4,351,000
Working Capital		15,865,000
Contingency Allowance		
TOTAL		$72,855,000
SCHEDULE		
Project Start		Jun, 1989
Mechanical Completion		Aug, 1990
Commercial Production		Nov, 1990
REVENUE		
Capacity Production Rate	(KG/HR)	18,889
Normal Production Rate	(KG/HR)	18,889
Initial Selling Price	($/KG)	0.2492*
COSTS		
Initial Cost Per Unit		
Operating Rate 100%	($/KG)	0.1862
Operating Rate 75%	($/KG)	0.2176
Operating Rate 50%	($/KG)	0.2804
DEBT		
Amount to be Financed	(50.00%)	$34,374,000
Interest Rate, Long-term		15.00%
PROFITABILITY		
Interest Rate of Return on Equity		15.00%
Payout Time -- Years		7.50
Return on Invest (ROI)		13.33%
Net Present Value		$-4,315,000
Venture Worth (Discount Rate = 17.0%)		$38,393,000
Initial Investment		$39,518,000
Venture Worth/Initial Investment		0.97
Break-Even Fraction		0.45
Break-Even Volume	(KG/HR)	7,695

* Calculated

432

TABLE 17.2

Annual Operating Cost

Production Capacity: 100.0%
Base Period: 1990, Quarter III
Principal Product: CMA
Plant Availability: 328/365.25 Days
Production Per year: 149023. (1000 KG)

Total Raw Material		$11,641,000
Total Utilities		$2,056,000

Labor
   Operators: 3 Per Shift, 3 Shifts
   Operating   (24000 HR)
   Maintenance
   Supervision
   Fringe Benefits
    Subtotal, Labor
	
	$438,000
	2,475,000
	583,000
	1,398,000
	
		$4,893,000

Supplies
   Operating
   Maintenance
    Subtotal, Supplies
	
	$44,000
	1,650,000
	
		$1,694,000

General Works
   General & Admin
   Property Tax
   Property Insurance
    Subtotal, General Works
	
	$2,097,000
	1,375,000
	550,000
	
		$4,022,000

Depreciation (20 Years, Straight Line)		$3,437,000
Gross Operating Cost Less: Byproduct Credit		$27,743,000 0
Net Operating Cost		$27,743,000

Summary:
  Fixed Cost
  Variable Cost
    Gross Cost Excl. Depreciation
    Gross Operating Cost
	
	$10,609,000
	$13,696,000
	
		$24,306,000
		$27,743,000

TABLE 17.3

Fixed Capital Estimate

** Project Completion: 1990, Quarter III **

	Material Cost	Labor Cost	Labor Hours
Process Units	$30,311,000	$0	
Utility Units	0	0	
Recvng, Shipping & Stor	0	0	
Service Building	2,380,000	1,864,000	70,000
Service Syst & Distrbutn	1,956,000	641,000	24,000
Additional Direct	0	0	
Subtotal	$34,647,000	$2,505,000	95,000
Site Development	$346,000	$184,000	7,000
Freight	693,000		
Sales Tax	1,039,000		
Total Direct Cost	$36,726,000	$2,689,000	102,000
Contractor Fld Indirects	$726,000	$1,963,000	62,000
Subtotal	$37,452,000	$4,652,000	$163,000
Total Direct & Field Indirect			$42,104,000
Contractor Engineering & Home Office		$5,052,000	
Owner's Cost		2,358,000	
Fees, Permits & Insurance		3,368,000	
Additional Depreciable		0	
Subtotal			10,779,000
Total Direct & Indirect			$52,882,000
Process Basis Contingency		$2,644,000	
Project Definition Contingency		13,221,000	
Total Contingency			15,865,000
Total Depreciable Capital			68,747,000
Land		$1,375,000	
Royalty & Expenses		0	
Additional Nondepreciable		0	
Subtotal (Nondepreciable Costs)			1,375,000
Total Fixed Capital			$70,122,000
Working Capital			4,351,000
Startup Cost			2,123,000
Total Investment			$76,597,000

TABLE 17.4

Cash Flow and Return Table

QUARTER	1989, II	1989, III	1989, IV
Capital Exp	106,000	6,358,000	21,785,000
Debt Rec'vd	53,000	3,179,000	10,893,000
Interest	1,000	59,000	350,000
Non-Depre	1,375,000	0	0
Net Cash	-1,429,000	-3,238,000	-11,243,000
Cum Cash	-1,429,000	-4,667,000	-15,910,000
Return	8,000	91,000	350,000
Cum Return	8,000	100,000	450,000
Balance	-1,437,000	-4,766,000	-16,360,000

QUARTER	2010, I	2010, II	2010, III
Sales Revn	34,730,000	35,322,000	23,950,000
Oper Cost	22,732,000	23,120,000	15,676,000
Sales & Admn	4,168,000	4,239,000	1,916,000
Debt Ret'rd	430,000	430,000	430,000
Interest	31,000	15,000	000
Income Tax	3,120,000	3,180,000	2,543,000
Working Cap	144,000	146,000	-10,771,000
Salvage Val			4,812,000
Net Cash	4,107,000	4,194,000	18,968,000
Cum Cash	121,203,000	125,396,000	144,365,000
Return	844,000	724,000	331,000
Cum Return	143,309,000	144,033,000	144,365,000
Balance	-22,106,000	-18,637,000	000

ACKNOWLEDGEMENT

This is to acknowledge the financial support of the New York State
Energy Research & Development Authority (NYSERDA) for this work, as carried
out on Contract Agreement No. "65-ERER-ER-89" with Northeastern University
entitled "Bioconversion of Woody Biomass to CMA." In particular, we wish to
thank Maria Caro, Ph.D. and Lawrence Hudson, Ph.D. of NYSERDA, as well as
Henry R. Bungay, Ph.D., Professor of Chemical Engineering, Rensselaer
Polytechnic Institute, Troy, New York.

REFERENCES

1   D.L. Wise, and D.C. Augstein, "An Evaluation of the Bioconversion
        of Woody Biomass to Calcium Acetate Deicing Salt" *Solar Energy*,
        41, 453--463 1988.
2   US patent No. 4,606,836; 4,636,467; 4,444,672; & 4,488,978.
3   M. Peters and Timmerhaus, *Plant Design and Economics for Chemical
        Engineers*, McGraw--Hill Book Co., third edition 1980.
4   R.H. Perry and D. Green, *Perry's Chemical Engineer's Handbook*,
        McGraw-Hill Book Co., sixth edition 1984.
5   L.R. Hudson, *Calcium Magnesium Acetate (CMA) from Low-grade
        Biomass*, Energy Biomas Systems, volume 11 1988.
6   J.M. Richardson, and J.F. Coulson, *Chemical Engineering, Vol 6,
        Design*, Pergamon Press., second edition 1985.
7   Manivannan et al *Desulfurization of Coal Using CMA* refer this
        text.

436

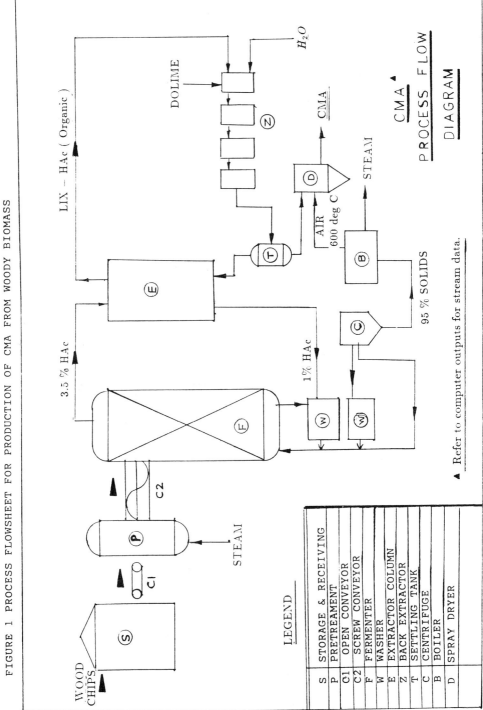

FIGURE 1 PROCESS FLOWSHEET FOR PRODUCTION OF CMA FROM WOODY BIOMASS

LEGEND

S	STORAGE & RECEIVING
P	PRETREAMENT
C1	OPEN CONVEYOR
C2	SCREW CONVEYOR
F	FERMENTER
W	WASHER
E	EXTRACTOR COLUMN
Z	BACK EXTRACTOR
T	SETTLING TANK
C	CENTRIFUGE
B	BOILER
D	SPRAY DRYER

▲ Refer to computer outputs for stream data.

CMA ▲
PROCESS FLOW
DIAGRAM

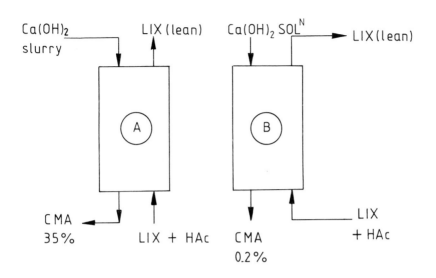

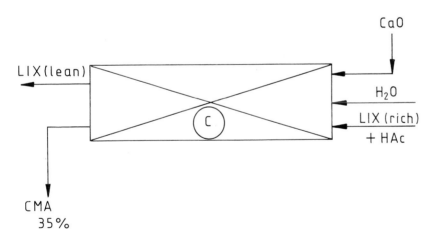

BACK EXTRACTOR FLOW MODELS

FIGURE 2   PROCESS OPTION FOR BACK-EXTRACTION UNIT

438

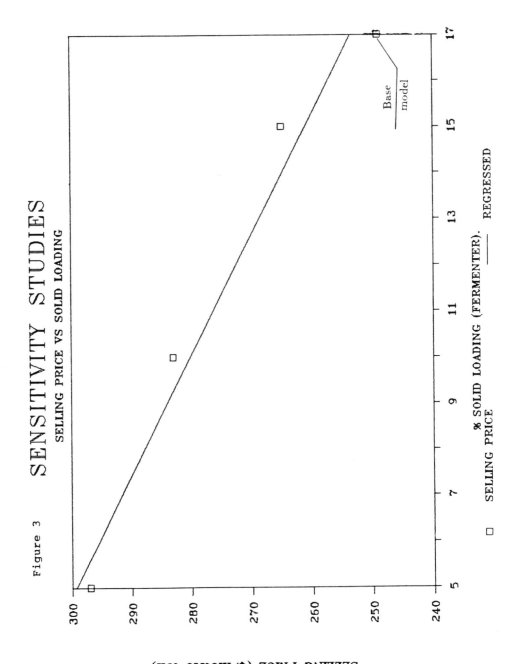

Figure 3

SENSITIVITY STUDIES

SELLING PRICE VS SOLID LOADING

439

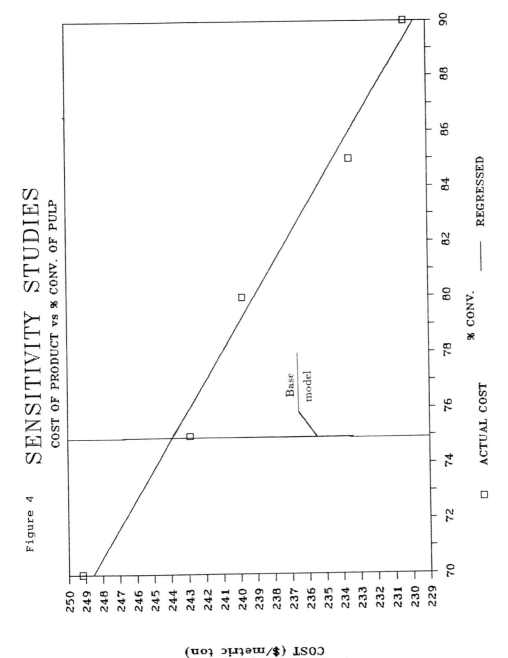

Figure 4   SENSITIVITY STUDIES

COST OF PRODUCT vs % CONV. OF PULP

440

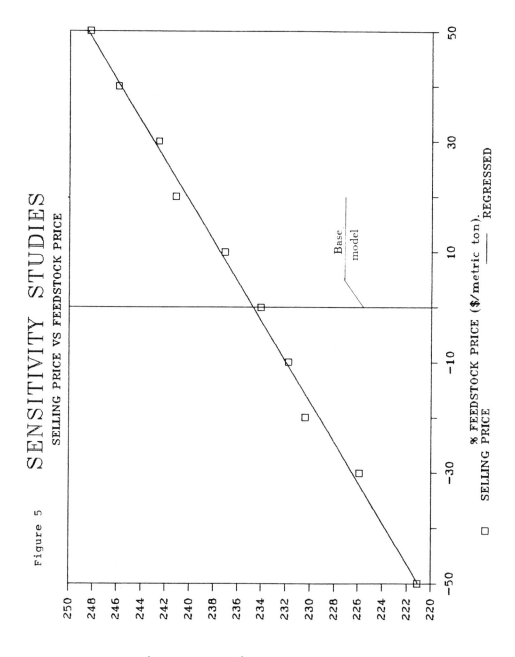

Figure 5

SENSITIVITY STUDIES

SELLING PRICE VS FEEDSTOCK PRICE

441

Figure 6   SENSITIVITY STUDIES

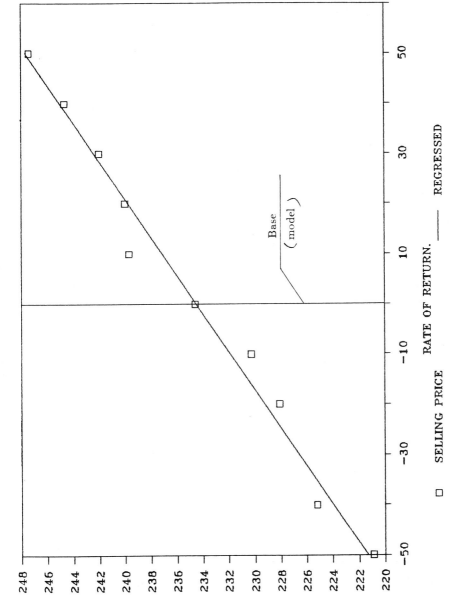

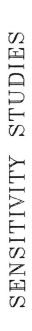

Figure 7   OVERALL SENSITIVITY ANALYSIS

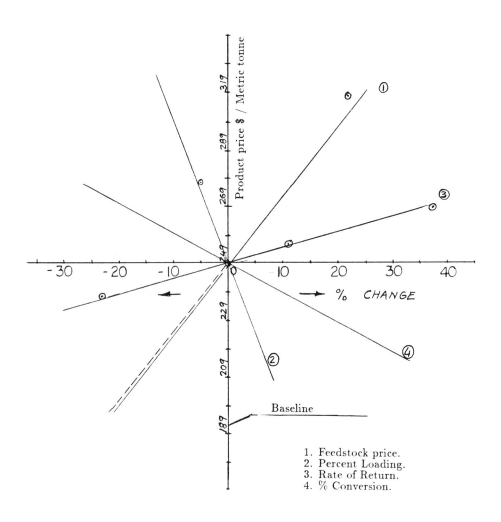

1. Feedstock price.
2. Percent Loading.
3. Rate of Return.
4. % Conversion.

– – – –   INTERPOLATED VALUES

Chapter 18

PILOT PLANT STUDIES AND PROCESS DESIGN FOR THE PRODUCTION OF CALCIUM-MAGNESIUM ACETATE.

R. Ostermann[1] and M.J. Economides[2]

[1]Petroleum Engineering Department, University of Alaska - Fairbanks, Fairbanks, AK 99775-1260

[2]Dowell-Schlumberger - London.

18.1 IMPLEMENTATION

The search for non-corrosive deicing chemicals for highway applications has taken on a national emphases as a part of the Strategic Highway Research Program. Calcium-Magnesium-Acetate (CMA) has been identified as the most promising alternative to the corrosive salts commonly in use for deicing.

Because of high transportation costs for shipping deicing chemicals to Alaska and because we felt that it may be economically feasible to produce CMA in Alaska, the Alaska Department of Transportation and Public Facilities has been investigating this product for use in Alaska.

This report describes research sponsored by the Alaska Department of Transportation and Public Facilities which has studied the manufacture of this non-corrosive deicing chemical and its application to Alaskan roads. Other reports concerning CMA studies are "Preliminary Design and Feasibility Study for Calcium-Magnesium-Acetate Unit", Report 83-16, "Corrosion of Steel in Calcium-Magnesium-Acetate (CMA) Deicer", Report 85-27, and a report on the environmental effects of the CMA produced in Alaska. The latter report is scheduled for completion in the spring of 1986 and is being coordinated with other federally sponsored environmental assessments of CMA.

In our studies University researchers and Department personnel have made and tested CMA in the laboratory and produced and field tested some 10,000 gallons of liquid CMA. Our field tests using a standard water truck and maintenance crews for application went smoothly and produced effective results.

As noted in the Conclusion, using the criteria discussed by the researchers, the cost to manufacture CMA in Alaska is between $415 and $442 per ton. This cost is highly dependent upon the cost of the acetic acid used in the manufacturing process; the costs above assume an acetic acid cost of $0.20 per pound.

If a decision is made to stop using the present corrosive chemicals and switch to CMA, the initial costs to the Department will increase.

444

However, as summarized in this report, the total costs including corrosion and environmental effects from our present practices are 10-15 times the first cost of purchasing sodium chloride, our most commonly used deicing chemical. If the Department decides to pursue implementation of CMA the next steps are to file an environmental impact statement using the results of the environmental study nearing completion and to conduct a full scale winter-long demonstration of CMA at a designated maintenance station.

18.2 ABSTRACT

Chloride salts are commonly used as deicing chemicals in many northern states. These chemicals are corrosive to automobiles, bridge decks and other public structures, and cause considerable
damage to plants found near salted roadways.

Calcium Magnesium Acetate (CMA) offers potential as an alternate deicing chemical. CMA is a generic term applied to the reaction product of acetic acid and limestone. It is non-corrosive and has no known potential to cause environmental damage.

During 1982-1984, over 12,000 gallons of saturated CMA solution were produced by researchers in the Petroleum Engineering Department of the University of Alaska - Fairbanks. The raw materials were acetic acid, hydrated lime and native limestone. The product CMA solution was used in road tests in Fairbanks as well as for environmental studies. The results of these pilot plant studies is presented in this report.

A process design for the production of 12,000 GPD of CMA from native limestone based on the results of the pilot plant studies is presented. For a grass-roots operation located in Fairbanks, the total capital cost is estimated at $215,000, including site and structure costs. Using current prices for acetic acid, limestone and hydrated lime, a product price of $413 per ton (dry CMA equivalent) is required for a 15% annual rate of return. It should be noted that raw materials costs amount to over 85% of the annual operating costs with capital cost amortization amounting to only 2% of the product cost. The process economics are thus relatively insensitive to changes in capital costs due to process design changes. Moreover, acetic acid cost alone accounts for 70% of the product price. The key to reducing the price of CMA lies in obtaining inexpensive acetic acid.

18.3 INTRODUCTION

Winter driving safety on icy roads is an issue of major concern to people living in most of the northern tier states and Canada. Currently, sodium chloride (common road salt) and calcium chloride are used extensively as deicing compounds throughout the country. The use of these chemicals has a definite positive value which must be measured in terms of human lives saved, as well as reduction in vehicular damages and lost productivity which may occur in ice related accidents.

There are also negative effects offsetting these benefits. With the use of salts, in addition to the costs of acquisition and application, secondary costs are incurred by associated corrosion and environmental damages. A commonly used estimate is that the total cost associated with salt usage, including damage costs, is roughly 10 to 15 times the cost of the applied salt itself.

A significant portion of these costs may be attributed to bridge deck corrosion, utility system corrosion, and in general, corrosion of susceptible public structures and equipment. However, the brunt of the damage costs are borne by the private sector, through the corrosion of personal automobiles. Additionally, salt is harmful to most plants found near the roadways. While it is difficult to estimate a dollar amount for this damage, it is very visible, and aggravating to the public. In response to "rusted out" automobiles and damaged lawns and landscaping, the public often offers potent criticism concerning salting practices, making the responsibility of the local DOT to assure the safety of the roadways doubly difficult.

The above concerns have triggered an earnest search for alternative methods of deicing roadways. Thermal, physical, and alternative chemical methods have been considered. F. Wood[1] in a paper dealing with salt damages to motor vehicles, presents a convenient summary of some of the physical and thermal deicing methods attempted. Zenewitz[2], has published a more extensive survey. While many different methods are technically feasible, he reports that the general feeling among highway personnel is that none of these methods are cost effective. However, it should be stressed that any economic assessment not taking into account the full spectrum of salt related damages should be questioned.

Several chemicals have been proposed as substitutes for road salt. Under the sponsorship of the Federal Highway Administration, researchers at the Bjorksten Research Laboratories in Madison, Wisconsin, conducted a major study assessing alternative deicing chemicals. In the final report on this study, Dunn and Schenk[3] report that only two chemicals, methanol and calcium magnesium acetate (CMA), have deicing properties similar to sodium chloride, are non-corrosive, and have the potential to be produced at reasonable costs. (CMA is in the simplest sense a salt, the reaction product of acetic acid and limestone.)

Salt can be obtained in most areas in the "lower 48" states at less than $20 per ton. If the damage costs are estimated at $200 per ton, it would appear that CMA may be a feasible substitute if it can be produced in the "lower 48" for less than $220/ton.

Considerable interest has been generated surrounding the properties and production of CMA in Alaska. The research section of the Alaska State Department of Transportation and Public Facilities (DOTPF), and the Petroleum Engineering Department at the University of Alaska - Fairbanks (UAF) have been investigating the production and use of CMA in Alaska. In the "lower 48" states, salt typically costs $10 to $20 per ton with damage costs of about $200/ton. In Alaska, since salt must be shipped by barge from the West Coast, the price is increased to over $100 per ton. It would be reasonable to assume that the damage costs in Alaska are also about $200/ton, adjusted for Alaskan prices. Repair and replacement costs are also high in Alaska. If damage repair costs were estimated at 25% greater than in the "lower 48", or $250 per ton, it would appear that the true "cost" for the use of road salt in Alaska could be nearly $350 per ton. Thus the benefits of using non-corrosive CMA, and hence the economically allowable price for CMA are higher in Alaska than in the "lower 48" states.

The obvious difficulty with the economic analysis of switching from salt to CMA lies in determining the "actual" cost of salt damage. While it is clear that the costs are substantial, and real, they are distributed over several sectors of the economy. Without a reasonably reliable estimate of the cost of salt damage, it will be difficult to evaluate what price the State could afford to pay for CMA and still achieve an economic benefit. Considerable damage to bridge decks and public structures is attributable to salt, but a large proportion of the damage is also borne by the private sector. Thus it may not be possible for the State DOTPF to show a "cost savings" through the use of CMA. Any price for CMA substantially above the current cost of salt, will result in increased operating costs for the DOTPF, even after deducting the potential repair costs and savings associated with decreased bridge deck corrosion and other damage to public structures.

It is anticipated that an election to use CMA will be more a policy decision (i.e. not to use corrosive salts) spurred on by adverse public opinion regarding salt usage, than an economic decision. Thus, the technical question may well be "How do we make CMA for the lowest possible cost?" as opposed to "Is that cost economical?"

For the last two years, researchers at UAF have been studying the production of CMA. Over 10,000 gallons of saturated CMA solution has been produced in a pilot plant operated at the local DOTPF facility in Fairbanks. This is the equivalent of roughly 12 tons of dry CMA product. The CMA has been used for limited road tests, and for environmental studies. In the current report, the results of our pilot plant and process design studies are presented.

18.4 BACKGROUND
18.4.1 Costs and Benefits of Deicing

Chloride salts have been used for many years for the control of ice and snow on the nation's streets, roads, and highways. Usage has risen to over 12 million tons annually. Even in Alaska, with relatively few miles of roadways, current usage is over 8,000 tons annually. As the public encounters "rusted out" automobiles, damaged lawns, and landscaping and the appearance of salt residue on shoes and carpets, controversy is raised regarding the use of these salts. The immediate question is, "Do the benefits outweigh the costs of salt usage?". If not, it may be appropriate to discontinue their use.

While the actual cost of salt and its application is easily obtained estimates of the benefits and damage costs are more elusive. The economic benefits of road deicing are primarily realized through the reductions in the number and severity of ice related accidents on treated roads as opposed to untreated roads. The cost of an accident is estimated in terms of productivity losses as well as actual structural damage.

One major benefit expected from deicing would be a reduction in deaths and injuries associated with ice related accidents. It is of course difficult to assign a dollar figure to this benefit.
Costs associated with the use of salts include vehicular and bridge deck corrosion, damages to structures and utility systems, and environmental damages. It is difficult to obtain reliable estimates of the magnitude of these damages since they are so pervasive and occur over extended periods of time.

The costs and benefits of deicing have been analyzed in some detail by Brenner and Moshman[4] and Murray and Ernest[5]. The results of the studies are summarized in Table 18.1.

Murray and Ernst[5] have estimated a total U.S. cost for salt use, including damages, at about $2.9 billion annually. Given an annual cost of $200 million for salt purchase and application, the total cost of salt would appear to be fifteen times greater than the cost of purchase and application alone. While the relationship may be used for the purposes of cost estimation in the "lower 48", the relationship would not apply for Alaska.

TABLE 18.1

Annual Costs of Road Salting
(In millions of dollars)

	Brenner and Moshman[4]	Murray and Ernst[5]
Utilities	2	10
Vehicle Corrosion	643	2000
Highway Bridge Decks	160	500
Trees and Vegetation	0	50
Water Supplies	10	150
Salt and Application	200	200
	1015	2910

In Alaska, salt costs nearly $100 per ton, or ten times as much as in the "lower 48". Using the ratio calculated above, damage costs for Alaska would be estimated at $1500 per ton. This figure is probably unrealistic.

If we assume that repair costs are roughly 25% higher in Alaska and that salt costs ten times as much as in Alaska as opposed to the "lower 48", the appropriate ratio would be only 3 to 1. That is, the total cost for salt usage is three times the cost of purchase and application, or $200-300 per ton. Considering the unique nature of Alaska, many other variables may be involved, making comparison to "lower 48" figures difficult. The above figures should be used with caution. However, in the absence of more accurate data, it may be used for the purpose of rough estimation.

The estimates of Brenner and Mosham[4] give a factor of three lower than Murray and Ernst[5]. The reason for this difference is unclear. However, the larger question as to whether the benefits of deicing outweigh the costs would appear to be clear in either case. Brenner and Mosham[4] (See Table 18.2) have estimated the total benefit from deicing to be $18.4 billion annually. Thus in the case of Brenner and Mosham the benefit cost ratio is roughly 18:1 while the ratio is 6:1 for Murray and Ernst[5]. Both studies indicate a substantial economic benefit to be derived from deicing. Moreover, the savings in human lives alone should be sufficient to justify the expense.

TABLE 18.2
Annual Economic Benefits of Deicing
(Brenner and Mosham[1])

	Annual Economic Savings In Millions of Dollars
Reduced Fuel Usage	200
Reduced Wage Losses	10,600
Reduced Production Losses	7,000
Reduced Losses in Goods Shipments	600
	18,400

18.4.2 Salt Uses in Alaska

Over 8,000 tons of road salts, including calcium and sodium salts are used annually in Alaska. In addition to deicing, calcium chloride is also used as a dust palliative. The vast majority of the salt used in Alaska is sodium chloride used in the Central region (Anchorage area). The current distribution of salt use in Alaska is presented in Table 18.3. (Data supplied by the State of Alaska DOTPF). About 90% of the salt is applied in the Central area with smaller amounts in Interior and Southeast.

TABLE 18.3
Current Salt Usage in Alaska

	Tons Per Year	
Region	NaCl	$CaCl_2$
Interior	500	0
Central	6950	440
Southeast	140	250
Total	7590	690

18.4.3 Alternate Deicing Chemicals

Dunn and Schenk[3] have compared alternate highway deicing chemicals. The primary requirements for a chemical to be a viable deicer substitute are: a) solubility of chemical in water, b) low rate of volatilization, c) minimal pollution effects, d) minimal corrosion effects, e) a non-hazardous nature, and f) low cost.

In the preliminary evaluation of the alternative chemicals, transuranium elements, actinide series, rare earth metals, noble gas, etc., are eliminated. Sodium salts are not considered because of the potential damage of sodium ions on plants, soils, and ground water contamination. All chlorides, nitrates, and sulphates are considered to be unsuitable. Dunn

and Schenk[3] suggested that potassium salts of carbonic acid (freezing point
- 36°C) might work. They also studied the possibility of using a eutectic
mixture of potassium salt of phosphoric acid, tetra potassium pyrophosphate,
ammonium salts of phosphoric acid, ammonium salts of carbonic acid, and
organic salts of Na, K, Mg, Ca and ammonium ions, glycine, methanol,
ethanol, isopropanol, acetone, urea formamide, dimethyl sulphoxide, and
urethane.

Dunn and Schenk[3] have reported that two chemical compounds, methanol
and CMA showed the most promise as deicing compounds. These chemicals were
evaluated on the basis of criteria such as: traction, skidding, friction,
field performance, compatibility with cement, asphalt and road paint, and
corrosion of the commonly used metals associated with vehicular and roadway
construction.

Dunn and Schenck[3] suggested that methanol should not pose any
significant environmental or corrosion problems. However, care should be
taken in the storage and use of methanol because of its flammability. CMA
is essentially nonflammable, nontoxic, and retards corrosion of most metals.

Both deicing agents can be produced from cellulosic and other organic
wastes by high pressure, high temperature technology with producer gas or
natural gas as an intermediate. Also, large deposits of dolomite are
available in many areas for the manufacture of CMA.

Methanol was found to be "less persistent" than CMA due to its high
volatility. CMA was selected as the preferred deicing substitute for sodium
and calcium chlorides. A detailed analysis is presented in the Dunn and
Schenk report[3].

18.4.4 Properties of CMA

The effective evaluation of CMA as a deicing compound will require
detailed knowledge of its physical and chemical properties. The primary
test of a chemical compound as an effective deicer is its ability to depress
the freezing point of water.

Figure 18-1 illustrates the freezing points of a solution of various
salts as a function of composition. Calcium acetate and magnesium acetate
are not quite as effective as the chloride salts in reducing the freezing
point of water. However, at higher concentrations, the acetates are
effective down to about 0°F.

However, the effectiveness of a deicing chemical depends on more than
just freezing point depression. There does not appear to be any firm,
quantitative measures of deicing effectiveness, although parameters such as
pounds of ice melted per pound of salt applied, and rate of melting are

451

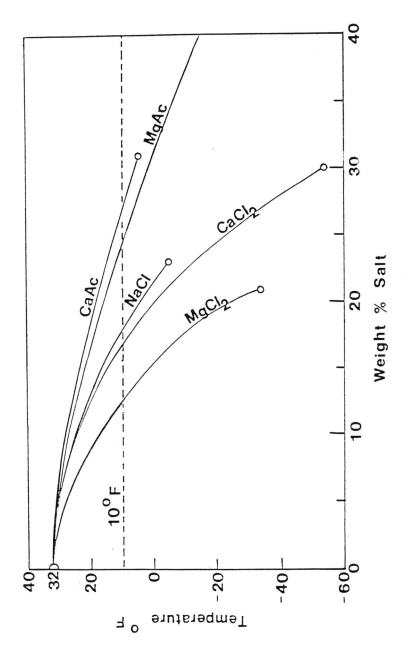

Figure 18-1. Phase relationships for various salts.

indicators.

Moreover, factors such as traction improvement and increased braking efficiency are important in determining the effectiveness of a deicer. Dunn and Schenk[3] have reported laboratory tests of braking traction and skidding friction for NaCl and CMA. The tests indicate that CMA should function well as a deicer.

In practice, however, the utility of CMA can best be assessed through large scale field tests. By the end of 1984 over 200 tons of CMA were field tested in Washington and Michigan. Within Alaska, about 5,000 gallons of saturated solution have been used on Fairbanks streets. The preliminary results of these tests indicate the CMA performs well as a deicer. In Fairbanks, a 50% increase in braking efficiency was reported within 15 minutes of CMA application. While these tests are encouraging, larger tests will have to be made before widespread application is prudent.

Repeated laboratory tests have indicated that CMA is not corrosive to steel, zinc or aluminum. Dunn and Schenk[3] present an extensive treatment of this topic. Additional tests are underway at the national level.

Calcium and Magnesium Acetates are generally regarded as environmentally benign. The Merck Index[6] lists the following lethal dose (LD) levels for mice and rats:

Calcium Acetate:	LD_{50} orally in rtes	4.28 g/kg
Magnesium Acetate:	LD_{50} intravenous in mice	18 mg/kg

By comparison, sodium chloride has a lethal level of LD_{50} (orally in rats) of 3.75 g/kg. In fact, calcium acetate is used commercially as a food stabilizer. Tests are currently underway in Alaska to assess the effects of CMA contamination on plant and animal life in ponds. At the present time, there are no indications of major problems.

18.5 PRODUCTION OF CMA
18.5.1 Basic Chemistry

CMA is a generic term applied to the reaction products of limestone or dolomite and acetic acid. Limestone is composed of calcium and magnesium carbonates ($CaCO_3$, $MgCO_3$) and varying amounts of impurities. Typical compositions of limestone and dolomite are presented in Table 18.4. The data were taken from a master's thesis by Sanusi[7]. The relative amounts of calcium and magnesium acetates found in the final product CMA will depend on the composition of the limestone or dolomite used.

TABLE 18.4

Typical Compositions from Dolomite and Limestone (After 7)

	Limestone[a]	Dolomite[b]
CaO	55.28	31.20
MgO	0.46	20.45
CO_2	43.73	47.87
SiO_2	.42	.11
Fe_2O_3	.05	.19
Other	.06	.18

a. Virginia High Calcium Limestone
b. Illinois Niagran Dolomite

   Magnesium acetate is actually a "better" deicing chemical than calcium acetate, showing a greater freezing point depression (See Figure 18-1). Most limestones are predominately composed of calcium carbonate, some having a purity of over 95%. Dolomites can have magnesium carbonate contents approaching 50%. Most of the potential limestone sources in Alaska are high in calcium content. Moreover, the chemistry for the production is largely the same in either case. However, limestone composition varies considerably, even among limestones having little magnesium content. For these reasons, this study focuses on the production of calcium acetate (CA) from pure calcium carbonate.

   Calcium acetate can be produced from limestone or hydrated lime.

18.5.2 Production From Limestone

   Calcium acetate is produced from the reaction of acetic acid (CH_3COOH or HAc) and calcium carbonate as follows:

$CaCO_3 + 2CH_3COOH \rightarrow Ca(CH_3COO)_2 + H_2O + CO_2$

The reaction is slightly exothermic (ΔH_r° = -4.6 Kcal/mol), and produces a considerable amount of carbon dioxide gas. Acetic acid will not react with calcium carbonate in the anhydrous state. It must be mixed with water to allow ionization. The solubility limit of CaAc in water is about 28% at 25°C. Hence, if a saturated solution of CA is desired, a simple material balance would indicate a water requirement of 388 g.H_2O/100 g.$CaCO_3$. However, the reaction does not go to completion under these conditions, making the reaction somewhat more complicated.

   The solubility of calcium carbonate in water is very low. Experiments have shown that if stoichiometric quantities of water, acetic acid, and calcium carbonate necessary to produce a saturated solution, are mixed, the reaction will not go to completion, but will stop at a pH of about 5.6 with only 86% of the acid reacted. Perhaps the easiest way to view the reaction is as a titration of an acetic acid solution with calcium carbonate.

Ideally, calcium carbonate would continue to be dissolved in the solution until neutrality was achieved. However, the solubility limit for calcium carbonate is reached before neutrality in all but solutions too dilute to be of value. Thus the normal product is a acidic solution containing unreacted acetic acid.

A series of experiments was conducted to determine the degree of completion achieved as a function of the composition of the original acid solution, and time. In general, the more dilute the solution, the greater the amount of carbonate which would react, and hence the more acid which would be neutralized. The results of these experiments are presented in Figure 18-2. The degree of completion is measured as percent of original acid consumed. In all reactions, an excess of calcium carbonate was used.

A cursory analysis of Figure 18-2 indicates that the rate of reaction and to a minor extent, the ultimate degree of completion depends on the original ratio of water to acetic acid. It should be noted that essentially no reaction continued beyond four hours. Figure 18-3 shows the relationship between reaction completion and initial water/acid ratio at various reaction times. In Figure 18-4, the time required to achieve 95% of maximum completion is plotted versus original water/acid ratio.

In general, CA can be produced by reacting acetic acid and calcium carbonate in any proportion in aqueous solution. While it may seem appropriate to aim at 100% acid utilization as a goal, it should be noted that this is not feasible. The maximum conversion obtained at a water/acid ratio of 4 is only 85%.

The goal of this study is to define a process for the production of saturated CA solution. At 25°C, a saturated solution of CA is 28% CA by weight. The stoichiometric proportions of water, acetic acid and calcium carbonate assuming 100% completion required are 388 g. H_2O, 120 g. HAc and 100 g. calcium carbonate. However, at a water/acid ratio of 3.88:1.20, the conversion is only 86%.

A material balance for the reaction starting with stoichiomectic proportions is given in Figure 18-5. The final product is a concentrated, but not saturated CA solution, with an excess of unreacted acetic acid. In a large scale plant, it my be feasible to recover the acid through any of a number of stock process. However, in a small or moderate sale plant, it will be more cost effective to use a neutralizing agent such as sodium or calcium hydroxide. In particular, the use of calcium hydroxide (hydrated lime) is recommended since the neutralization product is CA. The stoichiometric equation for this reaction is presented below.

$$Ca(OH)_2 + 2CH_3COOH \rightarrow Ca(CH_3COO)_2 + 2H_2O$$

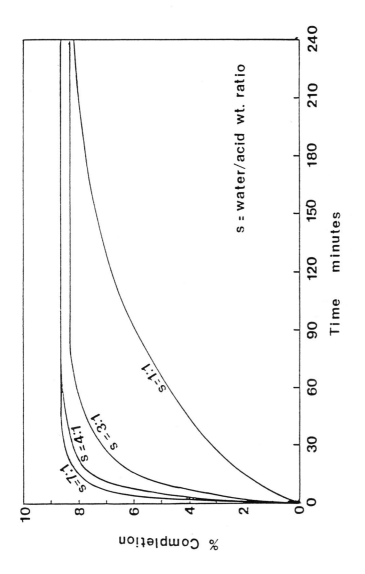

Figure 18-2. Reaction rates for CMA production.

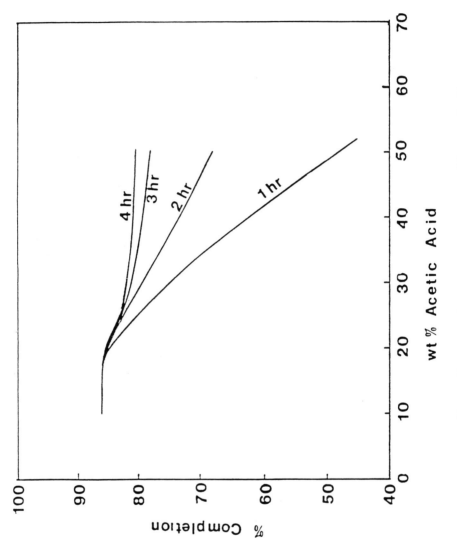

Figure 18-3. Reaction completion versus acid concentration.

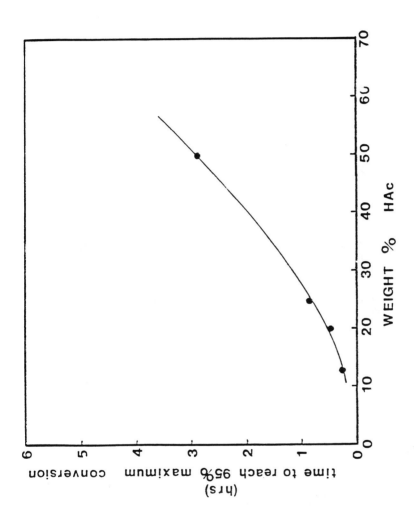

Figure 18-4. Reaction time versus acid concentration.

The amount of calcium hydroxide required for neutralization is dictated by the amount of unreacted acid remaining in solution, which in turn is directly related to the original acid concentration. This relationship is presented in Figure 18-6. The final solution CA concentration after neutralization may also be calculated from the original acid concentration. This relationship is presented in Figure 18-7.

Thus, a two step process is envisioned, wherein a solution of acetic acid and water is reacted with an excess of calcium carbonate. The resulting solution is decanted and neutralized with calcium hydroxide. A representative material balance is presented in Figure 18-8. This material balance is for the production of a 26 wt.% solution of CA, slightly less than saturated.

For original acid concentrations of less than 24 weight percent, the resulting product will be an undersaturated solution. At 24 weight percent acetic acid, a saturated solution results. For concentrations greater than 24 weight percent a saturated solution and solid CMA are produced. This relationship is a direct result of the stoichiometry of the reaction and the fact that a saturated solution contains 28 weight percent of CA. Solubility limits are influenced by temperature. The figures used here are at 25°C. It should be noted, however, that any solid product will be "soaked" with solution and would require considerable dewatering and drying to be marketable or useful as a "solid" product. The decision as to whether a saturated solution or solid product is more desirable will depend on the relative costs of lime (calcium hydroxide) and the preferred form of the product for application purposes. In this study, it is presumed that the desired product is a saturated or nearly saturated solution.

Finally, it should be noted that the above discussion is based upon reaction with <u>pure</u> calcium carbonate. If limestone is used, the reaction kinetics and ultimate degree of completion will vary with the nature of the limestone. The behavior of a representative dolomite is indicated on Figure 18-2 along with the data for pure calcium carbonate.

In general, it can be seen that the reaction with limestone is somewhat slower, and does not proceed to a great degree of completion as with pure calcium carbonate.

18.5.3 <u>Production From Lime</u>

Calcium Acetate (CA) can also be produced from the reaction of lime or hydrated lime with acetic acid as follows:

Hydrated Lime:

$Ca(OH)_2 + 2HAc \rightarrow CaAc_2 + 2H_2O$

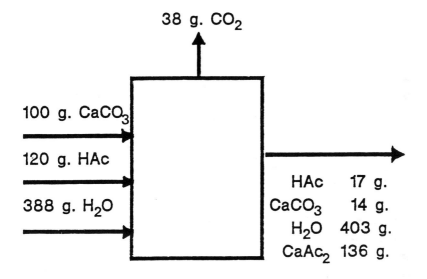

FIGURE 18-5

Material Balance with Stoichiometric
Feed

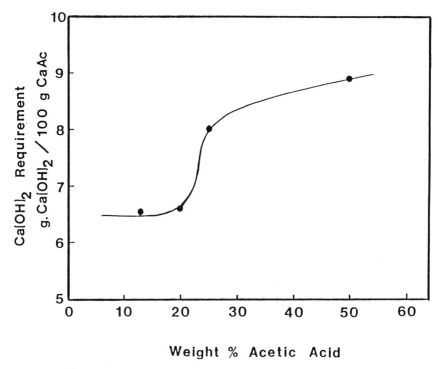

Weight % Acetic Acid

Figure 18-6. Calcium hydroxide versus acid concentration.

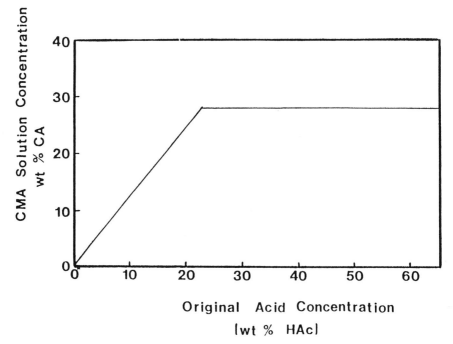

Original Acid Concentration
(wt % HAc)

Figure 18-7. Final solution strength versus acid concentration.

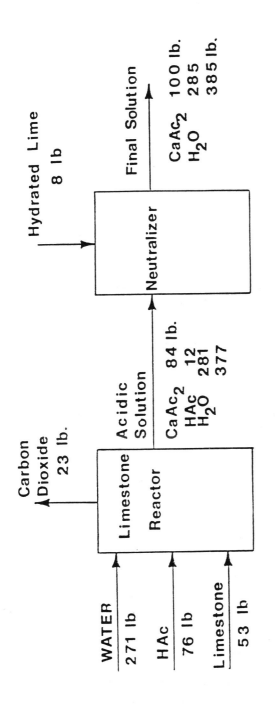

Material Balance for Two-Step Process

FIGURE 18–8

Lime:

$$CaO + 2HAc \rightarrow CaAc_2 + H_2O$$

This reaction is somewhat simpler than the reaction with limestone. No carbon dioxide is produced and no solubility barriers prevent the reaction from going to completion. As a result, the reaction may be conducted on a stoichiometric basis. A material balance for a representative reaction is presented in Figure 18-9.

18.5.4 Source of Raw Materials
Acetic Acid

There are currently no manufacturers of acetic acid in Alaska. Thus, all acetic acid must be obtained from outside Alaska. The current delivered cost of glacial acetic acid in small quantities in Alaska is about 48¢ per pound including shipping and handling. The current world market price for acetic acid is 25¢/lb (f.o.b Los Angeles). Shipping costs vary from 5-9¢/lb (Anchorage). This makes for a delivered price of about 30¢/lb or $600/ton. Since .76 lb of HAc are required per pound of CA product, the cost contribution of HAc is about $456/ton. Clearly, the key to economic CMA production is to control the HAc cost.

For CMA production, large quantities of acetic acid will be needed. Earlier calculations have shown that the cost of production of CMA can be significantly reduced if the cost of acetic acid is decreased. Moreover, it is not necessary to use glacial acetic acid. A 50-50 mixture (volume basis) of acetic acid-water will suffice. Since a large portion of the cost of glacial acetic acid manufacture is associated with purification of reaction products, it is reasonable to expect that a petroleum derived, low grade, low cost acetic acid could be made available within Alaska.

We have completed a literature search for various processes for the manufacture of acetic acid. Although acetic acid can be produced from various chemicals, the three major routes are as follows:
1.   The liquid phase oxidation of hydrocarbons.
2.   Carbonylation of methanol.
3.   Conversion of feedstocks derived from Biomass.
Of these three processes, the first is the most suitable for Alaska since light hydrocarbon fractions could be available from refineries within the State.

18.5.5 Liquid Phase Oxidation of Hydrocarbons

(i) Process Description. Liquid-phase calatytic oxidation (LPO) of n-butane was introduced by Celanese in 1952 at a large plant located near

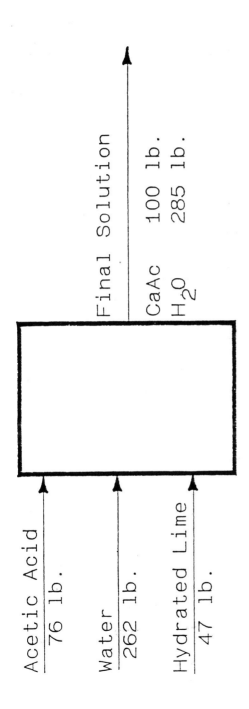

Material Balance Production from
Hydrated Lime

FIGURE 18-9

464

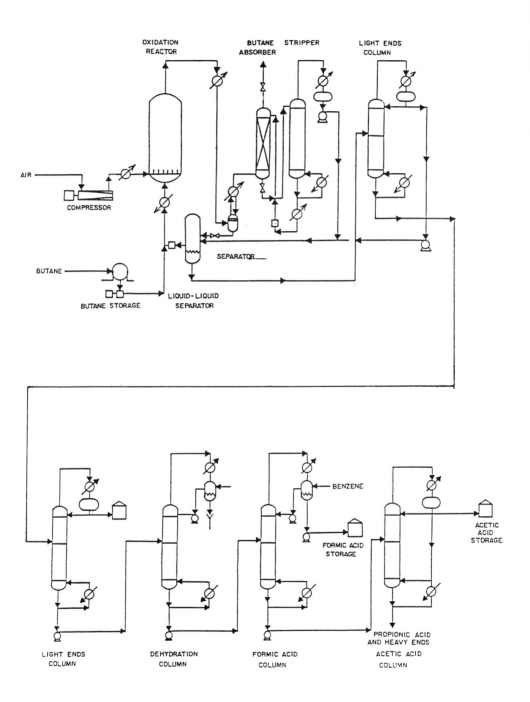

Figure 18-10. Process schematic for the production of acetic acid.

Pampa, Texas. Butane is dissolved in acetic acid along with a suitable
homogeneous catalyst (e.g. cobalt, chromium, vanadium or manganese acetate)
and sparged with air under high pressure. Pressures near the critical value
are used. Temperatures from 95-100°C and pressures from 10-54 atm are
usually required. Tubular reactors with a length to diameter ratio of 16-60
are required to obtain acceptable yields and conversions. The reaction
products contain methylethylketone, ethylacetate, methylvinylketone, acetic
acid, butyric acid, formic acid, propionic acid, and water. Figure 18-10
shows the process schematic. The acid-water product out of the light ends
column typically contains 70% HAc, 3% HBu, 4% HF, 2.5% HPr, with the rest
water. For CMA use, it may not be necessary to further purify this mixture,
as there is no reason to assume that formates, propionates, and butyrates
would not be similar to acetate as a deicer.

(ii) Laboratory Study. In our laboratories we are concentrating our
efforts on the process conditions for the manufacture of acetic acid from
liquid phase oxidation of propane and butane. Since propane can be made
available at the North Pole Refinery, it is desirable to ascertain whether
propane can be converted to acetic acid through LPO. The scope of this
study will be limited to determine process conditions such as pressure,
temperature, conversion, flow rate, etc., and no attempts will be made to
purify acetic acid. If and when need arises this aspect can be studied
further.

18.5.6 Limestone

To maximize the deicing characteristics of a CMA solution, the ratio of
magnesium to calcium ions should be as high as possible. Dolomite is a form
of limestone having a high percentage of magnesium ions. The chemical
formula for dolomite is $CaMg(CO_3)_2$. Typical compositions for dolomite and
limestone are given in Table 18.4.

A study was conducted to locate the Alaskan deposits of dolomitic
limestone most suitable for CMA production. The evaluation was accomplished
in three stages. The first stage was a thorough literature search. The
second stage was field work in which promising sites were examined in
detail. The third stage was analysis and evaluation of the data collected,
culminating in our recommendation of the most suitable sites for dolomitic
limestone recovery. The quantity and composition of the material,
accessibility and ease of transportation were considered.

18.5.7 Literature Search

The available literature concerning location, size, accessibility, and composition of limestone deposits within Alaska was examined. In addition, the Alaska Department of Geological and Geophysical Surveys and Mineral Industry Research Laboratory personnel at the University of Alaska provided up-to-date information and unpublished data. Dr. Gil Eakins, James Clough, and Dr. Florence Weber of the USGS, and Dr. P.D. Rao and Dr. Paul Metz of MIRL all made significant contributions.

A Master's Thesis by Sanusi[7] was the starting point for this study. The reader should consult this thesis for more detailed information, and an excellent treatment of limestone resources. A number of other references were obtained by cross referencing. The majority of the published data are related to either pure limestone ($CaCO_3$) for cement manufacture or marble for use as a building material. Many of the reports dealt with the marble deposits on the Southeastern panhandle. A map of carbonate deposits in Alaska excerpted from this study is given in Figure 18.11.

18.5.8 Field Study

From the list of limestone deposits, several locations were chosen for field study using the following criteria:

Composition: The deposit should approach the ideal composition of dolomite.

Accessibility: The deposit must be easily reached, also, it should not be in or close to Federally operated parks or preserves.

Location: The deposit should be relatively close to a major processing and distribution center.

Quantity: The deposit should be large enough for commercial use.

The Hoodoos deposit near Isabel Pass and several deposits along the Elliot Highway were examined in detail and samples of limestone were collected. The location of the field study sites are highlighted in Figure 18-11.

18.5.9   Data Analysis and Evaluation

The collected samples were analyzed for calcium and magnesium contents using a Direct Current Plasma Emission Spectrograph. Table 18.5 shows the results of the samples studied.

467

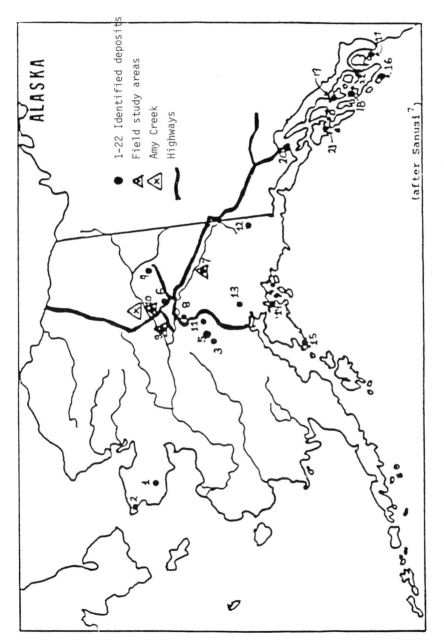

Figure 18-11. Location of carbonate deposits in Alaska.

(after Sanus1[7])

Figure 18.12 (continued)

Summary of Identified Limestone Deposits in Alaska (after Sanusi[7])

Name	Location	Mean % CaCO$_3$	Estimated Reserves (mil. tons)	Comments
		Western Region		
1) Mount Distin	Nome Area	N.A.	N.A.	Thickness ranges from 2,220 ft. to 3,200 ft. and of high purity. Further field work required.
2) Port Clarence	Lost River area Seward Peninsula	N.A.	N.A.	Several thousand feet thick with sequences of black limestone 500 to 1,000 thick exposed east of Lost River.
		Interior Region		
3) Chulitna River	West and southwest of Golden Zone mine	88.0	103	Two deposits at the heads of Long and Copeland Creeks. Easily accessible.
4) Crazy Mountains	Crazy Mts., 20 miles from Steese Highway	N.A.	N.A	Easily accessible but further field work required.
5) Foggy Pass	15 miles northwest of Cantwell at the entrance of Foggy Pass	89.0	100	Very close to McKinley Park. Consists of grey crystallized, folded and contorted limestone.
6) Fox	3/4 mile southwest of Fox	53.0	N.A.	Easily accessible. Too small for commercial development.
7) Hoo Doos	East of Isabel Pass	94.0	300	Highly fractured. Easily accessible.

Figure 18.12 (continued)
Summary of Identified Limestone Deposits in Alaska (after Sanusi[7])

Name	Location	Mean % CaCO$_3$	Estimated Reserves (mil. tons)	Comments
		Interior Region		
8) Nenana	Birch Creek Schist	95.0	N.A.	A blue-grey lens, 1 to 4 ft. thick and 400 ft. long. Too small for commercial development.
9) Rampart	15 miles north of Manley Hot Springs, North Fork of Baker Creek	90.0	N.A.	Easily accessible. Further exploration work needed.
10) Tolovana	40 miles northwest of Fairbanks, Minto Flats-Dugan Hills	99.0	N.A.	A very large deposit which needs evaluation.
11) Windy Creek	Windy Creek	92.0	180	Easily accessible and approximately 4 miles from Cantwell.
		Southcentral Region		
12) Chitistone & Nizina	Wrangell Mtns.	N.A.	N.A	Sporadically distributed along the southern flank of the Wrangell Mountains. Very little is known.
13) Kings River	Matanuska Valley, Kings River drainage. North of Glenn Highway	97.0	33	Consists of several large steeply dipping masses.
14) Potter	1/2 mile northeast of Potter	96.0	N.A.	Consists of several lenses. Needs further exploration work.

Figure 18.12 (continued)
Summary of Identified Limestone Deposits in Alaska (after Sanusi[7])

Name	Location	Mean % CaCO$_3$	Estimated Reserves (mil. tons)	Comments
		South Central Region		
15) Seldovia	Seldovia, Kenai Peninsula	89.0	0.2	Transportation limited to barge.
16) Dall & Long Island	Waterfall Bay and Gleva Bay	95.0	200	Massive and extensively folded. Accessibility almost impossible.
17) Glacier Bay	Willoughby, north and south Marble Islands	97.0	N.A.	Accessible with difficulties.
18) Heceta-Tuxekan	Heceta and Tuxekan Islands	94	N.A.	High purity, massive and thickness extremely variable.
19) Mud Bay	Northwest Shrubby Island	N.A.	15	Exposed as a 1500 foot beach out crop.
20) Pleasant Camp	Haines cut-off international boundary	N.A.	N.A.	Further field work required.
21) Saginaw Bay	Kui Island	N.A.	N.A.	Needs further exploration.
22) San Alberto Bay	Wadleigh Island	N.A.	40	Accessible by waterway.

NOTE: N.A. - not available

TABLE 18.5

Calcium and Magnesium Content of Selected Samples
(Percentage of whole rock)

Location	Calcium	Magnesium	Mg/Ca
Hoodoos 1	38.45	0.168	0.0044
Hoodoos 2	33.47	1.490	0.0445
Hoodoos 3	37.95	0.317	0.0083
Hoodoos 4	33.93	0.389	0.0114
Amy Creek 1	22.27	12.020	0.5393
Amy Creek 2	22.98	13.000	0.5657
Livengood 1	22.75	0.280	0.0123
Livengood 2	21.20	12.186	0.5748
Livengood 3	36.29	0.303	0.0083
Livengood 4	22.39	12.490	0.5578

18.6   CONCLUSIONS

The Livengood #4 location was selected as the most promising site for future study and utilization. This deposit is part of a major outcropping. High grade dolomite is present in quantities far beyond that necessary for CMA production. The current projected CMA requirements is about 9,000 T/yr. This would require about 9,000 T of limestone, or about 4,500 yards of limestone/year. Although the extent of this deposit is not clearly defined, the deposit is several hundred feet wide, at least 40 feet in height and extends (judging from outcroppings), for several miles. This deposit is located approximately 40 miles north of Fox, Alaska on the Elliot Highway. The dolomite is exposed behind a roadcut on the east side of the road.

The dolomite is medium gray in color on a fresh surface and weathers to light brown on exposed areas. It is finely crystalline, and highly fractured. No fossils were found as it has been recrystallized. A number of the fractures contain calcite.

The site was chosen because of its relatively high percentage of magnesium, proximity to Fairbanks along a major roadway, and the large volume of material present. Another important aspect is the fact that the deposit has already been used by the Department of Transportation for roadwork. Therefore, the state already has rights to the deposit and no environmental impacts are apparent.

18.6.1 Lime

Hydrated lime or calcium hydroxide is a common compound often used as a soil conditioner. It is produced from limestone in the form of calcium oxide (CaO). The addition of water converts CaO to calcium hydroxide through the following reaction:

$$CaO + H_2O \rightarrow Ca(OH)_2$$

Lime is a by-product of the production of acetylene gas from calcium carbide and water, and is used in the manufacture of cement. There are currently no cement plants in Alaska, but there are at least two acetylene plants, one in Palmer (Big-Three Lincoln) and one in Anchorage (Liquid Air Corporation).

The plant in Palmer produces about 200 tons/year of dry calcium hydroxide. Table 18.6 presents the analysis of the calcium hydroxide produced at Big Three Lincoln in Palmer, Alaska. This product is currently classified as a "waste" and poses a disposal problem for the manufacturer. The product is in the form of a white powder. Roughly 5,000 cubic yards have accumulated at the plant site. The facility in Anchorage produces a smaller amount (exact production not known). To produce 8,000 ton/year of CA (current projected demand), roughly 3,700 ton/year of calcium hydroxide would be required. Current calcium hydroxide production rates are not sufficient to serve as the raw material for CA production. However, if limestone is the base calcium source, the amount of calcium hydroxide required for neutralization is only 800 ton/year.

Thus one quarter to one half of the required lime could be obtained in Alaska. The lime is currently a disposal problem, and could probably be obtained at a relatively low cost.

TABLE 18.6

Hydrated Lime Analyses
(Dry Basis)

	Acetylene By-Product From Generator	Generator Hydrate from Settling Pond	Commercial Hydrate Sample 1
$Ca(OH)_2$	96.50	95.00	96.44
Available CaO	(73.00)	(69.80)	(72.50)
$CaCO_3$	1.25	2.00	1.76
SiO_2	1.10	0.83	0.38
$R_2O_3 (Al_2O_3, Fe_2O_3)$	0.50	0.20	0.57
$Mg(OH)_2$	0.25	0.20	0.57
S	0.15	0.16	0.03
P	–	0.01	0.01
Free Carbon	–	0.40	–

18.7   PILOT PLANT STUDIES

In the past three years, we have produced CMA successfully from lime and
limestone, in lab scale, bench scale and pilot plant scale processes. In
this section of the report, our experiences are summarized.

18.7.1 <u>Bulk Production of CMA From Hydrate Lime (Calcium Hydroxide)</u>

As discussed above, the reaction of acetic acid with hydrated lime is
much easier to conduct than that with limestone. In many areas, hydrated
lime is readily available for purchase as a common agricultural soil
conditioner. In Alaska, significant quantities are available as a
by-product of acetylene manufacture. At the present time, this material is
a nuisance, posing a disposal problem. As a result, it may be available as
a raw material for CA manufacture at little or no cost. However, the
current level of lime production from this source is far less than would be
required if all salt usage in the state were replaced by CA. Thus, it is
envisioned that if CA is produced in Alaska, it would be produced from
limestone, and not from lime. However, as mentioned above, there is a need
in the limestone process for a neutralizing agent such as hydrated lime
(calcium hydroxide).

In the summers of 1983 and 1984, 5,000 and 7,000 gallons respectively of
saturated CA solution were produced. At this juncture, we were in need of
a significant quantity of CMA for road and environmental tests. Since
hydrated lime was available at no cost in sufficient quantity for our
purposes, this reaction route was chosen. A pilot plant was constructed at
Alaska State DOTPF facilities in Fairbanks by personnel of the Petroleum
Engineering Department at UAF. In the sections that follow, this plant,
and our experiences in operating it are detailed.

18.7.2 <u>Equipment</u>

A schematic diagram of the pilot plant is presented in Figure 18-13.
The plant consisted essentially of two 500 gallon stainless steel reactors,
associated feed tanks, product tanks, and transfer pumps and hoses. The
key elements are described below:

1. Reactors (2) - 500 gallon stainless steel tank with a loosely fitted top
    closure, and a bottom drain (1.5" pipe). A mounting bracket for an
    agitator was attached in the top center of the reactor (see Figure
    18-13).
2. Agitators (2) - 1.5 hp electrically driven, single shaft, twin propeller
    agitator.
3. Acid Pump - Wilden air operated Teflon diaphragm pump, 5-20 gpm

474

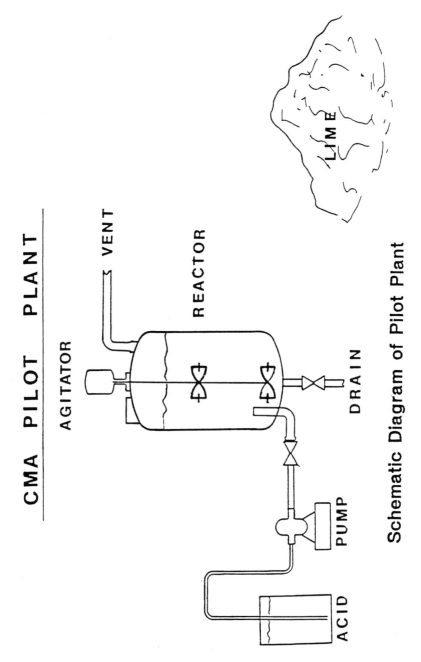

Schematic Diagram of Pilot Plant

FIGURE 18-13

capacity.

4. Air Compressor - 1.5 hp air compressor to drive acid pump.
5. Generator - 6kW generator to power the air compressor and agitators.
6. Miscellaneous Tubing and Piping - Two 25 foot sections of Teflon hose
   (1" ID) were used for transfer of acid and products.

Many modifications to the above equipment would be acceptable. The
important factor to consider, is that the reactor and any pumps or tubing
coming in contact directly with the acid should be of a corrosion resistant
material. As described below, it is possible to conduct the reaction so
that the reactor itself never encounters an acidic solution. In this event,
all that is required is an agitated vessel capable of withstanding exposure
to caustic lime. The only equipment coming into contact with the acid
would be the transfer lines and pump.

18.7.3 Raw Materials

Glacial acetic acid was shipped in 55 gallon drums from bulk chemical
suppliers in the "lower 48" states. It was of research purity, although
this is certainly not required. Virtually any grade of acid would suffice.
As the pilot plant was set up in a "remote" location without water,
electricity, or sewer facilities, water was hauled in and stored in a 1000
gallon tank. Hydrated lime was donated by Big Three - Lincoln in Palmer,
Alaska. An analysis of the lime is presented in Table 18.6. The lime had
been exposed to the elements for several years, and had "acquired" some
impurities, mostly inerts. The most significant problem with the lime was
a varying water content throughout the stockpile, and over the course of
the summer. This uncertainty made quality control measurements a
necessity.

18.7.4 Operating Procedures

The reactions were conducted in the batch mode. Little attempt was made
to minimize the reaction and turnaround times. In 1983, the plan called for
the production of 5,000 gallons of saturated solution. Ten batches were
prepared. In 1984, an additional 7,000 gallons were produced in fourteen
batches. The procedure was the same for each batch, although the final
adjustment of pH and density for each batch was unique. The following
procedure was followed:

1. Reactor was filled with 300 gallons of fresh water.
2. Roughly 22 cubic feet, or 1200 lbs. of lime were added to the tank,
   while agitating. The lime was added manually using 2.5 gallon buckets,

which were filled by hand from the storage pile. Initially, the buckets were weighed, in an attempt to more accurately control the amount of lime added. However, the variation in water content of the lime was significant enough to make exact knowledge of the appropriate amount of lime to add unknown. Thus, the amount of lime added varied from one batch to the next. The quality control procedure described below assured a constant product quality.

3. While agitating, two barrels of acetic acid were pumped into the bottom of the tank. The acid was added slowly to avoid contacting the tanks directly with undiluted acid. The reaction is exothermic, and the solution temperature rose to nearly 180°F during the mixing process. If all of the acid were added suddenly, it may be possible to approach the boiling point of the solution.

4. The tank was agitated for up to 8 hours, or until the pH became stable. The solution was then allowed to settle for several hours.

5. A sample of clear liquid was withdrawn from the tank and the gravity and pH were measured. Typically, the solution was less than saturated with CA. The CA concentration was estimated from the specific gravity of the solution. A computation was made to determine the amount of lime required to bring the density up to the desired mark. This amount of lime was then added to the solution.

6. The density having been adjusted, acetic acid was added to bring the pH to approximately eight. The amount of additional acetic acid was determined by scaling up a simple titration of the solution liquid. It was also possible to monitor the tank pH directly while adding acid.

7. The tanks were allowed to settle for several more hours. The clear solution was then pumped to storage facilities. In 1983, the solution was pumped directly into a 3000 gallon tank truck for storage and application. In 1984, the solution was in a variety of drums and tanks.

18.7.5 Quality Control

The desired product for the pilot plant was saturated CMA solution. The inherent uncertainty in the water content and composition of the lime made it impossible to assure product quality through metering of the reactants. It was necessary to adjust both the pH and the gravity of the product solution to the desired values. The desired specifications for the product were: Specific Gravity = 1.14; Weight Percent = 25; pH = 8.0.

Density was determined by weighing a known volume of fluid. Using published data on density of CA solutions versus weight fraction CA, the CA concentration can be determined from density. Care was required to see that the fluid sample did not contain appreciable amounts of suspended solids. Filtration is recommended.

The concentration of the solution can be increased to the desired value (25 weight percent) by adding calcium hydroxide and acid. A simple material balance calculation yields the following relationship for calculating the pounds of calcium hydroxide required to increase the concentration to 25 weight percent:

$$S = (W + A + L)/1.48 * (.25-X)$$

where W = original amount of water used lb., L = original amount of lime used (lb), A = original acid used (lb), and X = actual weight fraction CA.

The adjustment of pH is somewhat easier. A small sample (200 cc.) of solution is titrated to a pH of eight with acetic acid. The amount of acid required is then divided by 200 to obtain a ratio of volume acid required per volume of solution. The volume of solution in the tank is roughly 400 gallons at this point. The ratio is simply multiplied by 400 to obtain the estimate of the gallons of acid required to neutralize the tank. In the final adjustment, the pH of the tank is monitored directly, and acid or lime is added manually to bring the pH to eight.

18.7.6 Material Balances

A material balance representing a typical batch is presented in Figure 18-14. It should be recognized that in reality, some water and inert material will be present, in the "Lime", necessitating the use of slightly more "lime" (by weight) then indicated in the Figure. In general, the correct amount of lime required can be determined by dividing the pure lime requirement by the weight fraction of pure lime in the actual lime used. If the lime is particularly wet, the amount of water added with the lime should be subtracted from the total water required.

18.7.7 Observations

Little difficulty was encountered with the pilot plant operations. Once operating procedures were established, the process worked quite well. It was possible to complete the production of a 500 gallon batch in under six hours. Two people could produce two parallel batches in an eight hour shift (2000 gallons).

This particular pilot plant design was one of expediency. The process was labor intensive, and could certainly be improved upon. The purpose of this pilot plant was two-fold. First, a supply of CA was necessary for environmental studies and field tests. Second, it was demonstrated that CA can be produced simply and efficiently with a minimal amount of equipment.

478

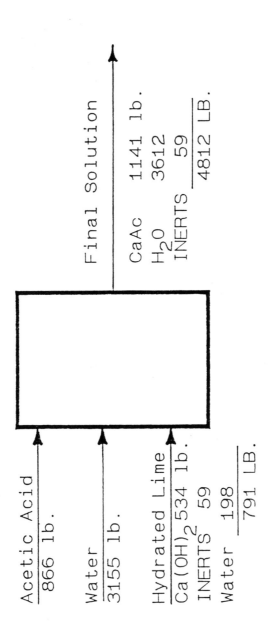

FIGURE 18-14

Material Balance For Pilot
Plant Production

18.7.8 Problems

Acetic acid in pure form freezes at about 60°F. We encountered difficulty with the acid freezing routinely. It will be necessary to provide heated storage facilities for the acid. The problem can be alleviated somewhat by mixing water in with the acid. The heat of mixing upon contact with water tends to warm the mixture, and the freezing point of aqueous acetic acid solutions is much lower than pure acetic acid.

The reaction of acetic acid with lime is exothermic. Although no gas is produced, there is a tendency for the mixture to foam at high temperature. Care must be taken regarding the speed with which the acid is added to avoid overheating the solution.

Acetic acid vapor is a health hazard as well as being flammable. In general, any personnel involved in this process should be instructed on proper safety procedures regarding the use of acetic acid. More importantly, the reaction area should be adequately ventilated. While the fumes given off from the tanks themselves are minimal, some small amounts of acid may escape during transfer operations. Extreme caution should be exercised at all times in handling the acid.

It is possible to use too little water in the reactors and to produce more CA than can be held in solution. On two occasions this occurred, once by accident, and once intentionally. When CA precipitates, it forms a hydrated crystal, thus pulling additional water from the solution. As a result, the solution in effect "gels" suddenly, to a consistency not unlike thick oatmeal. With the proper equipment, this paste could be collected and further dried to a powdered form. However, for the purposes of this study, and with the equipment at hand, the paste was unmanageable. However, the addition of a small amount of water was sufficient to redissolve the entire mixture. In one case, as little as five gallons of water added to the 400 gallons of paste was sufficient to redissolve the entire mixture. Care should be taken to see that this precipitation does not occur, as it may damage the agitators.

18.7.9 The Production of CMA From Limestone

The production of CMA from limestone was somewhat more complicated than described above. A two step process is required. First, calcium carbonate is dissolved in a solution with an excess of acetic acid. In the second step, the excess acid is neutralized with calcium hydroxide. The evolution of carbon dioxide is a further complication.

Numerous lab scale studies were undertaken to determine the

stoichiometry and kinetics of the reaction of acetic acid with limestone.
A limited number of bench scale studies were also performed. In the
previous section of this report dealing with CMA chemistry, the basics of
the reaction and results of these studies are summarized. A detailed
report covering the results of these experiments will be released at a
later date. However, the information essential for a process design is
presented here.

18.8   PROCESS DESIGN

Process designs are presented below for the production of CMA from
limestone and lime. These designs are based on the results of our
experiences with the pilot plant, and numerous lab scale studies of the
reactions involved. The designs have been verified in our laboratories, and
should provide a valid basis for specific equipment design and cost
analysis. The design basis is 12,000 GPD production with a stream factor
of .95, for an annual production of 5140 TPY of dry CA equivalent. The
economics of CMA production will be site specific, and will depend heavily
on the costs and properties of the raw materials. The final desired form
for the product i.e., solid or saturated solution, may also influence the
final design.

18.8.1 Production of CMA From Limestone

(i) Material Balance. The production of CMA from limestone requires
a two step process. In the initial step, a solution of acetic acid is
allowed to react with excess calcium carbonate (or limestone). The liquid
product is then separated from the solid, unreacted carbonate, and
neutralized with calcium hydroxide. The concentration of the original
acetic acid solution will determine to some extent the concentration of the
final product. In general, if the acid concentration exceeds 24 weight
percent, the product will be a saturated solution with a solid precipitate.
As described in the discussion of the pilot plant above, this precipitation
will result in a thick paste-like product, which would require special
handling. If the concentration of the acid is exactly 24 weight percent, a
saturated solution will result after neutralization. If the acid
concentration is less than 24 weight percent, an undersaturated solution
will result. The relationship between original acid concentration and
product concentration is presented graphically in Figure 18-7.

While it may appear that the best course of action is to use a high
original acid concentration, it should be noted that the amount of calcium
hydroxide required per pound of CA increases with increasing acid

concentration (See Figure 18-6). Since calcium hydroxide is normally more
expensive than limestone, it may be best to minimize calcium hydroxide
consumption. However, if calcium hydroxide consumption is eliminated
through the use of a dilute solution of the acid, the resulting product
would be too dilute to be useful directly, and would require expensive
concentration. The final decision as to how to proceed will be an
optimization question involving the prices of limestone and lime, and the
expense of concentrating the solution.

A material balance for the production of saturated solution is presented
in Figure 18-8. A similar material balance for the production of saturated
solution and solid CMA is presented in Figure 18-15.

The reaction represented in Figure 18-15 is valid, but would result in
the production of solid CMA along with a saturated solution. While it may
be desirable to produce a solid product the resulting mixture is too thick
to be handled in a stirred tank reactor, and contains unreacted HAc and
solid limestone in addition to the CA product.

The process design specified in this report is for the production of a
nearly saturated solution of CA. The material balance presented in Figure
18-16 is the basis for this design.

The first reactor, in which the acetic acid and limestone react, may be
operated either in the batch or continuous mode. A residence time of about
four hours is required. If the reaction is to be conducted in the
continuous mode, more detailed information regarding the kinetics of the
reaction will be required for proper sizing. The degree of agitation and
particle size will also be important parameters.

Operation in the batch mode is more forgiving of incomplete kinetics
characterization. Moreover, in the continuous mode any impurities in the
limestone would build up in the reactor. The continuous mode would require
continuous cleaning, or the availability of a duplicate reactor for
operation during the cleaning cycle for the first reactor.

The process design presented in this report is based on the material
balance given in Figure 18-16. For the purpose of standardization, the
design is based on the reaction of pure calcium carbonate. If limestone is
used, slight changes are experienced due to the presence of varying amounts
of inert materials, and somewhat decreased conversion versus that obtained
with pure carbonate. These factors have very little effect on the process
design, but do affect the material balance, and hence the process
economics.

The major effect of inerts is to require correspondingly more
"limestone" to obtain the proper amount of carbonate. If the limestone

482

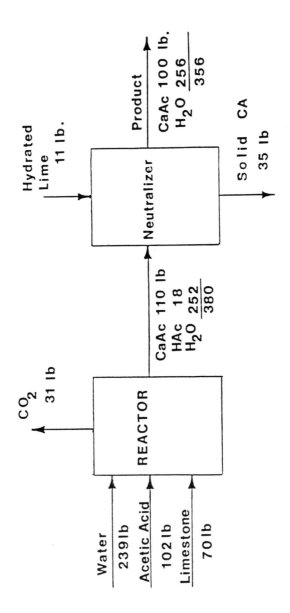

Figure 18-15. Material balance for production of saturated solution and solid CA.

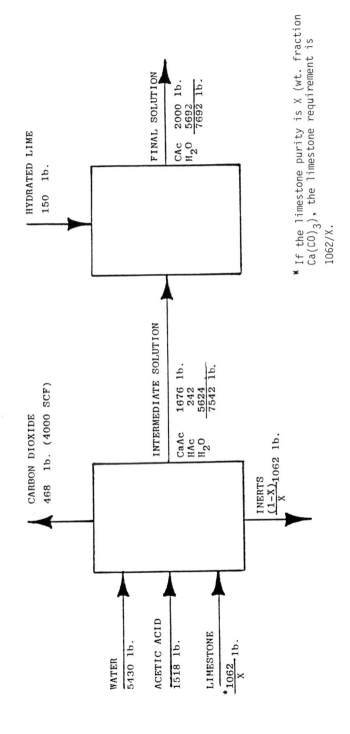

Figure 18-16. Material balance for process design - Production from limestone (based on a production of 1 ton of dry CA in a 26 wt.% solution).

contains X (weight fraction) calcium carbonate, then based on the material
balance presented in Figure 18-16, 53.1/X pounds of limestone would be
required per 100 lbs. of $CaAc_2$ as opposed to 53.1 lbs. If an excess of
limestone, Y, is used (Y = weight fraction excess calcium carbonate). The
total limestone feed would be 53.1(1 + Y)/X. The resulting waste stream of
inerts would be (1 + Y)(53.1)(1-X)/X and the amount of unreacted limestone
would be 53.1 Y.

The effect of changes in conversion is a change in the amount of $Ca(OH)_2$
required for neutralization. This effect is shown in Figure 18-17. The
equipment design and operation procedures would be unaffected.

   (ii) _Process Description._ The process could be operated in either a
batch or continuous mode. Continuous mode operation is normally more
economical for large scale operations. However, a large investment in real
time control equipment is required. This equipment would include flow
controllers, level controllers, pH controllers, density measurement
equipment, solid metering systems, etc. The presence of this equipment
would require a highly trained staff of at least two.

The same process can be conducted in the batch mode by a two-man crew,
using manual controls. In this case, the extra expenditure necessary to
"automate" the process cannot be justified. If a design were contemplated
for a much larger plant, (greater than 100,000 GPD) a continuous process
might be justified.

While the reaction must be conducted in two steps, reaction and then
neutralization, the same reactor can be used for both reactions. The basic
process design in this case consists of three 4,000 gallon reactors and the
associated transfer equipment. A detailed equipment list is given below.
A process schematic is presented in Figure 18-18.

A complete batch can be produced in one 8-hour shift. Thus it is
possible to achieve a 12,000 GPD production rate with a single reactor and
three shift operation. However, as is demonstrated below, the reactor cost
is a small factor in determining the overall process cost. Thus it is more
desirable to install parallel reactors and conduct the entire daily
production in one shift. This approach greatly simplifies the logistics of
operation, and allows the flexibility to increase production to 36,000 GPD
if desired. In particular, the production could be performed seasonally if
desired, requiring four months to produce the annual equivalent of 12,000
GPD production. In this case, however, large storage capacity (over 4
million gallons) would be required.

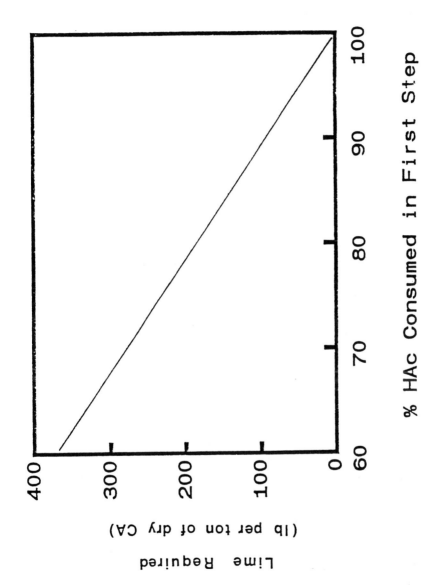

Figure 18-17. Effect of conversion on $Ca(OH)_2$ consumption.

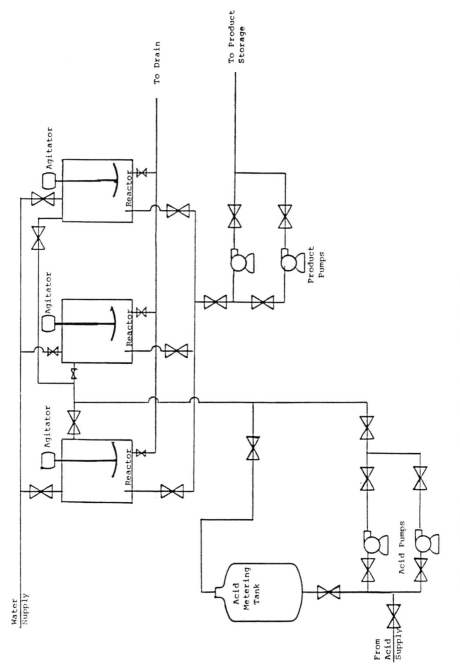

Figure 18-18. Process schematic – Production from limestone.

(iii) Equipment List

Reactors (3) - 4,000 gallon stainless steel (316) (any other material resistant to acetic acid would suffice). Reactor should be fitted with top closure to accommodate CO_2 venting. Since foaming can occur, at least 20% freeboard should be allowed. Approximate dimensions 8' diameter - 9' height. Unit should be equipped for top entering turbine agitator (20 hp).

Unit Cost - $5,200

Acid Pump (2) - 100 GPM pump rated for acetic acid (316 ss).

Unit Cost - $2,000

Solids Conveyor (2) Small conveyor system capable of moving limestone and lime from storage to tank. Must be moveable to service all tanks.

Unit Cost - $8,100

Product Pump (3) - 300 GPM pump rated for mild slurry service.

Unit Cost - $3,300

Agitator (3) - 20 hp top entering, turbine agitator. All immersed parts -316ss.

Unit Cost - $7,000

Acid Storage/Metering Tank - 1,000 gallon stainless steel tank with top closure for venting. Tank is used to measure acid quantity before addition to reactors.

Unit Cost - $7,000

Piping and Vent Lines - Miscellaneous transfer lines.

Unit Cost - $4,000

Front End Loader - For moving limestone.

Unit Cost - $13,000

Analytical Equipment - pH meters (3), one ton scales for weighing lime and limestone, laboratory balances, glassware, furniture, etc.

$8,000

Building - 50 x 60 metal building with 20' ceiling clearance, heated with office and laboratory (500 feet) space.

$70,000

Land -   Suitably zoned site which can accommodate
                          excavation   to bury 50,000 gallon storage
                          tanks.
                                    (Extremely Variable - est. $20,000)

(iv) Utilities. Electricity and water are required for the process.
Heating oil is required for the building. For the building described
above, fuel distributors in the Fairbanks area estimate the total annual
fuel requirement at 6,000 gallons of No. 2 Diesel Heating Oil. Current
prices in Fairbanks are about $1.00 per gallon, for an annual requirement
of about $6,000.

About 7,000 kwh/month of electricity is required. Most of the power is
for the operation of the agitators (3-20 hp motors). Current rates in
Fairbanks are about $.11 per kwh, for a total annual cost of about $9,230.
Rates may vary around the state.

About 15,000 GPD, or 5.5 million gallons of water on an annual basis are
required. It may be possible to use river or well water at a very low cost.
However, the current cost in Fairbanks for city water service would amount
to about $13,570 per year.

(v) Labor. The manufacturing operation can be handled easily by two
laborers. The major operations involved are loading the reactors with
limestone, water and acetic acid. Additional time will also be required
for quality control measurements and adjustments. A supervisory position
with responsibility for the entire operation is also advised. With added
experience and training, it should be possible to reduce the staff to two,
however, the design is based on a three man operation.

Labor costs will vary widely depending on location and other factors.
For the purposes of this study, it was assumed that manual labor would cost
$30,000 per year plus benefits, and that the supervisory position could be
filled for about $40,000 plus benefits. The total benefit rate was
estimated at 50%.

(vi) Raw Materials. The basic raw materials for this process are
acetic acid, limestone, and calcium hydroxide. The base prices assumed for
these materials delivered to Fairbanks were $.20/lb., $75/ton, and $200/ton
respectively. The raw materials are required in the following amounts
relative to tons of CA produced:

acetic acid	0.760 ton/ton CA
limestone	0.530 ton/ton CA
calcium hydroxide	0.075 ton/ton CA

(vii) Economic Analysis. With a very small operation such as this,
many economies may be realized through integration of the process with some

existing facilities. As detailed below, for a stand alone facility, the building cost alone amounts to over half of the total capital expense. If a building were already in use with 2-3000 square feet of extra space, considerable savings could be realized.

This analysis is based on a stand-alone grass-roots operation.

Capital Cost

	#	Unit Cost	Source	Total
Conveyor/Elevator	1	$8,100	Goodman, Inc.	$8,100
Reactors	3	5,100	Greer Tanks	15,300
Acid Tank	1	3,000	Greer Tanks	3,000
Acid Pump	2	2,000	Durco	4,000
Product Pump	2	3,300	Gallagher	6,600
Agitator	3	7,000	Chemineer	21,000
Piping & Valves	–	4,000	misc. (est.)	4,000
Storage Tanks	4	8,000	Greer Tanks	42,000
Front End Loader	1	13,000	Yukon Equipment	13,000
Analytical Equip.	–	8,000	misc. (est.)	8,000
				$125,000
Building (50x60 metal on slab)				70,000
Land (will vary greatly)				20,000
				$215,000

The total capital cost of $215,000 could be amortized over a 20 year period at 15% for $34,349/year. If a building is available, and if storage facilities for the product liquid are available (or unnecessary... as for an on demand production basis), the total capital cost would drop to $73,000.

Annual Operating Cost:

As is to be expected, the operating costs are totally controlled by the raw material costs. A breakdown is given below.

	Annual Cost
UTILITIES:	
Heating (6,000 gal. @ $1/gal.)	$ 6,000
Electricity (7,000 kwh/mo. @ $.11)	9,230
Water (5.5 mil. gal. @ $2.25/1000 ga. + $100/mo.)	13,570
	$28,800
LABOR:	
Supervisor (1 @ $40,000)	$40,000
Laborers (3 @ $30,000)	90,000
	$130,000
BENEFITS AT 50%	65,000
TOTAL	$195,000
RAW MATERIALS:	
Acetic Acid (3,900 ton @ $400/ton)	$1,560,000
Limestone (2,730 ton @ $75/ton)	204,750
Calcium Hydroxide (380 ton @ $200/ton)	76,000
	$1,840,750
SUBTOTAL - Operating Costs	$2,064,550
CONTINGENCY (10% of Capital)	$22,000
	$2,086,550

The process is fairly easy to analyze, consisting of a cash flow and one initial investment of $215,000. The gross cash flow (before tax) can be computed as: $I = 5,140 \ S - 2,086,550$

where S is the selling price in dollars per ton of dry CA equivalent and I is annual before tax net income. Based on an initial investment of $215,000, the required selling price for several rates of return on investment (ROI) is listed in Table 18.7, and plotted in Figure 18-19.

TABLE 18.7

CA Price Required for Specific ROI
(Limestone - 20 year life)

Annual ROI	Net Cash Flow	CA Price $/ton
0%	10,750	408
10	25,253	411
15	34,349	413
20	44,152	415
25	54,377	417
30	64,841	419

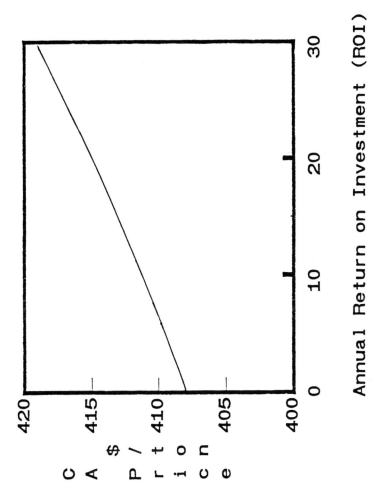

Figure 18-19. Effect of ROI on selling price for production from limestone.

It is useful to examine the effects of raw material prices on the required selling price for a 15% ROI. The total annual operation cost exclusive of raw materials is $245,800 giving a per ton cost (based on 5,140 ton/year CA) of $48/ton CA. The total cash flow required for 15% ROI is $34,349, or $6.40/ton. It is clear that capital cost amortization is a minor part of the required selling price of CA. Furthermore, labor and utility costs represent less than 12% of the selling price.

The total annual acetic acid requirement is 7,800,000 pounds. The total limestone requirement is 2,730 tons (based on 100% purity). The total lime requirement is 380 tons/year. These figures are based on an annual production rate of 5,140 ton/year of CA. The total required revenues from CA at 15% ROI can then be calculated as:

Annual Revenue =
245,800 (op. cost) + 34,349 (ROI) + 7,800,000 C (acid cost) + 2,730 PL (limestone) + 380 PC (Calcium hydroxide)

where C is the per pound cost of acetic acid delivered to Fairbanks, PL is the cost of ground (1-1.5 mm) limestone delivered to Fairbanks, and PD is the cost of powdered or pelletized calcium hydroxide delivered to Fairbanks. (If the limestone and lime are less than 100% pure, the PL and PC terms should be divided by the weight fraction of limestone and lime.)

For an annual production rate of 5,400 tons, the above equation can be reduced to: P = 54.5 + 1,518 C + .53 PL + .074 PC

where P is the required selling price for CA on a dollar per dry ton equivalent basis, for an annual ROI of 15%. For the design case where C = $.20/lb, PL = $75/ton, and PC = $200/ton, the required price is $415/ton. The relationship between required selling price and raw material costs are plotted in Figures 18-20 and 18-21.

18.8.2 Production of CA From Hydrated Lime

CMA can be produced more easily from hydrated lime (calcium hydroxide). The chemistry of the reaction is discussed in section 18.5.3 and a material balance is given in Figure 18-22.

Pilot plant studies indicate that the reaction with lime requires only about 1 hour to go to completion, with about an hour of settling time and one hour for pH adjustment required. Thus it should be possible to complete two batches in an eight hour shift. Hence two - three thousand gallon reactors should be sufficient.

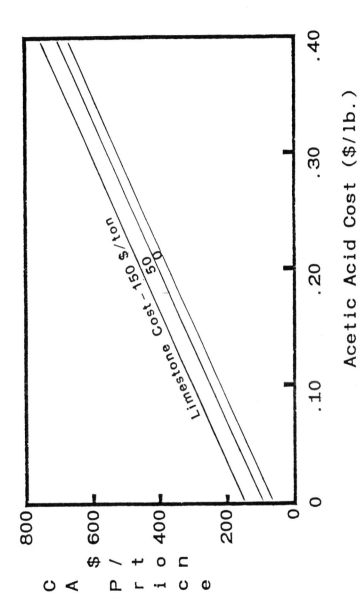

Figure 18-20. Effect of acid and limestone cost on selling price of CA for production for limestone.

494

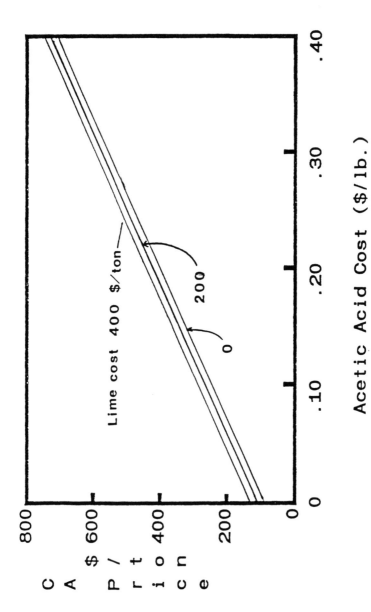

Figure 18-21. Effect of hydrated lime costs on CA selling price for production from limestone.

495

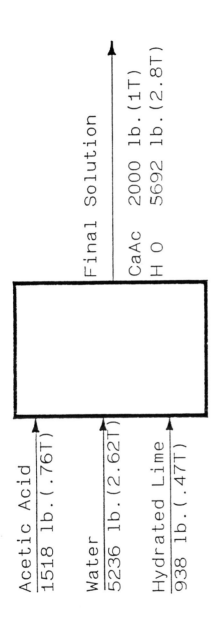

Acetic Acid
1518 lb. (.76T)

Water
5236 lb. (2.62T)

Hydrated Lime
938 lb. (.47T)

Final Solution

CaAc   2000 lb. (1T)
H O   5692 lb. (2.8T)

Material Balance for Process Design
Production from Hydrated Lime

FIGURE 18-22

496

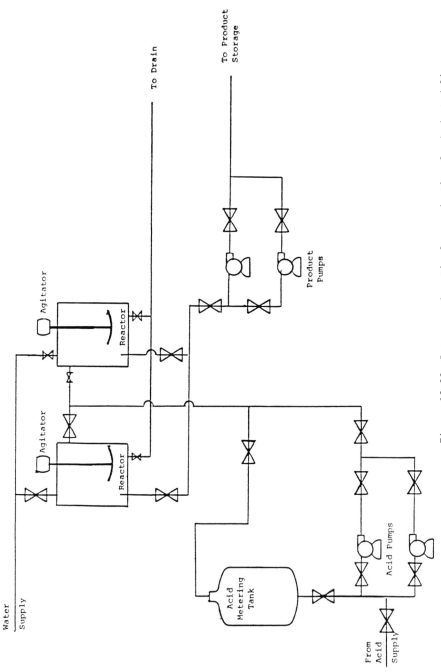

Figure 18-23. Process schematic for production for hydrated lime.

(i) <u>Material Balance</u>. The material balance for this process design is presented in Figure 18-22. The reaction proceeds stoichiometrically with no surprises. It should be noted however, that the material balance is based on pure $Ca(OH)_2$.

Thus for the production of one ton of CA, 940 lb of lime is required. For a lime purity of X (wt fraction $Ca(OH)_2$), 940/X lbs of impure lime would be required.

(ii) <u>Process Description</u>. The process consists of a simple acid-base titration. The reactor is first filled with the appropriate amount of lime and water. Then sufficient acetic acid is added to bring the pH to about 8. After about one hour of agitation, the pH and density are adjusted through the addition of additional lime and acid as described in section 18.7.5. A schematic of the process is presented in Figure 18-23.

(iii) <u>Equipment List</u>. The equipment descriptions for this process are exactly the same as those presented in section 18.8.1 (ii) for the design for production from limestone, with the following changes:

reactors -   two 3,000 gal. reactors of the same type described in 18.8.1;

agitators - two of the same type described in 18.8.1;

(iv) <u>Utilities</u>. The utility requirements for this process are essentially the same as for the limestone process. About 6,000 gal/year of No. 2 heating oil is required for building heat, at a cost of about $6,000.

About 15,000 GPD, or 5.5 million gallons per year of water are required for the process and cleanup. It may be possible to use well or river water at a very low cost. However, the cost for "city" water would be about $13,570/year.

Since this process only requires two agitators instead of three, the electrical usage would be correspondingly lower than for production from limestone. The total electrical requirement is estimated at $6,153/year (4660 kwh/month at $.11 per).

(iv) <u>Labor</u>. Since this is a one step reaction, a single laborer should be able to handle loading and operating the reactors. A supervisory manager could handle the quality control measurements and oversee the operation.

Labor costs will vary widely. For this analysis, it was assumed that manual labor would cost about $30,000/year plus benefits, and that the supervisory position could be filled for $40,000/year plus benefits. The benefit rate was estimated at 50%.

(v) <u>Raw Materials</u>. The base raw materials for this process are hydrated lime (calcium hydroxide), acetic acid, and water. The hydrated

lime would be shipped in bags or by bulk shipment form the lower 40 states.
Acetic acid was assumed to be pure (100%). The estimated costs of lime and
acid delivered to Fairbanks, were $200/ton and $.20/lb respectively.

The raw materials are required in the following amounts relative to tons
of CA produced:

acetic acid -   0.76 ton/ton CA

hydrated lime -   0.47 ton/ton CA

(vi) Economic Analysis. The economics of production from hydrated lime
is very similar to that for production from limestone. Capital costs
represent a very small portion of the annual cash flow. Over 80% of the
operating costs are direct raw materials costs.

The economic analysis presented below is for a 12,000 GPD (5,400
ton/year) grass-roots facility.

Capital Costs

	#	Unit Cost	Source	Total
Reactors	2	$5,100	Greer Tanks	$10,200
Acid Tank	1	3,100	Greer Tanks	3,000
Acid Pumps	2	2,000	Durco	4,000
Product Pump	2	3,300	Gallagher	6,600
Conveyor/Elevator	1	8,100	Goodman, Inc.	8,100
Agitators	2	7,000	Chemineer	14,000
Analytical Equip.	–	8,000	–	8,000
Piping	–	4,000	–	4,100
Storage Tanks	4	8,000	–	42,000
Front End Loader	1	13,000	–	13,000
				$113,000
Building (50x60 metal on slab)				70,000
Land (will vary)				20,000
			TOTAL	$203,000

The total capital cost of $203,000 could be amortized over a 20 year
period at 15% for $32,432/year. If a building and storage facilities are
available, the total capital cost drops to $71,000, which could be
amortized at $11,343/year.

(vii) Annual Operating Cost. Operating costs are almost totally
controlled by raw material costs. Over 90% of the operating costs are
attributed to direct raw material costs. A breakdown is give below:

	Annual Cost
Utilities:	
• Heating Oil (6,000 gal. @ 1 $/gal)	$6,000
• Electricity (4,670 kwh/mo @ 11¢ per kwh)	6,150
• Water (5.5 mil gal @ $2.25/1,000 gal + 100 $/mo)	13,570
	$25,720
Labor:	
• Supervisor (1 @ $40,000)	$40,000
• Laborer (1 @ $30,000)	60,000
Subtotal	**$100,000**
Benefits	50,000
	$150,000
Raw Materials:	
• Acetic Acid (3,900 ton @ $400/ton)	$1,560,000
• Hydrated Lime (2,410 ton at $200/ton)	482,000
	2,042,600
Subtotal Operating Costs	2,217,720
Contingency	20,000
Total Operating Costs	**$2,237,720**

The process is fairly easy to analyze, consisting of a cash flow and single initial investment of $203,000. (It should be noted that the total operating cost is almost identical with the limestone case at $2,086,550). Gross (before tax) cash flow can be computed as:

$$I = 5,140 \ S - 2,237,720$$

where S is the selling price for CA in dollars per dry ton equivalent, and I is annual before tax, net income. Based on an initial investment of $203,000, the required selling price for several rates of return are listed in Table 18.8 and plotted in Figure 18-24.

TABLE 18.8

CA Price Required for Specific ROI
(Hydrated Lime - 20 year life)

Annual ROI	Net Cash Flow $/yr	CA Price
0%	10,203	437
10	23,844	440
15	32,431	442
20	41,687	443
25	51,342	445
30	61,222	447

The effects of raw material costs on required selling price at 15% ROI has been investigated. The total operating cost, exclusive of raw materials is $195,720 for a per ton cost (based on 5,140 ton/year production) of $38/ton. For a 15% ROI, the capital cost contribution is only $6/ton of CMA. Clearly, the economics of the process depends mostly on raw material costs.

The total acetic acid requirement is 7,800,000 pounds. The total hydrated lime requirement is 2,410 tons (based on 100% purity). The total annual revenue required from CA sales can be computed as a function of acid and hydrated lime price as:

Annual Revenue + $195,720 (Op. Cost) + 32,431 (ROI)
+ 7,800,000 C (Acid Cost) + 2,410 PC (Hydrated Lime)

where C is the per pound cost of acetic acid delivered to Fairbanks, and PC is the delivered cost of hydrated lime (calcium hydroxide). If the lime purity is less than 100%, PC should be divided by the weight fraction calcium hydroxide.   For an annual rate of production of 5,140 tons, the above equation can be reduced to:

P = 44.4 + 1,518C + .47 PC

where P is the required CA selling price per ton of dry CA equivalent for an annual ROI of 15%.

For the design case where C = 0.20 $/ton and PC = 200 $/ton, the required selling price is 4342 $/ton. The relationship between required selling price and raw material costs are presented in Figure 18-25.

18.9   CONCLUSIONS

CMA can be produced in a fairly simple process from either limestone or hydrated lime. The capital cost are roughly the same in either case, and

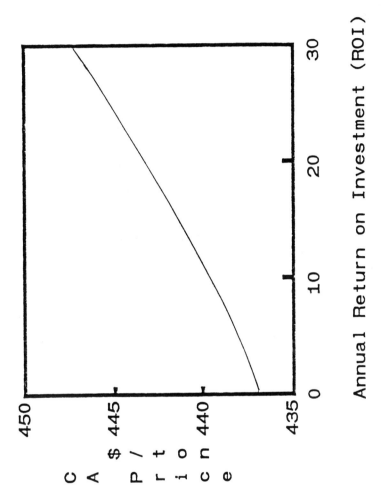

Figure 18-24. Effect of ROI on selling price.

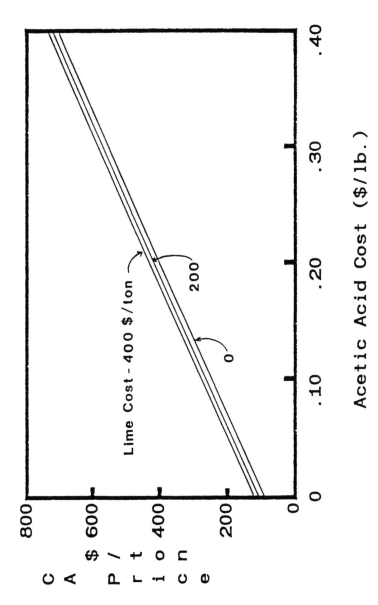

Figure 18-25. Effect of acid and hydrated lime cost for production for hydrated lime.

operating costs are comparable. For production from limestone, raw
material costs are somewhat lower while capital costs are higher. The
required selling price for 15% ROI is $415/ton. For production from lime,
capital costs are slightly lower ($12,200 less), and operating costs are
slightly lower, but raw materials costs are higher. The required selling
price for production from lime at 15% ROI is $442/ton.

In either case, the controlling factor in the economics is the price of
acetic acid. About.76 tons of acetic acid are required per ton of dry CA.
Since acetic acid costs about $.2/lb or $400/ton, the contribution of acid
to total cost in $304/ton. A substantial discount in acetic acid price to
say $.10/lb would bring the acetic acid contribution down to $152/ton, or a
total cost of 250 $/ton.

With current prices, acetic acid is available on the West Coast at a
delivered price of $.25/lb. Shipping costs are roughly $.05/lb.
Therefore, the current contribution of acetic acid would be $456/ton.
However, with a long term purchase agreement and the possibility of using
"off-spec" acetic acid, it may be possible for the price to be reduced to
the neighborhood of $.20/lb delivered in Anchorage.

For production from limestone, the capital cost contributes only $6/ton
to the selling price of CA. Hence considerable latitude in equipment
design is possible. The operating costs for utilities and labor amounts to
about $48/ton of CA produced. The total non-raw material cost is then only
$54/ton. Furthermore, the contribution of limestone costs to the selling
price is only about $40/ton. Thus variations in limestone cost have a
small effect (less than 10%) on the total cost.

In general, capital cost, operating cost and limestone and hydrated lime
costs are minor factors is the total CA price. The major factor is acid
cost. While the product may be attractive at $400/ton, the key to reducing
this cost significantly lies in the reduction of acid cost.

504

REFERENCES

1   F.O. Wood, "Motor-Vehicle Corrosion from Deicing Salt", Transportation Research Record, 762, pp. 32-36.
2   J.A. Zenewitz, "Survey of Alternatives to the Use of Chlorides for Highway Deicing," Federal Highway Administration, U.S. Department of Transportation, Rept. FHWA-RD-77-52, May 1977.
3   S.A. Dunn, and R.V. Schenk, "Alternative Highway Deicing Chemicals," In Snow Removal and Ice Control Research, TRB, Special Rept. 185, pp. 261-269, 1979. (Also see FHWA-RD-79-08.)
4   R. Brenner, and J. Moshman, "Benefits and Costs in the Use of Salt to Deice Highways", Institute for Safety Analysis, Washington, D.C., November 1976.
5   D.M. Murray, and U.F.W. Ernst, "An Economic Analysis of the Environmental Impact of Highway Deicing," U.S. E.P.A., Rept. EPA-60012-76105, May 1976.
6   I. Merck, and M. Windholz, Ed., 9th Edition, Merck and Co., Inc., Rahway, N.J., 1976.
7   A.C. Sanusi, "Agricultural Limestone Demand Requirements and Supply Production in Alaska," M.S. Thesis, University of Alaska, 1983 (MIRL Rept. No. 67).
8   G. Arce, "The Distribution of Carbonate Deposits Along Alaska's Roads," Petroleum Engineering Department, University of Alaska - Fairbanks, 1983.
9   R. Moell, Private Communication, Big Three Lincoln Alaska, Inc., 1983.

INDEX